AF612203

Mathematics and Its Applications (*East European Series*)

Introduction to the Theory of Games

J. Szép and F. Forgó

Institute of Mathematics and Computer Science
Karl Marx University of Economics, Budapest

Introduction to the Theory of Games

Springer-Science+Business Media, B.V.

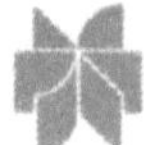

Library of Congress Cataloging in Publication Data

Szép, J.
Introduction to the theory of games.

(Mathematics and its applications (East European series) ; v. 3)
Translation of: Bevezetés a játékelméletbe.
Bibliography: p.
1. Game theory. I. Forgó, Ferenc. II. Title. III. Series: Mathematics and its applications (D. Reidel Publishing Company). East European series ; v. 3.
QA269.S94 1985 519.3 84-13457

DOI 10.1007/978-94-009-5193-8

Originally published by Akadémiai Kiadó, Budapest, Hungary in 1985
MyCopy version of the original edition 1985

www.springer.com/mycopy

Contents

Editor's Preface

Approach your problems from the right end and begin with the answers. Then one day, perhaps you will find the final question.

'The Hermit Clad in Crane Feathers' in R. van Gulik's *The Chinese Maze Murders*.

It isn't that they can't see the solution. It is that they can't see the problem.

G. K. Chesterton. The Scandal of Father Brown 'The Point of a Pin'.

Growing specialization and diversification have brought a host of monographs and textbooks on increasingly specialized topics. However, the "tree" of knowledge of mathematics and related fields does not grow only by putting forth new branches. It also happens, quite often in fact, that branches which were thought to be completely disparate are suddenly seen to be related.

Further, the kind and level of sophistication of mathematics applied in various sciences has changed drastically in recent years: measure theory is used (non-trivially) in regional and theoretical economics; algebraic geometry interacts with physics; the Minkowsky lemma, coding theory and the structure of water meet one another in packing and covering theory: quantum fields, crystal defects and mathematical programming profit from homotopy theory; Lie algebras are relevant to filtering; and prediction and electrical engineering can use Stein spaces.

This program, Mathematics and Its Applications, is devoted to such (new) interrelations as exempli gratia:

— a central concept which plays an important role in several different mathematical and/or scientific specialized areas;

— new applications of the results and ideas from one area of scientific endeavor into another;
— influences which the results, problems and concepts of one field of enquiry have and have had on the development of another.

The Mathematics and Its Applications programme tries to make available a careful selection of books which fit the philosophy outlined above. With such books, which are stimulating rather than definitive, intriguing rather than encyclopaedic, we hope to contribute something towards better communication among the practitioners in diversified fields.

Because of the wealth of scholarly research being undertaken in the Soviet Union, Eastern Europe, and Japan, it was decided to devote special attention to the work emanating from these particular regions.

Thus, it was decided to start three regional series under the umbrella of the main MIA programme.

The present volume is devoted to game theory and several of its many applications. More or less recent developments, both theoretical (Shapley value and related concepts) and practical (algorithms for calculating equilibrium points), have vastly extended the potential applicability of game theory, thus making it desirable that these ideas are readily accessible to those interested in the applications of game theory. This book meets this demand.

The unreasonable effectiveness of mathematics in science...

Eugene Wigner

Well, if you knows of a better 'ole, go to it.

Bruce Bairnsfather

What is now proved was once only imagined.

William Blake

As long as algebra and geometry proceeded along separate paths, their advance was slow and their applications limited.
But when these sciences joined company they drew from each other fresh vitality and thenceforward marched on at a rapid pace towards perfection.

Joseph Louis Lagrange

Preface

Since 1944, when the famous book *Theory of Games and Economic Behavior* by von Neumann and Morgenstern was published, several books devoted to the subject have appeared. Although most of them intended to give a general overview of the actual state of the theory their approach has been quite different. Some try to make the subject popular, others are concerned primarily with behavioural-economic-interpretational problems or specialize on the mathematics of the theory etc. In addition to keeping to a theoretical-mathematical presentation we pay more attention to methods for determining "solutions" of games than it is done in other textbooks.

Although we looked through much of the relevant literature, with special emphasis on the vintage of the last ten years, we do not pretend to have written a monograph. So much the more because the size of the book had to be kept within limits. The length of a chapter does not necessarily indicate its importance, either. As an example we mention cooperative games which are discussed more briefly than the abundance of papers recently published on this field would justify it. We will touch upon subjects covered in great detail in other books very concisely (e.g. games played over the unit square [79] and stable sets [126] which are discussed thoroughly by Karlin and Von Neumann–Morgenstern resp.). Thus the material selected bears marks of the subjective judgment of the authors. In addition, numerous significant publications have been left out of consideration because they were not available for us or escaped our attention. For these shortcomings the reader must have our apologies.

Two main topics are entirely omitted. We do not deal with differential games. There are good books solely devoted to the subject (see e.g. [76]). Games with an infinite number of players are also left out since generality assured by n-person games seems to meet most of the demand of practice and theory of other sciences alike.

Only part of the abundant bibliography has been directly used for the preparation of the book. The rest is aimed at providing information on the subject covered by the book and may help to encourage further studies.

We hope that our book will be of use for those interested in game theory and who have the not too high-level mathematical background to understand it.

ACKNOWLEDGEMENTS

We would like to express our deep thanks to Dr. Ferenc Szidarovszky whose contribution to the material of this book is very significant. His suggestions and comments are an integral part of the final product. Special thanks go to Dr. N. N. Vorobyev whose constructive criticism helped a great deal to improve the original manuscript. We are also grateful to Mrs. Éva Németh for her patient and very competent secretarial assistance.

Introduction

Basic notions and problems of the theory of games – just as in other branches of mathematics – are deeply rooted in everyday life and in the theory and practice of various sciences. These fundamental notions are rather complex. They are not as easy to interpret as, e.g., those of classical geometry: point, straight line, plane etc. It is therefore necessary to deal with the fundamentals of game theory in a "qualitative way" before giving the exact mathematical definitions.

We shall speak about two kinds of games: games of chance and strategic games. In games of chance none of the players can control the outcome of the game (e.g., dice). In strategic games the players can influence the result of the game. Examples of strategic games are: chess, most card games and business games.

The distinction between games of chance and strategic games seem to be necessary in spite of the fact that in the theory of games "random mechanisms" can be incorporated as "players". In special cases each player is allowed to be a "random mechanism" with no strategic choice whatever. However, theory of games is concerned with strategic games.

Every game is, in fact, a conflict situation. The interests of the players are generally different. Any player is supposed to pursue his own goals in the game. Since the players have more or less opposing interests, a conflict does arise.

Depending on the number of players we speak of two, three or n-person games ($n \geqq 3$). Even games with an infinite number of players can be treated by suitable mathematical tools. By "person" we do not necessarily mean an "individual": it can be a group of people, a company or even a "random mechanism". This way – in a sense – we can consider strategic games where man plays against nature.

In an actual game a number of situations arise and the players should decide how to proceed. E.g., in chess, any time a player is to make a move, a concrete situation realizes. In all possible situations the behaviour and actions of a player are determined by his strategy. Thus strategy is a plan of action telling the player in any conceivable situation what to do. A strategy also refers to situations which actually never come up in a particular play of the game but which theoretically can occur. Generally the players are supposed to choose their strategies before starting to play. The strategy choice is, of course, affected by any prior knowledge acquired about the game.

A player can usually choose from several strategies. The set of all possible strategies available for him is called the *strategy set*. If the strategies chosen by the players determine the outcome of the game uniquely the game is said to be *deterministic*. If the outcome is determined only up to a probability distribution (which is dependent on the strategies chosen), then the game is called *stochastic*.

The players generally are interested in the outcome of the game. Primarily this fact motivated the development of the mathematical theory of games since in order to assure themselves favourable outcomes the players should apply scientific methods when confronting the problem of strategy selection. We shall always assume that the possible outcomes of a game form an ordered set for each player. In other words, for each pair of outcomes K, K' exactly one of the relations $K<K'$, $K=K'$, $K>K'$ holds for any player. $K<K'$ is usually interpreted as: K' is more "favourable" than K for a particular player whereas $K=K'$ means that K and K' are "equally favourable". It should be kept in mind that notations $<$, $=$, $>$ refer only to an ordering which is not necessarily numerical.

Although some of the basic problems and general theorems of the theory of games can be explicated by assuming only that the set of outcomes is completely ordered for each player, we will confine ourselves to a lower level of generality in our book. In particular, each outcome of a game will be characterized by a real number.

The real-valued function defining an ordering on the set of outcomes is called a *pay-off function.* Since the outcome of a game is supposed to depend (in a deterministic or stochastic way) on the strategy choice of the players we denote the pay-off functions by $K_i(\sigma_1, \ldots \sigma_n)$ where $\sigma_i \in \Sigma_i$, $(i=1, \ldots n)$ is the strategy used by the ith player and Σ_i is the set of his strategies.

Each player is assumed to have partial or complete information (at least in principle) about the strategy sets and pay-off functions of all players. Except for cooperative games the players pick their strategies independently, i.e., none of them knows beforehand what the others are up to.

If the players are assumed to be "rational" (this term also deserves some explanation but we accept its everyday meaning) and they know that all the others behave rationally, too, then they have to seek "security" as to the outcome of the game. This requires the selection of a "safety" strategy. To move away from this strategy is not advantageous for any of the players provided the other stick to their strategies. The collection of safety strategies is called *the equilibrium point of the game.*

Of course things are different if some of the players do not have complete information about the game (about the strategy sets, pay-off functions). In this case, those having full information about the game can deviate from their "safety" strategies hoping to reach higher pay-off (of course, taking the risk of being worse off with a certain probability compared to the outcome assured by the "safety" strategy).

In an important class of games – we call them *cooperative games* – groups of players are allowed to form coalitions. Then the coalition acts as a single player in the game. The strategy set of a coalition is generally the Cartesian product of the strategy sets of the participants and the pay-off function is the sum of the individual pay-off functions. In cooperative games the problem of coalition formation and fair division of gains are the most interesting and intriguing questions.

While games of chance are treated primarily by methods of probability theory, the tools used for the analysis of strategic games come from several branches of mathematics. To be able to read the book some knowledge of probability theory, calculus, linear algebra and mathematical programming is needed. These disciplines have developed independently from the theory of games and are subjects of standard university courses. Only the average university level is required to understand the book.

Throughout the text we use standard mathematical notation denoting vectors and matrices by bold type letters **a**, **A**. The source (not necessarily the original) of each theorem is indicated by reference to an item in the bibliography.

1. On equilibrium of systems

In the past 10–15 years a new branch of mathematics, mostly referred to as mathematical systems theory, has been developing very fast. It aims at integrating several special mathematical disciplines such as automata theory, theory of dynamic systems, control theory and most recently theory of games, as well. Though these fields have a wide range of applications and their own well established theory, yet a comprehensive treatment might reveal new aspects of the subject matter. As an effort to do this the first chapter of our book is intended to put game theory in a system theoretical framework.

1.1. BASIC IDEAS

Let us define a system S by the n-tuple

$$S=\{S_1, \dots, S_n\}$$

where the S_i $(i=1, \dots, n)$ are called *organs*. Denote by X_i, $(i=1, \dots, n)$ metric spaces with denumerable bases and distance functions denoted by ρ_i. The set

$$\Sigma_i \subseteqq X_i$$

is called the state set of organ S_i $(i=1, \dots, n)$. The product set

$$\Sigma=\Sigma_1 \times \dots \times \Sigma_n$$

$(\Sigma \subseteqq X_1 \times \dots \times X_n)$ is the *state set of system* S. The metric on $X=$

$=X_1 \times X_2 \times \ldots \times X_n$ is defined by the distance function ρ which in addition to the usual axioms satisfies the following condition: if

$$\mathbf{x}=(x_1, \ldots, x_{i-1}, x_i, x_{i+1}, \ldots, x_n)$$

$$\mathbf{x}'=(x_1, \ldots, x_{i-1}, y_i, x_{i+1}, \ldots, x_n),$$

then

$$\rho(\mathbf{x}, \mathbf{x}')=\rho_i(x_i, y_i), \quad (i=1, \ldots, n)$$

A set L is also given

$$L \subseteqq \Sigma$$

which we call the *set of admissible states* for system S.

The functions

$$F_1, \ldots, F_n$$

mapping the set Σ into a well-ordered set Y are called *preference functions of the organs.* (In most cases Y is the real line. Unless otherwise stated $Y=\mathbb{R}$.) The composite function

$$F(F_1, \ldots, F_n)$$

mapping Σ into Y is said to be the preference function of the system S.

The point-to-set mappings

$$\Phi_1, \ldots, \Phi_n$$

where $\Phi_i(\mathbf{x})$ is a nonempty subset of Σ_i for each $\mathbf{x}=(x_1, \ldots, x_n) \in \Sigma$, are called the *neighbourhood functions* of organs $S_1, \ldots, S_n$. The neighbourhood functions are assumed to satisfy the following conditions:

(a)

$$x_i \in \Phi_i(\mathbf{x}); \ \mathbf{x}=(x_1, \ldots, x_n) \in L, \quad (i=1, \ldots, n),$$

(b)

$$\text{if} \quad \mathbf{x}=(x_1, \ldots, x_{i-1}, x_i, x_{i+1}, \ldots, x_n) \in L$$

$$\text{and} \quad \mathbf{x}'=(x_1, \ldots, x_{i-1}, x_i', x_{i+1}, \ldots, x_n) \in \Sigma ,$$

then $x_i' \in \Phi_i(\mathbf{x})$ implies $\mathbf{x}' \in L$, $(i=1, \ldots, n)$.

Let

$$\Phi(\mathbf{x})=\Phi_1(\mathbf{x}) \times \ldots \times \Phi_n(\mathbf{x}), \quad (\mathbf{x} \in \Sigma) .$$

$\Phi(\mathbf{x})$ is said to be the neighbourhood function of system S.

By properties (a) and (b) $\mathbf{x} \in \Phi(\mathbf{x})$ for each $\mathbf{x} \in L$and if $\mathbf{x} \in L$ and $\mathbf{x}' \in \Sigma$ differ only in their ith components, then $\mathbf{x}' \in \Phi(\mathbf{x})$ implies $\mathbf{x}' \in L$.

To each organ we assign *cost functions*

$$C_1, \ldots, C_n$$

where C_i maps $\Sigma \times \Sigma$ into Y, it is nonnegative and $C_i(\mathbf{x}, \mathbf{x})=0$ for each $\mathbf{x} \in \Sigma$, $(i=1, \ldots, n)$. The composite function $C(C_1, \ldots, C_n)$ mapping $\Sigma \times \Sigma$ into $Y$is called the *cost function of system S*.

To sum up, system S is defined by

1. state sets $\Sigma_1, \ldots, \Sigma_n$,
2. preference functions $F_1, \ldots, F_n, F$,
3. neighbourhood functions $\Phi_1, \ldots, \Phi_n$,
4. cost functions $C_1, \ldots, C_n, C$,
5. set of admissible states L.

Thus we shall denote system S by the symbol

$$S=\{\Sigma_1, \ldots, \Sigma_n; \quad F_1, \ldots, F_n, F;$$
$$\Phi_1, \ldots, \Phi_n, C_1, \ldots, C_n, C, L\} .$$

The functions F_i, Φ_i, C_i are not necessarily independent. Nevertheless to keep things relatively simple we assume them to be independent.

To illustrate the above ideas we give an economic interpretation. Let the economic system S consist of companies S_i, $(i=1, \ldots, n)$. Each company can be characterized by a state vector at any moment of time. (The components of this vector can be: labour force, fixed assets, turnover capital, profit etc.). There is a set of feasible (at least in principle) states for each company Σ_i. Out of these theoretically feasible states only those can come true which are in compliance with the states attained by the other companies at the given time. This way we get the set of admissible states L. The companies evaluate their state by the preference functions F_i. The evaluation can depend on the other companies' states, as well. This is why F_i is a function of $\mathbf{x} \in \Sigma$ not only of x_i. The dependence of a company on the others is not absolute, thus within certain bounds they can change their states. This is described by the neighbourhood functions Φ_i i.e. given $\mathbf{x} \in L$, company S_i can "move about" only in $\Phi(\mathbf{x})$. The cost functions measure the cost incurred when changing a state.

1.2. CHAINS AND TRAVERSABLE REGIONS

Before dealing with equilibrium points and equilibrium sets we have to give a few definitions. Let $\mathbf{x}, \mathbf{y} \in \Sigma$. If $\mathbf{x}$ and $\mathbf{y}$ differ in at most one component (say in the ith one, $(i = 1, \ldots, n)$), they are said to be *comparable*. This relation is denoted by

$$\mathbf{x} \overset{i}{\sim} \mathbf{y} .$$

Clearly, this is not an equivalence relation. If $\mathbf{x} \overset{i}{\sim} \mathbf{y}$ and $y_i \in \Phi_i(\mathbf{x})$ (or equivalently $\mathbf{y} \in \Phi(\mathbf{x})$), then $\mathbf{y}$ is said to be *directly accessible* from $\mathbf{x}$ and is denoted by

$$\mathbf{x} \overset{i}{\to} \mathbf{y} .$$

For any $\mathbf{x}$ and i we have $\mathbf{x} \overset{i}{\to} \mathbf{x}$ but $\mathbf{x} \overset{i}{\to} \mathbf{y}$ does not necessarily imply $\mathbf{y} \overset{i}{\to} \mathbf{x}$ and the relation is not transitive. If $\mathbf{x} \in L$ and $\mathbf{x} \overset{i}{\to} \mathbf{y}$, then by the definition of the neighbourhood function $\mathbf{y} \in L$ follows.

Let a finite or infinite sequence of the elements of Σ be given

$$\alpha : \mathbf{x}_1, \mathbf{x}_2, \ldots, \mathbf{x}_n, \ldots .$$

This sequence is called a *chain* if any member of it is directly accessible from its predecessor i.e. there is a sequence $\{i_m\}$, $(1 \leqq i_m \leqq n)$ such that

$$\mathbf{x}_m \overset{i_m}{\rightarrow} \mathbf{x}_{m+1}, \qquad (m = 1, 2, \ldots).$$

If $\mathbf{y}$ is a member of a chain α emanating from $\mathbf{x}$, then $\mathbf{y}$ is said to be accessible from $\mathbf{x}$ by the chain α and is denoted by

$$\mathbf{x} \overset{\alpha}{\rightarrow} \mathbf{y}.$$

Apparently, this relation is not reversible either, but it is transitive. Furthermore, if $\mathbf{x} \in L$, then any $\mathbf{y}$ accessible by a chain from $\mathbf{x}$ also belongs to L.

Let $\mathbf{x}, \mathbf{y} \in \Sigma$ and $\mathbf{x} \overset{i}{\rightarrow} \mathbf{y}$. If

$$F_i(\mathbf{x}) \leqq F_i(\mathbf{y}) - C_i(\mathbf{x}, \mathbf{y}),$$

then $\mathbf{y}$ is said to be an improvement of $\mathbf{x}$ with respect to i. If

$$F_i(\mathbf{x}) < F_i(\mathbf{y}) - C_i(\mathbf{x}, \mathbf{y}),$$

then $\mathbf{y}$ is a strict improvement of $\mathbf{x}$ with respect to i. These relations will be denoted by

$$\mathbf{x} \overset{i}{\leqq} \mathbf{y}$$

and

$$\mathbf{x} \overset{i}{<} \mathbf{y}$$

respectively. Relations $\mathbf{x} \overset{i}{\leqq} \mathbf{y}$ and $\mathbf{x} \overset{i}{<} \mathbf{y}$ are not transitive either.

Let the sequence $\{\mathbf{x}_m\}$ be a chain in Σ. If $\mathbf{x}_m$ is an improvement (strict improvement) of $\mathbf{x}_{m-1}$ for $m = 2, 3, \ldots$, then the chain is said to be monotone (strictly monotone). Given a point $\mathbf{x} \in \Sigma$, denote by $\tau_{\leqq}(\mathbf{x})$ $(\tau_{<}(\mathbf{x}))$ the set of points accessible by monotone (strictly monotone) chains from $\mathbf{x}$. The set $\tau_{\leqq}(\mathbf{x})$ is called the *traversable region* while the set $\tau_{<}(\mathbf{x})$ is called the *strict traversable region* of $\mathbf{x}$.

The traversable regions $\tau_{\leqq}(\mathbf{x}), \tau_{<}(\mathbf{x})$ are said to be proper if they are proper subsets of Σ.

It follows from the previous definitions that if $\mathbf{x} \in L$, then

$$\tau_{\leqq}(\mathbf{x}) \subseteqq L$$

$$\tau_{<}(\mathbf{x}) \subset L.$$

Furthermore, if $\mathbf{y} \in \tau_{\leqq}(\mathbf{x})$ and $\mathbf{y}' \in \tau_{<}(\mathbf{x})$, then

$$\tau_{\leqq}(\mathbf{y}) \subseteqq \tau_{\leqq}(\mathbf{x})$$

$$\tau_{<}(\mathbf{y}') \subset \tau_{<}(\mathbf{x}).$$

Obviously

$$\tau_{<}(\mathbf{x}) \subset \tau_{\leqq}(\mathbf{x})$$

for any $\mathbf{x} \in \Sigma$.

Finally, the neighbourhood function Φ_i is said to be *uniformly connected* on the set Σ if there exists a positive δ such that if $\mathbf{x} \overset{i}{\sim} \mathbf{y}$ and $\rho(\mathbf{x}, \mathbf{y}) < \delta$, then $x_i \in \Phi_i(\mathbf{y})$ and $y_i \in \Phi(\mathbf{x})$.

1.3. EQUILIBRIUM POINT, STABILITY SET, EQUILIBRIUM SET

First we define an equilibrium point which plays central role in the theory of games.

A point $\mathbf{x}^* \in L$ is said to be a (Nash) *equilibrium point* if its strict traversable region is the empty set.

A set $H_{\leqq} \subseteqq L$ is called a *stability set* of system S if for any $\mathbf{x} \in H_{\leqq}$

$$[\tau_{\leqq}(\mathbf{x})] = H_{\leqq}.$$ [1]

Analogously, the set $H_{<} \subseteqq L$ is defined to be a *strict stability set* of S if for any $\mathbf{x} \in H_{<}$

$$[\tau_{<}(\mathbf{x})] = H_{<}.$$

[1] $[A]$ denotes the closure of set A.

A set $H^*_{\leqq} \subsetneqq \Sigma$ is said to be an *equilibrium set* of system S if for any $\mathbf{x} \in H^*_{\leqq}$

$$\tau_{\leqq}(\mathbf{x}) = H^*_{\leqq} ,$$

whereas $H^*_{<} \subsetneqq \Sigma$ is called a *strict equilibrium set* of system S if for any $\mathbf{x} \in H^*_{<}$ we have

$$\tau_{<}(\mathbf{x}) = H^*_{<} .$$

The stability and equilibrium sets $H_{\leqq}$, $H_{<}$, $H^*_{\leqq}$, $H^*_{<}$ are called *proper* if $H_{\leqq}$, $H_{<}$ and $H^*_{\leqq}$, $H^*_{<}$ are proper subsets of L and Σ resp. Stability sets are closed by definition but this does not necessarily hold for equilibrium sets. It also follows from the definitions that if for any $\mathbf{x}$ the traversable and strict traversable regions are closed, then any stability and strict stability set is an equilibrium or strict equilibrium set resp. Moreover, if a stability or strict stability set contains a point whose traversable or strict traversable region is closed, then the corresponding stability or strict stability set is an equilibrium (strict equilibrium) set, as well.

1.4. "EQUILIBRIUM PROPERTIES" OF EQUILIBRIUM POINTS AND EQUILIBRIUM SETS

It is obvious that an equilibrium point, consisting of one element, is an equilibrium set, as well. If for some $\mathbf{x} \in H_{\leqq}$, $H_{<}$ the traversable and strict traversable sets of $\mathbf{x}$ are closed, then stability sets and strict stability sets are equilibrium and strict equilibrium sets resp.

It can easily be seen that

(a) the stability and equilibrium sets are "minimal" in the sense that none of their proper subsets is a stability or equilibrium set resp.,

(b) if A and B are stability (equilibrium) sets such that $A \neq B$, then $A \cap B = \emptyset$.

Let now $\mathbf{x}^* = (x_1^*, \ldots, x_n^*)$ be an equilibrium point of S. It follows from our definition that for any i, $(i = 1, \ldots, n)$ and for any

$$\mathbf{x}^{(i)} = (x_1^*, \ldots, x_{i-1}^*, x_i, x_{i+1}^*, \ldots, x_n^*)$$

where $x_i \in \Phi_i(\mathbf{x}^*)$ we have

$$F_i(\mathbf{x}^*) \geqq F_i(\mathbf{x}_i^{(i)}) - C_i(\mathbf{x}^*, \mathbf{x}^{(i)}) .$$

In other words if $n-1$ members of system S are already "in an equilibrium state", then the remaining organ cannot do better but also attain state x_i^*.

We shall show that this property carries over to equilibrium sets as well. Let $H_{\leqq}^*$ be an equilibrium set of system S

$$H_{\leqq}^* = \{H_1^*, \ldots, H_n^*\} .$$

Let $x_i^* \in H_i^*$, $(i = 1, \ldots, n)$ which means that the system has reached an equilibrium state. The ith organ is allowed to move about only in $\Phi_i(\mathbf{x}^*)$. Since $\tau_{\leqq}(\mathbf{x}^*) = H_{\leqq}^*$, any change in x_i^* resulting in a state not worse for the ith organ than x_i^* is necessarily an element of $\Phi_i(\mathbf{x}^*) \subseteqq H_i^*$. This means that once the system got somehow into an equilibrium set, "individually profitable" moves will not lead out of this set. Thus system S can just as well be regarded as having reached "equilibrium".

1.5. ON THE EXISTENCE OF AN EQUILIBRIUM POINT

Before dwelling upon the questions concerning the existence of stability and equilibrium sets we give a sufficient condition for the existence of an equilibrium point.

THEOREM 1. [177] If the following six conditions are satisfied, then system $S=\{\Sigma_1, \ldots, \Sigma_n; F_1, \ldots, F_n; \Phi_1, \ldots, \Phi_n; C_1, \ldots, C_n; L\}$[1] has at least one equilibrium point.

(a) The sets Σ_i are bounded subsets of the m_i-dimensional Euclidean space R^{m_i}, $(i=1, \ldots, n)$.

(b) The functions F_i are concave functions of $\mathbf{x}_i$ on Σ_i for each fixed vector $(\mathbf{x}_1, \ldots, \mathbf{x}_{i-1}, \mathbf{x}_{i+1}, \ldots, \mathbf{x}_n)$, $(i=1, \ldots, n)$.

(c) The functions F_i are continuous functions of $\mathbf{x}=(\mathbf{x}_1, \ldots, \mathbf{x}_n)$ on Σ, $(i=1, \ldots, n)$.

(d) L is a closed, convex subset of Σ.

(e) The cost functions $C_i(\mathbf{x}, \mathbf{y})$ are continuous, nonnegative convex functions of $\mathbf{y}_i$ for any fixed $\mathbf{x}$ and $(\mathbf{y}_1, \ldots, \mathbf{y}_{i-1}, \mathbf{y}_{i+1}, \ldots, \mathbf{y}_n)$, $(i=1, \ldots, n)$.

(f) For any $\mathbf{x} \in L$ the set $T_{\mathbf{x}}=\{\mathbf{y} \in L \mid \mathbf{x} \in \Phi(\mathbf{y})\}$ is either the whole set L or an open subset of L.

Proof.[2] Let $\mathbf{x}=(\mathbf{x}_1, \ldots, \mathbf{x}_n)$, $\mathbf{y}=(\mathbf{y}_1, \ldots, \mathbf{y}_n)$ vary independently in L. By our assumptions

$$G(\mathbf{x}, \mathbf{y})=\sum_{i=1}^{n} [F_i(\mathbf{y}_1, \ldots, \mathbf{x}_i, \ldots, \mathbf{y}_n) - C_i(\mathbf{y}, \mathbf{y}_1, \ldots, \mathbf{x}_i, \ldots, \mathbf{y}_n)]$$

is a continuous function of $\mathbf{x}$, $\mathbf{y}$ and is concave in $\mathbf{x}_i$.

We shall first show that if there exists a vector $\mathbf{y}^*=(\mathbf{y}_1^*, \ldots, \mathbf{y}_n^*) \in L$ such that for any $\mathbf{x} \in \Phi(\mathbf{y}^*)$ the inequality

$$(1) \qquad G(\mathbf{y}^*, \mathbf{y}^*) \geqq G(\mathbf{x}, \mathbf{y}^*)$$

holds, then $\mathbf{y}^*$ is an equilibrium point of S. Inequality (1) should hold for any $\bar{\mathbf{y}}^{(i)}=(\mathbf{y}_1^*, \ldots, \mathbf{x}_i, \ldots, \mathbf{y}_n^*) \in \Phi(\mathbf{y}^*)$, $(i=1, \ldots, n)$.

[1] We assume here that the preference and cost functions F, C of the whole system S are identically 0.

[2] The proof goes along the lines of the proof for the Nikaido-Isoda theorem in [31].

Putting $\mathbf{x}=\bar{\mathbf{y}}^{(i)}$ into (1) we have for any $\mathbf{x} \in \Phi(\mathbf{y}^*)$

$$F_i(\mathbf{y}_1^*, \ldots, \mathbf{y}_n^*) \geqq F_i(\mathbf{y}_1^*, \ldots, \mathbf{x}_i, \ldots, \mathbf{y}_n^*) - $$
$$- C_i(\mathbf{y}^*, \mathbf{y}_1^*, \ldots, \mathbf{x}_i, \ldots, \mathbf{y}_n^*),$$
$$(i=1, \ldots, n)$$

which means that $\mathbf{y}^*$ is really an equilibrium point of S.[1]

Now we proceed by using an indirect proof, i.e., we suppose that there is no vector $\mathbf{y}^* \in L$ satisfying (1). In other words, to any $\mathbf{y} \in L$ there can be found an $\mathbf{x} \in \Phi(\mathbf{y})$ for which $G(\mathbf{y}, \mathbf{y}) < G(\mathbf{x}, \mathbf{y})$. Consider now the set

$$(2) \qquad H_{\mathbf{x}} = \{\mathbf{y} \in L \mid G(\mathbf{y}, \mathbf{y}) < G(\mathbf{x}, \mathbf{y})\}\,.$$

The sets $H_{\mathbf{x}}$ cover L entirely i.e.

$$L = \bigcup_{\mathbf{x} \in L} H_{\mathbf{x}}\,.$$

Since the function $G(\mathbf{x}, \mathbf{y})$ is continuous, the sets $H_{\mathbf{x}}$ are open, relative to L. Applying Borel's covering theorem, there exist finitely many vectors $\mathbf{x}^{(1)}, \ldots, \mathbf{x}^{(q)}$ such that $L = \bigcup_{j=1}^{q} H_{\mathbf{x}^{(j)}}$. Consider now the nonnegative functions

$$g_j(\mathbf{y}) = \max\{G(\mathbf{x}^{(j)}, \mathbf{y}) - G(\mathbf{y}, \mathbf{y}), 0\} \quad (j=1, \ldots, q)\,.$$

Since to any $\mathbf{y} \in L$ there is an $\mathbf{x} \in \Phi(\mathbf{y})$ such that $G(\mathbf{x}, \mathbf{y}) > G(\mathbf{y}, \mathbf{y})$, therefore to any $\mathbf{y} \in L$ there exists an index j, $(1 \leqq j \leqq q)$ for which $G(\mathbf{x}^{(j)}, \mathbf{y}) > G(\mathbf{y}, \mathbf{y})$. Thus the function $g(\mathbf{y}) = \sum_{j=1}^{q} g_j(\mathbf{y})$ is positive for any $\mathbf{y} \in L$. Hence the vector

$$\sum_{j=1}^{q} \frac{g_j(\mathbf{y})}{g(\mathbf{y})} \mathbf{x}^{(j)}$$

[1] $\mathbf{y}^*$ is a "local equilibrium point" i.e. (1) holds for any $\mathbf{x} \in \Phi(\mathbf{y}^*)$ but there might be a $\mathbf{z} \notin \Phi(\mathbf{y}^*)$ for which $G(\mathbf{y}^*, \mathbf{y}^*) \ngeq G(\mathbf{z}, \mathbf{y}^*)$.

belongs to L for any $\mathbf{y} \in L$ by the convexity of L. The functions $g(\mathbf{y})$, $g_j(\mathbf{y})$, $(j=1, \ldots, q)$ are obviously continuous; therefore the mapping

$$\mathbf{y} \to \sum_{j=1}^{q} \frac{g_j(\mathbf{y})}{g(\mathbf{y})} \mathbf{x}^{(j)}$$

is a continuous mapping of the closed, bounded convex set L into itself. By Brouwer's fixed point theorem there is a $\bar{\mathbf{y}} \in L$ for which

$$\bar{\mathbf{y}} = \sum_{j=1}^{q} \frac{g_j(\bar{\mathbf{y}})}{g(\bar{\mathbf{y}})} \mathbf{x}^{(j)}.$$

Since $G(\mathbf{x}, \bar{\mathbf{y}})$ is concave in $\mathbf{x}$, therefore

$$G(\bar{\mathbf{y}}, \bar{\mathbf{y}}) \geqq \sum_{j=1}^{q} \frac{g_j(\bar{\mathbf{y}})}{g(\bar{\mathbf{y}})} G(\mathbf{x}^{(j)}, \bar{\mathbf{y}}). \tag{3}$$

But this inequality does not hold since

(i) there is an index j, for which

$$G(\mathbf{x}^{(j)}, \bar{\mathbf{y}}) > G(\bar{\mathbf{y}}, \bar{\mathbf{y}}),$$

(ii) if $G(\mathbf{x}^{(j)}, \bar{\mathbf{y}}) < G(\bar{\mathbf{y}}, \bar{\mathbf{y}})$, then $g_j(\bar{\mathbf{y}}) = 0$ by definition.

(i) and (ii) together imply

$$\sum_{j=1}^{q} \frac{g_j(\bar{\mathbf{y}})}{g(\bar{\mathbf{y}})} G(\mathbf{x}^{(j)}, \bar{\mathbf{y}}) > \sum_{j=1}^{q} \frac{g_j(\bar{\mathbf{y}})}{g(\bar{\mathbf{y}})} G(\bar{\mathbf{y}}, \bar{\mathbf{y}}) = G(\bar{\mathbf{y}}, \bar{\mathbf{y}})$$

contradicting (3). ∎

Theorem 1 does assure the existence of an equilibrium point but the characterization of the points from which at least one monotone chain leads to an equilibrium point is still an unsolved problem.

1.6. ON THE EXISTENCE OF STABILITY SETS

As we saw in the previous section we had to impose rather strict conditions on the sets and functions defining a system in order to be able to guarantee the existence of an equilibrium point. It seems reasonable to expect that milder conditions are needed if we only want to get existence theorems on equilibrium and stability sets. In the following we give a sufficient condition for the existence of a strict stability set. If one ever tried to realize our model computationally only strict stability and equilibrium sets would be important because, by using floating point arithmetics equalities (in the strict mathematical sense) are defined only up to positive precision. Treating equilibrium sets involves more difficulties of a mathematical nature therefore we only deal with stability sets here.

First we give a few definitions needed for the theorems on the existence of strict stability sets.

Let $\mathbf{x} \overset{i}{\leqq} \mathbf{y}$ and

$$\mathbf{x}=(x_1, \ldots, x_{i-1}, x_i, x_{i+1}, \ldots, x_n)$$

$$\mathbf{y}=(x_1, \ldots, x_{i-1}, y_i, x_{i+1}, \ldots, x_n).$$

If there is a continuous function $t \to x(t)$, $(t \in [\alpha, \beta], x(t) \in \Sigma_i)$ of a real variable t such that $x(\alpha)=x_i$, $x(\beta)=y_i$ and for any $\alpha \leqq t_1 < t_2 \leqq \beta$ the inequality

$$\mathbf{x}(t_1) \overset{i}{\leqq} \mathbf{x}(t_2),$$

holds where

$$\mathbf{x}(t_1)=(x_1, \ldots, x_{i-1}, x(t_1), x_{i+1}, \ldots, x_n)$$

$$\mathbf{x}(t_2)=(x_1, \ldots, x_{i-1}, x(t_2), x_{i+1}, \ldots, x_n),$$

then we say that $\mathbf{y}$ is accessible from $\mathbf{x}$ by continuous improvement.

Analogously, if $\mathbf{x} \overset{i}{<} \mathbf{y}$ and

$$\mathbf{x}(t_1) \overset{i}{<} \mathbf{x}(t_2),$$

then $\mathbf{y}$ is accessible from $\mathbf{x}$ by continuous strict improvement.

We now prove the following:

THEOREM 2. [177] Let the sets Σ_i be nonempty bounded and closed, the functions F_i, C_i continuous and Φ_i uniformly connected for any i.

If for each $\mathbf{x} \in \Sigma$ any of its strict improvement is a continuous strict improvement, as well, then for all $\mathbf{y} \in [\tau_<(\mathbf{x})]$ the relation $\tau_<(\mathbf{y}) \subseteqq [\tau_<(\mathbf{x})]$ holds.

Proof. Let $\mathbf{x} \in \Sigma$ be arbitrary and $\mathbf{x}^* \in [\tau_<(\mathbf{x})]$. If $\mathbf{x}^* \in \tau_<(\mathbf{x})$, then the assertion of the theorem is obvious. Let therefore $\mathbf{x}^*$ be a boundary cluster point of $\tau_<(\mathbf{x})$ not belonging to $\tau_<(\mathbf{x})$. We have to show that

$$\tau_<(\mathbf{x}^*) \subseteqq [\tau_<(\mathbf{x})] .$$

Let us suppose on the contrary that there is a $\mathbf{y}^*$ satisfying

$$\mathbf{y}^* \in \tau_<(\mathbf{x}^*) \quad \text{and} \quad \mathbf{y}^* \notin [\tau_<(\mathbf{x})] .$$

This means that $\mathbf{y}^*$ is accessible from $\mathbf{x}^*$ by a strict monotone chain. Let $\mathbf{y}^*$ be the first member of the chain not belonging to $[\tau_<(\mathbf{x})]$. Let further

$$\mathbf{x}^* \overset{i_1}{<} \mathbf{y}_1 \overset{i_2}{<} \mathbf{y}_2 \overset{i_3}{<} \dots \overset{i_k}{<} \mathbf{y}_k \overset{i_*}{<} \mathbf{y}^*.$$

Since $\mathbf{y}^* \notin [\tau_<(\mathbf{x})]$, therefore $\mathbf{y}_i \notin \tau_<(\mathbf{x})$, $(i=1, \dots, k)$. Since $\mathbf{y}^*$ is the first member of the above chain not belonging to $[\tau_<(\mathbf{x})]$, all $\mathbf{y}_i$ are boundary cluster points of $[\tau_<(\mathbf{x})]$, $(i=1, \dots, k)$.

Let $\mathbf{y} = \mathbf{y}_k$. The point $\mathbf{y}$ is a boundary cluster point of $\tau_<(\mathbf{x})$ whose improvement $\mathbf{y}^*$ is not in $[\tau_<(\mathbf{x})]$. Now we are going to show that there exists a boundary cluster point $\mathbf{u}$ of $\tau_<(\mathbf{x})$ from which an improvement arbitrarily close to $\mathbf{u}$ and not belonging to $[\tau_<(\mathbf{x})]$ can be found.

Consider now the points $\mathbf{y}$ and $\mathbf{y}^*$ defined above. We have already seen $\mathbf{y} \overset{i_*}{<} \mathbf{y}^*$. By assumption, there exists a continuous function

$$t \to y_{i^*}(t), \quad (t \in [\alpha, \beta]; \quad y_{i^*}(t) \in \Sigma_i)$$

such that $\mathbf{y}(\alpha) = \mathbf{y}$, $\mathbf{y}(\beta) = \mathbf{y}^*$ and for any $\alpha \leqq t_1 < t_2 \leqq \beta$

$$\mathbf{y}(t_1) \overset{i_*}{<} \mathbf{y}(t_2) .$$

By definition

$$\mathbf{y}(\alpha)=\mathbf{y} \in [\tau_<(\mathbf{x})]$$

$$\mathbf{y}(\beta)=\mathbf{y}^* \notin [\tau_<(\mathbf{x})] .$$

Let

$$t^* = \sup \{t \mid t \in [\alpha, \beta] \quad \text{and} \quad \mathbf{y}(t) \in [\tau_<(\mathbf{x})]\} .$$

Since $[\tau_<(\mathbf{x})]$ is closed, therefore $\mathbf{y}(t^*) \in [\tau_<(\mathbf{x})]$ and $\mathbf{y}(t^*)$ is a boundary cluster point of $\tau_<(\mathbf{x})$. By the continuity of $\mathbf{y}(t)$ to any positive number δ^* there can be found a positive δ such that for any t satisfying $t > t^*$ and $t - t^* < \delta$ the following relations hold

$$\rho(\mathbf{y}(t), \mathbf{y}(t^*)) < \delta^*$$

$$\mathbf{y}(t) \notin [\tau_<(\mathbf{x})] .$$

By assumption, the functions Φ_i are uniformly connected, i.e., there exists a positive δ_1 such that $\mathbf{x} \xrightarrow{i} \mathbf{y}$ and $\rho(\mathbf{x}, \mathbf{y}) < \delta_1$ implies $y_i \in \Phi_i(\mathbf{x})$, $(i = 1, \dots, n)$.

As we showed previously there exists a boundary cluster point $\mathbf{z}$ of $\tau_<(\mathbf{x})$ (in fact $\mathbf{z} = \mathbf{y}(t^*)$) and a $\mathbf{z}^* \notin [\tau_<(\mathbf{x})]$ such that $\mathbf{z} \overset{i_*}{<} \mathbf{z}^*$ and $\rho(\mathbf{z}, \mathbf{z}^*) < \dfrac{\delta_1}{2}$. Since $\mathbf{z}$ is a cluster point of $\tau_<(\mathbf{x})$ there exists a sequence $\{\mathbf{z}^{(m)}\}$ such that $\mathbf{z}^{(m)} \to \mathbf{z}$ and $\mathbf{z}^{(m)} \in \tau_<(\mathbf{x})$, $(m = 1, 2, \dots)$. Let us introduce the notations[1]

$$\mathbf{z} = (z_1, \dots, z_{i-1}, z_i, z_{i+1}, \dots, z_n) ,$$

$$\mathbf{z}^* = (z_1, \dots, z_{i-1}, z_i^*, z_{i+1}, \dots, z_n) ,$$

$$\mathbf{z}^{(m)} = (z_1^{(m)}, \dots, z_{i-1}^{(m)}, z_i^{(m)}, z_{i+1}^{(m)}, \dots, z_n^{(m)}) ,$$

$$\mathbf{z}^{*(m)} = (z_1^{(m)}, \dots, z_{i-1}^{(m)}, z_i^*, z_{i+1}^{(m)}, \dots, z_n^{(m)}) .$$

[1] For the sake of simplicity we denote i^* simply by i.

Clearly, $\lim_{m\to\infty} z^{*(m)} = z^*$. For sufficiently large m we get the estimation

$$\rho(\mathbf{z}^{(m)}, \mathbf{z}^{*(m)}) = \rho_i(z_i^{(m)}, z_i^*) \leqq \rho_i(z_i^{(m)}, z_i) + \\ + \rho_i(z_i, z_i^*) < \frac{\delta_1}{2} + \frac{\delta_1}{2} = \delta_1 .$$

On the other hand $\mathbf{z}^{(m)} \not\stackrel{i}{\to} \mathbf{z}^{*(m)}$ for any m. Thus by the uniform connectedness of the neighbourhood functions Φ_i we have

$$\mathbf{z}^{(m)} \stackrel{i}{\to} \mathbf{z}^{*(m)}.$$

Now we are going to prove that $\mathbf{z}^{(m)} \stackrel{i}{<} \mathbf{z}^{*(m)}$ also holds. Let

$$F_i(\mathbf{z}^*) - F_i(\mathbf{z}) - C_i(\mathbf{z}, \mathbf{z}^*) = \delta_2 > 0 .$$

By the continuity of F_i, C_i, for sufficiently large m we get

$$| F_i(\mathbf{z}^{*(m)}) - F_i(\mathbf{z}^*) | < \frac{\delta_2}{4},$$

$$| C_i(\mathbf{z}^{(m)}, \mathbf{z}^{*(m)}) - C_i(\mathbf{z}, \mathbf{z}^*) | < \frac{\delta_2}{4} \quad (i = 1, \ldots, m).$$

$$| F_i(\mathbf{z}^{(m)}) - F_i(\mathbf{z}) | < \frac{\delta_2}{4}.$$

Now we have the estimation

$$F_i(\mathbf{z}^{*(m)}) - F_i(\mathbf{z}^{(m)}) - C_i(\mathbf{z}^{(m)}, \mathbf{z}^{*(m)}) = F_i(\mathbf{z}^{*(m)}) - \\ - C_i(\mathbf{z}^{(m)}, \mathbf{z}^{*(m)}) - [F_i(\mathbf{z}^*) - C_i(\mathbf{z}, \mathbf{z}^*)] + \\ + [F_i(\mathbf{z}^*) - F_i(\mathbf{z}) - C_i(\mathbf{z}, \mathbf{z}^*)] + \\ + [F_i(\mathbf{z}) - F_i(\mathbf{z}^{(m)})] = [F_i(\mathbf{z}^{*(m)}) - F_i(\mathbf{z}^*)] - \\ - [C_i(\mathbf{z}^{(m)}, \mathbf{z}^{*(m)}) - C_i(\mathbf{z}, \mathbf{z}^*)] + \\ + [F_i(\mathbf{z}^*) - F_i(\mathbf{z}) - C_i(\mathbf{z}, \mathbf{z}^*)] + \\ + [F_i(\mathbf{z}) - F_i(\mathbf{z}^{(m)})] > -\frac{3}{4}\delta_2 + \delta_2 = \frac{\delta_2}{4} > 0 .$$

Thus we have proved that $\mathbf{z}^{(m)} \overset{i}{<} \mathbf{z}^{*(m)}$ for sufficiently large m. This means that

$$\mathbf{z}^{*(m)} \in \tau_{<}(\mathbf{z}^{(m)}) \subseteqq \tau_{<}(\mathbf{x}) \subseteqq [\tau_{<}(\mathbf{x})] .$$

Since $[\tau_{<}(\mathbf{x})]$ is closed, we have

$$\lim_{m \to \infty} \mathbf{z}^{*(m)} = \mathbf{z}^* \in [\tau_{<}(\mathbf{x})]$$

which contradicts the assumption $\mathbf{z}^* \notin [\tau_{<}(\mathbf{x})]$.

It should be noted that the continuity of the functions $C_i(\mathbf{x}, \mathbf{y})$ was only used at points $\mathbf{x} \neq \mathbf{y}$. ▮

Now Theorem 2 enables us to prove the existence theorem of strict stability sets.

THEOREM 3. [177] Let the functions $F_i(\mathbf{x})$, $C_i(\mathbf{x}, \mathbf{y})$ be continuous $(\mathbf{x} \neq \mathbf{y})$; $\Phi_i(\mathbf{x})$ be uniformly connected $(i = 1, \ldots, n)$; and the sets Σ_i be nonempty, closed and bounded. If any strict improvement of each $\mathbf{x} \in \Sigma$ is a continuous strict improvement as well, then there exists at least one strict stability set in each $[\tau_{<}(\mathbf{x})]$ of system S.

Proof. Let $\mathbf{x} \in \Sigma$ be arbitrary. We shall show that among the subsets of $[\tau_{<}(\mathbf{x})]$ there exists a stability set. If $[\tau_{<}(\mathbf{x})]$ itself is not a stability set, then there is an $\mathbf{x}_1 \in [\tau_{<}(\mathbf{x})]$ such that

$$[\tau_{<}(\mathbf{x}_1)] \neq [\tau_{<}(\mathbf{x})] .$$

By Theorem 2, $\tau_{<}(\mathbf{x}_1) \subseteqq [\tau_{<}(\mathbf{x})]$ which implies $[\tau_{<}(\mathbf{x}_1)] \subseteqq [\tau_{<}(\mathbf{x})]$, since $[\tau_{<}(\mathbf{x})]$ is closed. Thus

$$[\tau_{<}(\mathbf{x}_1)] \subset [\tau_{<}(\mathbf{x})] .$$

If $[\tau_{<}(\mathbf{x}_1)]$ is not a strict stability set, then there is an $\mathbf{x}_2 \in [\tau_{<}(\mathbf{x}_1)]$ such that

$$[\tau_{<}(\mathbf{x}_2)] \subset [\tau_{<}(\mathbf{x}_1)] .$$

If after a finite number of steps we do not obtain a strict stability set, then we get an infinite sequence of nonempty closed, bounded sets

$$[\tau_<(\mathbf{x})] \supset [\tau_<(\mathbf{x}_1)] \supset [\tau_<(\mathbf{x}_2)] \supset \dots$$

If the intersection of these sets (the intersection is nonempty by Cantor's theorem) is not a strict stability set, then starting from an element $\mathbf{x}'_1$ of this set we repeat the above procedure. Thus we get the sequence

$$[\tau_<(\mathbf{x})] \supset [\tau_<(\mathbf{x}_1)] \supset \dots \supset [\tau_<(\mathbf{x}'_1)] \supset [\tau_<(\mathbf{x}'_2)] \supset \dots$$

Applying the Baire–Hausdorff theorem (see [3]) for this sequence there is an index r such that

$$[\tau_<(\mathbf{x}_r)] = [\tau_<(\mathbf{x}_{r+1})] = \dots,$$

which is a contradiction. Thus after a finite number of steps we have to reach a strict stability set. ∎

2. The n-person game

As we have seen in Chapter 1 several monotone chains may start out of each point of the state set Σ_i of system S. A monotone chain can just as well be regarded as a "state trajectory".

The n-person game can be conceived as a system consisting of n organs – now called players. A strategy σ_i of the ith player is a subset of state trajectories amenating from a particular point of Σ_i. The strategy set of the ith player is the set of all strategies belonging to any starting point of Σ_i.

In the following, instead of system we shall speak about *game*, the organs will be called *players* and we shall use the term *strategy* in the sense specified above. These terms are most commonly used in the literature of game theory.

In "classical" game theory only the strategy sets and preference functions (possibly the neighbourhood functions) play an important role. Therefore, movement along a trajectory may be regarded as a "one-step move" from the starting point to the final state. If we fix the starting point, then choosing a strategy amounts to picking an element σ_i of Σ_i for each player in an n-person game. Thus the preference functions F_i depend only on $\sigma_1, \ldots, \sigma_n$. Since we are not interested now how the players have reached states σ_i, the cost functions $C_1, \ldots, C_n$, C and the preference function of the whole system F are taken identically 0. Unless otherwise stated the strategy choices of the players are independent, therefore the set of admissible states (strategies) $L = \Sigma = \Sigma_1 \times \ldots \times \Sigma_n$.

To emphasize that we are dealing with games instead of systems, from now on we change the notation of preference functions F_i to K_i

and call them *pay-off functions* $(i=1, \ldots, n)$. Thus an n-person game Γ is given by the strategy sets $\Sigma_1, \ldots, \Sigma_n$, the pay-off functions $K_1, \ldots, K_n$ defined on Σ and by the neighbourhood functions $\Phi_1, \ldots, \Phi_n$ also defined on Σ. Formally

$$(1) \qquad \Gamma=\{\Sigma_1, \ldots, \Sigma_n; K_1, \ldots, K_n; \Phi_1, \ldots, \Phi_n\}.$$

We say that Γ is given now in *normal form* (as opposed to the extensive form we are going to deal with later on).

One of the most important problems in game theory is finding an equilibrium point. Just as for systems an n-tuple $(\sigma_1^*, \ldots, \sigma_n^*) \in \Sigma$ is said to be an *equilibrium point* if

$$K_i(\sigma_1^*, \ldots, \sigma_n^*) \geqq K_i(\sigma_1^*, \ldots, \sigma_{i-1}^*, \sigma_i, \sigma_{i+1}^*, \ldots, \sigma_n^*)$$

holds for any $\sigma_i \in \Phi_i(\sigma_1^*, \ldots, \sigma_n^*)$, $(i=1, \ldots, n)$.

If the strategy sets $\Sigma_1, \ldots, \Sigma_n$ are finite we speak of a *finite game*. If $\Sigma_1, \ldots, \Sigma_n$ are closed, bounded subsets of a finite dimensional Euclidean space, the game is said to be *continuous*. (The pay-off functions $K_1, \ldots, K_n$ are not necessarily continuous). The game $\Gamma=\{\Sigma_1, \ldots, \Sigma_n; K_1, \ldots, K_n; \Phi_1, \ldots, \Phi_n\}$ is called *zero-sum* if

$$\sum_{i=1}^{n} K_i(\sigma_1, \ldots, \sigma_n)=0$$

for each $\sigma_i \in \Sigma_i$, $(i=1, \ldots, n)$. In case of two-person zero-sum games the interests of the players are antagonistic, i.e.

$$K_2(\sigma_1, \sigma_2)=-K_1(\sigma_1, \sigma_2)$$

for any $\sigma_1 \in \Sigma$, $\sigma_2 \in \Sigma_2$.

In the following if we do not write out explicitly the neighbourhood functions $\Phi_1, \ldots, \Phi_n$ when defining a game Γ we automatically assume that $\Phi_i \equiv \Sigma_i$, $(i=1, \ldots, n)$.

Especially from a mathematical point of view, *mixed strategies*, first introduced by von Neumann, are of great importance. We give the definition of a mixed strategy for two important special cases.

1. The game $\Gamma = \{\Sigma_1, \ldots, \Sigma_n; K_1, \ldots, K_n\}$ is finite, i.e. $\Sigma_i = \{\sigma_i^{(1)}, \ldots, \sigma_i^{(N_i)}\}$, $(i = 1, \ldots, n)$. Let $\mathbf{x}_i = (x_i^{(1)}, \ldots, x_i^{(N_i)})$ be a probability vector[1], i.e.

$$\sum_{k=1}^{N_i} x_i^{(k)} = 1, \quad x_i^{(k)} \geqq 0, \qquad (i = 1, \ldots, n).$$

The vector $\mathbf{x}_i$ is called a *mixed strategy* of the ith player. This means that he plays strategy $\sigma_i^{(k)}$ with probability $x_i^{(k)}$. In this context the original strategies $\sigma_i^{(k)}$ are called *pure strategies.* Let $\bar{\Sigma}_i$ denote the set of all mixed strategies of the ith player. $\bar{\Sigma}_i$ is in fact a simplex in $\mathbb{R}^{N_i}$, $(i = 1, \ldots, n)$.

If the players are allowed to apply mixed strategies (they pick strategies randomly according to the probability distribution given by $\mathbf{x}_i$), then it seems reasonable to redefine the pay-off function as the expected value of the pay-offs $K_i(\sigma_1^{(k_1)}, \ldots, \sigma_n^{(k_n)})$, $(i = 1, \ldots, n)$. Thus by playing the mixed strategies $\mathbf{x}_1, \ldots, \mathbf{x}_n$ the new pay-off functions are

$$\bar{K}_i(\mathbf{x}_1, \ldots, \mathbf{x}_n) = \sum_{k_n=1}^{N_n} \cdots \sum_{k_n=1}^{N_1} K_i(\sigma_1^{(k_1)}, \ldots, \sigma_n^{(k_n)}) \, x_1^{(k_1)} \ldots x_n^{(k_n)}, \quad (i = 1, \ldots, n).$$

The game

$$\bar{\Gamma} = \{\bar{\Sigma}_1, \ldots, \bar{\Sigma}_n; \bar{K}_1, \ldots, \bar{K}_n\}$$

stemming from the original game $\Gamma = \{\Sigma_1, \ldots, \Sigma_n, K_1, \ldots, K_n\}$ is usually called the *mixed extension* of Γ.

2. The strategy sets $\Sigma_i = [a_i, b_i]$ are closed intervals of $\mathbb{R}$. Let $F_i(x)$ be a distribution function defined on $[a_i, b_i]$, i.e.

$$F_i(x) = \begin{cases} 0, & \text{if} \quad x \leqq a_i \\ 1, & \text{if} \quad x > b_i \end{cases} \qquad (i = 1, \ldots, n)$$

[1] We remark that stochastic interpretation of mixed strategies is only one (though the most usual) possibility.

$F_i(\sigma_i)$ is the probability of the ith player's choosing a strategy in the interval $[a_i, \sigma_i)$. Now a mixed strategy of the ith player is a distribution function $F_i(x)$, the strategy set $\bar{\Sigma}_i$ is the family of all distribution functions defined on $[a_i, b_i]$, $(i=1, \ldots, n)$. Accordingly, the pay-off function is given by the Stieltjes-integral

$$\bar{K}_i(F_1, \ldots, F_n) = \int_{a_n}^{b_n} \ldots \int_{a_1}^{b_1} K_i(\sigma_1, \ldots, \sigma_n)\, dF_1, \ldots, dF_n,$$

$$(i=1, \ldots, n).$$

The game $\bar{\Gamma} = \{\bar{\Sigma}_1, \ldots, \bar{\Sigma}_n; \bar{K}_1, \ldots, \bar{K}_n\}$ thus obtained is called the (continuous) *mixed extension of the original game* Γ.

The mixed extension of games, where the strategy sets are intervals of the $m_1, \ldots, m_n$ dimensional Euclidean space can be defined analogously.

Finally we define *strategic equivalence*. Games $\Gamma = \{\Sigma_1, \ldots, \Sigma_n; K_1, \ldots, K_n\}$ and $\Gamma' = \{\Sigma_1, \ldots, \Sigma_n; K'_1, \ldots, K'_n\}$ are said to be strategically equivalent if any equilibrium point of Γ is an equilibrium point of Γ', as well as vice versa.

3. Existence theorems of equilibrium points

Let a game Γ be given in normal form

$$\Gamma = \{\Sigma_1, \ldots, \Sigma_n; K_1, \ldots, K_n; \Phi_1, \ldots, \Phi_n\}.$$

As we have already seen an equilibrium point of the game Γ is an n-tuple of strategies $(\sigma_1^*, \ldots, \sigma_n^*)$, $(\sigma_i^* \in \Sigma_i, i=1, \ldots, n)$ for which

(a) $\sigma_i^* \in \Phi_i(\sigma_1^*, \ldots, \sigma_n^*)$, $\quad (i=1, \ldots, n)$,
(b) $K_i(\sigma_1^*, \ldots, \sigma_{i-1}^*, \sigma_i, \sigma_{i+1}^*, \ldots, \sigma_n^*) \leqq K_i(\sigma_1^*, \ldots, \sigma_n^*)$

for each $\sigma_i \in \Phi_i(\sigma_1^*, \ldots, \sigma_n^*)$, $(i=1, \ldots, n)$.

The equilibrium point thus defined is generally referred to as a (generalized) *Nash-equilibrium point*.

Very often we have to deal with games where the condensed information supplied by the normal form is not enough to describe the behaviour of the players, the rules of the game etc. in detail. As a first step to formalizing an actual game the so-called *extensive form* seems to be more adequate. An n-person game Γ is said to be given in extensive form if a directed rooted tree (finite or infinite) can be associated with it[1] by the following properties:

(a) The game starts at the root of the tree.

(b) To any node of the tree a player is assigned and the game proceeds to the vertex chosen by that particular player.

(c) Each player knows the vertices at which he has to make a decision.

[1] In the following – unless otherwise stated – we shall deal only with extensive games given by finite trees.

(d) There are given functions $f_1, \ldots, f_n$ assigning pay-offs to any player at each terminal node of the tree.

The game is said to be of *complete information* if each player when making his move knows exactly at which node he is and remembers perfectly how the game got to that particular node.

An additional "player", a "random mechanism" called "*chance*" may also participate in the game. If it is chance's turn to move, then the game gets to the next node according to a probability distribution. Chance is not a conscious player thus its behaviour is not governed by a strategy but by a probability distribution. For player i (except for "chance"), a strategy σ_i is a function defined on the nodes where he is to move and tells him how to proceed if the game ever gets to the particular node. Thus if chance is not involved in the game, then an n-tuple of strategies $(\sigma_1, \ldots, \sigma_n)$ uniquely determines a terminal node of the tree where each player gets $f_i(\sigma_1, \ldots, \sigma_n)$ as pay-offs. If chance is one of the players, then each terminal node V_j $(j=1, \ldots, N)$ is reached with some probability $p_j(\sigma_1, \ldots, \sigma_n)$. Obviously

$$(1) \qquad \sum_{j=1}^{N} p_j(\sigma_1, \ldots, \sigma_n) = 1 .$$

In this case the expected pay-offs

$$K_i(\sigma_1, \ldots, \sigma_n) = \sum_{j=1}^{N} f_i^{(j)} p_j(\sigma_1, \ldots, \sigma_n), \quad (i=1, \ldots, n)$$

are considered as pay-off functions, where $f_i^{(j)}$ denotes the pay-off of player i at the jth terminal node.

Let us now form subsets $H_1, \ldots, H_s$ of players in such a way that each player belongs to at least one H_i. Denoting by n_i the cardinality of H_i, $(i=1, \ldots, s)$ we have $\sum_{i=1}^{n} n_i \geqq n$. If $H_i \cap H_j = \emptyset$, $(i \neq j)$, then we speak about a *partition*. An n-tuple of strategies $(\sigma_1^*, \ldots, \sigma_n^*)$ is said to be a *group equilibrium point* if

$$(2)\qquad \sum_{j=1}^{n_i} K_{i_j}(\sigma_1^*, \ldots, \sigma_n^*) \geqq \sum_{j=1}^{n_i} K_{i_j}(\sigma_1^*, \ldots,$$
$$\ldots, \sigma_{i_1-1}^*, \sigma_{i_1}, \sigma_{i_1+1}^*, \ldots, \sigma_{i_j-1}^*, \sigma_{i_j}, \sigma_{i_j+1}^*, \ldots$$
$$\ldots, \sigma_{i_{n_i}-1}^*, \sigma_{i_{n_i}}, \sigma_{i_{n_i}+1}^*, \ldots, \sigma_n^*), \quad (i=1, \ldots, s).$$

If the players form disjoint groups, then finding a group equilibrium-point seems to be a more rational objective for them, than pursuing their goals independently. These groups can be considered as coalitions though the conditions and incentives of coalition forming are not subjects of our discussion now.

It is obvious that if all players form a group (coalition), then finding an equilibrium point amounts to maximizing the sum of the pay-off functions.

Generally if an n-person game Γ has an equilibrium point, then it does not necessarily have a group-equilibrium point for any disjoint partition of the players under the same conditions. However this situation cannot occur in case of n-person games with complete information given in extensive form.

THEOREM 1. [85], [178] Any n-person game with complete information representable by a finite rooted tree in extensive form has at least one group-equilibrium point for any disjoint partition $H_1, \ldots, H_s$ of the players.

Proof. From the root I of the tree F there leads exactly one route to any terminal node V_j, $(j=1, \ldots, N)$. Let $|I, V_j|$ denote the number of edges between I and V_j and define the length of the game $\Gamma=\{\Sigma_1, \ldots, \Sigma_n; K_1, \ldots, K_n\}$ by $h(\Gamma)=\max_j |I, V_j|$. We are going to prove our theorem by induction on $h(\Gamma)$.

In the case of $h(\Gamma)=0$, the assertion of the theorem is trivial, the root is the only vertex of the tree, and the "optimal" strategy of all players is "doing nothing".

Consider now the case $h(\Gamma)\geqq 1$. Let the vertices of the tree connected to the root I by an edge be $I_1, \ldots, I_m$. The vertices

$I_1, \ldots, I_m$ can be considered as roots of "subtrees" whose terminal vertices are some of those of the tree F. Consider now these subtrees $F_1, \ldots, F_m$ and the "subgames" $\Gamma^{(k)} = \{\Sigma_1^{(k)}, \ldots, \Sigma_n^{(k)}; K_1^{(k)} \ldots, K_n^{(k)}\}$, $(k=1, \ldots, m)$ determined by them. The pay-offs at terminal points of the subtrees F_k are those realizable in F and the players assigned to nonterminal vertices are the same in F_k $(k=1, \ldots, m)$ as in F.

Obviously $h(\Gamma^{(k)}) < h(\Gamma)$, $(k=1, \ldots, m)$ and by induction each game $\Gamma^{(k)}$, $(k=1, \ldots, m)$ has at least one group-equilibrium point. Let $\sigma_1^{*(k)}, \ldots, \sigma_n^{*(k)}$ be a a group-equilibrium point in game $\Gamma^{(k)}$. Thus by (2) we have

$$\sum_{j=1}^{n_i} K_{i_j}^{(k)}(\sigma_1^{*(k)}, \ldots, \sigma_n^{*(k)}) \geq \sum_{j=1}^{n_i} K_{i_j}^{(k)}(*, \ldots, \sigma_{i_1}^{(k)}, \ldots$$
$$\ldots, \sigma_{i_2}^{(k)}, *, \ldots, \sigma_{i_{n_i}}^{(k)}, \ldots, *)$$

for $i=1, \ldots, s$; $k=1, \ldots, m$. (For notational simplicity we have written out only the strategies without the asterisk $*$, the strategies with superscript $*$ are briefly denoted by $*$.)

If the player assigned to root I is "chance", then strategies $\sigma_i^* = \bigcup_{k=1}^{m} \sigma_i^{*(k)}$, $(i=1, \ldots, n)$ obviously form an equilibrium point in Γ.

Thus we have only to consider the case when a "real" player belongs to I. Let this player be i_0. His strategy set Σ_{i_0} in game Γ is the following: $\bigcup_{k=1}^{m} [\Sigma_{i_0}^{(k)} \cup \{\text{choose the } I\text{st edge emanating from } I\}$ $\cup \ldots \cup \{\text{choose the } m\text{th edge emanating from } I\}]$.

For the other players $\Sigma_i = \bigcup_{k=1}^{m} \Sigma_i^{(k)}$, $(i \neq i_0)$.

We are now able to construct equilibrium strategies for the game Γ. Let

$$\sigma_i^* = \bigcup_{k=1}^{m} \sigma_i^{*(k)}, \quad (i \neq i_0),$$

and let $\sigma_{i_0}^*$ be that strategy of the player i_0 (i.e. he chooses that edge emanating from I, say edge k_0,) for which $\sum_{i \in H} K_i^{(k_0)}(\sigma_1^*, \ldots, \sigma_n^*)$ is maximal. (H is the index set of the group containing i_0). By choosing the k_0-th edge player i_0 has passed over to subtree F_{k_0} and subgame $\Gamma^{(k_0)}$. Since there is an equilibrium point in Γ_{k_0} the way player i_0 chose edge k_0 assures that $\sigma_1^*, \ldots, \sigma_n^*$ is an equilibrium point of the game Γ. ∎

To games in an extensive form, with complete information which are representable only by an infinite tree the above existence theorem does not carry over as shown by Gale and Stewart [85].

Now we have to say a few words about games with *incomplete information.* Most card-games fall into this category.

The game Γ to be considered is given in extensive form represented by a finite tree–just as for games with complete information we partition the vertices of the tree into disjoint subsets $B_0, B_1, \ldots, B_n$ called *player sets.* At vertices belonging to B_0 it is chance's turn to move while at vertices of B_i, ($i=1, \ldots, n$) "real" players are to make a move. The player set B_i, ($i=1, \ldots, n$) is further partitioned into sets $B_i^{(j)}\left(\bigcup_j B_i^{(j)}=B_i\right)$ called *information sets.* The number of edges going out of each vertex of an information set is assumed to be the same and there is no route connecting two vertices within an information set. To each information set $B_i^{(j)}$ belongs an index set $I_i^{(j)}$. (If the number of edges going out of each vertex of $B_i^{(j)}$ is k, then $I_i^{(j)}$ consists of k of the numbers 0, 1, 2, $\ldots$, n allowing repetitions). From each vertex of $B_i^{(j)}$ the index set of players directly reachable by an edge is the same, i.e. $I_i^{(j)}$.

In a game with incomplete information every player knows in which information set he is but he does not know at which vertex. Of course, we obtain a game with complete information as a special case if each information set consists of exactly one vertex.

EXAMPLE. Consider the following two-person game given by the tree on Figure 1. Both players choose either number 1 or 2 at each vertex and move along the edge going to the left or to the right accordingly.

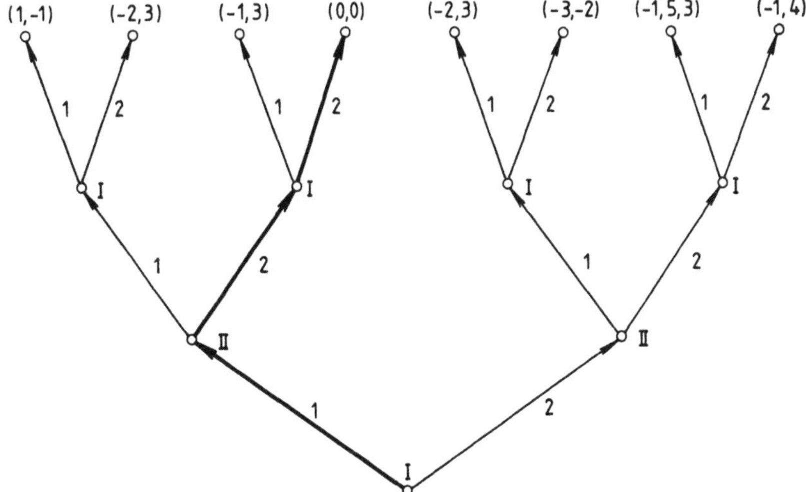

Figure 1

This game being of complete information has an equilibrium point by Theorem 1. It is easy to see that if both players apply their equilibrium strategies, then the game arrives at the terminal vertex with pay-off vector (0, 0). (The path leading from the root to this terminal vertex is marked by thick lines).

Suppose now that player I has forgotten his and player II's first move by the time he is to make his second move. Thus, he does not know at which of the encircled nodes of the tree he is. (See Figure 2. The encircled nodes form an information set.) Now player I has two alternatives: he either goes to the left or to the right.

The pay-off of which he can assure himself by going to the right is -3; while going to the left guarantees him -2. Thus player I's optimal strategy is to go to the left. Player II's optimal strategy is to move to the right no matter what player I's first move was and thus he gets at least 0. But provided player II goes to the right, player I is better off going to the right also, which means that the game does not have an equilibrium point. Writing the game as a bimatrix game (see Chapter 8) leads to the same conclusion.

More can be found about games in extensive form in [126] and [205].

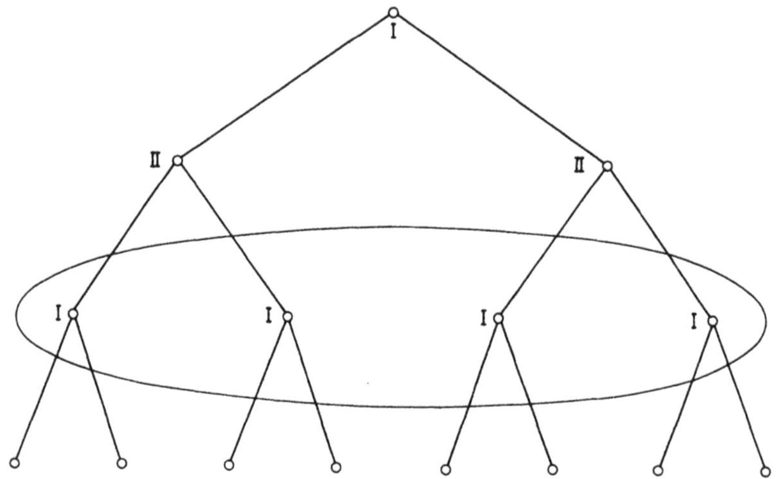

Figure 2

From now on we will study games given in normal form and establish existence theorems under various conditions imposed on the strategy sets and pay-off functions.

THEOREM 2. [31] Game $\Gamma = \{\Sigma_1, \ldots, \Sigma_n;\ K_1, \ldots, K_n;\ \Phi_1, \ldots, \Phi_n\}$ has an equilibrium point if the following conditions are satisfied:

(a) The strategy sets Σ_i are nonempty closed, bounded convex subsets of $\mathbb{R}^{n_i}$, $(i = 1, \ldots, n)$.

(b) K_i is a concave function of $\boldsymbol{\sigma}_i$ for fixed

$$\boldsymbol{\sigma}_1, \ldots, \boldsymbol{\sigma}_{i-1}, \boldsymbol{\sigma}_{i+1}, \ldots, \boldsymbol{\sigma}_n, \quad (i = 1, \ldots, n).$$

(c) K_i is a continuous function of $(\boldsymbol{\sigma}_1, \ldots, \boldsymbol{\sigma}_n)$, $(i = 1, \ldots, n)$.

(d) For any $\boldsymbol{\sigma} = (\boldsymbol{\sigma}_1, \ldots, \boldsymbol{\sigma}_n)$ the set $T_{\boldsymbol{\sigma}} = \{\boldsymbol{\sigma}' | \boldsymbol{\sigma} \in \Phi(\boldsymbol{\sigma}')$ is an open subset in $\Sigma = \Sigma_1 x \ldots x\Sigma_n$.

Proof. This theorem can be proved in exactly the same way as Theorem 1 of Chapter 1. ∎

REMARK. If $\Phi_i \equiv \Sigma_i$, then $\Gamma = \{\Sigma_1, \ldots, \Sigma_n; K_1 \ldots, K_n\}$ is a game in the "classical sense" and reduces to Theorem 1, the famous Nikaido–Isoda Theorem [31].

THEOREM 3. [31] The mixed extension of any finite n-person game has at least one equilibrium point.

Proof. As we have seen in Chapter 2, the pay-off functions in the mixed extension of a finite n-person game are multilinear, therefore continuous and concave in each variable. The strategy sets are bounded polyhedrons of finite dimensional Euclidean spaces. Thus all the conditions of Theorem 2 are met ($\Phi_i \equiv \Sigma_i$, $(i=1, \ldots, n)$) and therefore Theorem 3 holds. ∎

THEOREM 4. [178] The n-person game satisfying the conditions of Theorem 2 has a group-equilibrium point for any disjoint partition of the players if the pay-off function of each group is concave in the variables controlled by the members of the group.

Proof. Let $H_1, \ldots, H_s$ be a disjoint partition of the players. Denote the index of the players in H_i by $i_1, \ldots, i_{r_i}$ where r_i is the number of players in H_i.

Let $\Sigma_{H_i} = \mathop{\times}\limits_{k=1}^{r_i} \Sigma_{i_k}$ be the Cartesian product of the strategy sets of players in H_i and

$$K_{H_i}(\tilde{\sigma}_1, \ldots, \tilde{\sigma}_s) = \sum_{k=1}^{r_i} K_{i_k}(\sigma_1, \ldots, \sigma_n), \quad (i=1, \ldots, s)$$

the pay-off function. Here $\tilde{\sigma}_i = (\sigma_{i_1}, \ldots, \sigma_{i_{r_i}})$ is the joint strategy vector of the players in H_i. The assertions below readily follow from our assumptions.

(a) The sets Σ_{H_i}, $(i=1, \ldots, s)$ are nonempty closed, bounded and convex. (By condition (a) of Theorem 2.)

(b) The pay-off functions K_{H_i} are concave in $\tilde{\boldsymbol{\sigma}}_i$. (This has been assumed in our theorem we are proving now since condition (b) of Theorem 2 does not render it valid.)

(c) The functions K_{H_i} are continuous. (Implied by condition (c) of Theorem 2.)

(d) As (d) in Theorem 2 for the neighborhood functions $\bar{\Phi}_i = \bigcup_{k \in H_i} \Phi_k$, $(i=1, \ldots, s)$.

Then applying Theorem 2 to the s-person game

$$\Gamma' = \{\Sigma_{H_1}, \ldots, \Sigma_{H_s}; K_{H_1}, \ldots, K_{H_s}; \bar{\Phi}_1, \ldots, \bar{\Phi}_s\}$$

we can conclude that Γ' has an equilibrium point which is a group equilibrium point of the original game Γ. ∎

Now we are turning our attention to continuous games, in particular to mixed extensions of games where the pure strategies are closed intervals of $\mathbb{R}^1$. Investigation of concave games of a very general nature (concave n-person games defined over linear metric spaces) seems to be the most appropriate way to get existence theorems for mixed extensions of continuous games. Because of its generality the theorem stated and proved below is worthy[1] of attention.

Let $\Gamma = \{\Sigma_1, \ldots, \Sigma_n; K_1, \ldots, K_n\}$ be an n-person game, where the strategy sets Σ_j, $(j=1, \ldots, n)$ are convex subsets of linear, metric spaces L_j, the pay-off functions $K_j(\sigma_1, \ldots, \sigma_n)$ are concave and continuous in σ_j for each fixed n-1-tuple of strategies $(\sigma_1, \ldots, \sigma_{j-1}, \sigma_{j+1}, \ldots, \sigma_n)$, $\sigma_k \in \Sigma_k$, $(k=1, \ldots, n)$.

The strategy set Σ_k, $(k=1, \ldots, n)$ is said to be *weakly compact* with respect to the functions $K_1, \ldots, K_n$ if for any sequence $\sigma_k^{(1)}, \sigma_k^{(2)}, \ldots,$ of Σ_k there exists a subsequence $\sigma_k^{(i_1)}, \sigma_k^{(i_2)}, \ldots$ and a strategy $\sigma_k^{(0)} \in \Sigma_k$ such that

[1] This theorem can be generalized somewhat allowing K_i to be quasiconcave functions in condition (b) (see e.g. [58]).

$$K_l(\sigma_1, \ldots, \sigma_{k-1}, \sigma_k^{(i_j)}, \sigma_{k+1}, \ldots, \sigma_n) \to$$
$$\to K_l(\sigma_1, \ldots, \sigma_{k-1}, \sigma_k^{(0)}, \sigma_{k+1}, \ldots, \sigma_n)$$

uniformly on the set $\bigtimes_{\substack{t=1 \\ t \neq k}}^{n} \Sigma_t$ as $j \to \infty$ for each $l = 1, \ldots, n$.

THEOREM 5. [31], [179] If all strategy sets Σ_k, $(k=1, \ldots, n)$ are weakly compact, then the n-person game Γ with the properties specified above has at least one (Nash) equilibrium point.

Proof. Define the real-valued function Φ by

$$\Phi(\boldsymbol{\sigma}, \boldsymbol{\tau}) = \sum_{k=1}^{n} K_k(\sigma_1, \ldots, \sigma_{k-1}, \tau_k, \sigma_{k+1}, \ldots, \sigma_n)$$

for any $\sigma_k, \tau_k \in \Sigma_k$, where $\boldsymbol{\sigma} = (\sigma_1, \ldots, \sigma_n)$, $\boldsymbol{\tau} = (\tau_1, \ldots, \tau_n)$. We saw in the proof of Theorem 1 of Chapter 1 that if there exists an n-tuple of strategies $\boldsymbol{\sigma}^*$ for which $\Phi(\boldsymbol{\sigma}^*, \boldsymbol{\sigma}^*) \geqq \Phi(\boldsymbol{\sigma}^*, \boldsymbol{\tau})$ holds for any $\boldsymbol{\tau} \in \Sigma$, then $\boldsymbol{\sigma}^*$ is an equilibrium point of the game Γ. (Here Σ denotes the Cartesian product $\Sigma_1 \times \ldots \times \Sigma_n$.)

Suppose now that our theorem is not true. This means that to any vector $\boldsymbol{\sigma} \in \Sigma$ there exists a $\boldsymbol{\tau} \in \Sigma$ for which $\Phi(\boldsymbol{\sigma}, \boldsymbol{\sigma}) < \Phi(\boldsymbol{\sigma}, \boldsymbol{\tau})$. Define for any $\boldsymbol{\tau} \in \Sigma$ the set

$$H_{\boldsymbol{\tau}} = \{\boldsymbol{\sigma} \mid \boldsymbol{\sigma} \in \Sigma, \Phi(\boldsymbol{\sigma}, \boldsymbol{\sigma}) < \Phi(\boldsymbol{\sigma}, \boldsymbol{\tau})\}.$$

Clearly the sets $H_{\boldsymbol{\tau}}$ cover entirely Σ and by the continuity of the payoff functions they are open. We shall now show that finitely many out of the sets $H_{\boldsymbol{\tau}}$ also cover Σ. The proof of this assertion consists of two parts.

(a) Let $\boldsymbol{\sigma}^{(0)} = (\sigma_1^{(0)}, \ldots, \sigma_n^{(0)}) \in \Sigma$ be arbitrary and $\varepsilon > 0$ a fixed number. Define $E_\varepsilon(\boldsymbol{\sigma}^{(0)})$ to be the set of vectors $\boldsymbol{\tau} = (\tau_1, \ldots, \tau_n) \in \Sigma$ for which the inequality

$$|K_k(\sigma_1, \ldots, \sigma_{i-1}, \tau_i, \sigma_{i+1}, \ldots, \sigma_n) -$$
$$- K_k(\sigma_1, \ldots, \sigma_{i-1}, \sigma_i^{(0)}, \sigma_{i+1}, \ldots, \sigma_n)| < \varepsilon$$

holds for any $(\sigma_1, \ldots, \sigma_n) \in \Sigma$ and $i = 1, \ldots, n$; $k = 1, \ldots, n$.

We first show that for any $\varepsilon>0$ there exists a finite number of vectors $\boldsymbol{\sigma}^{(1)}, \ldots, \boldsymbol{\sigma}^{(r(\varepsilon))} \in \Sigma$ such that the sets $E_\varepsilon(\boldsymbol{\sigma}^{(l)})$, $(l=1, \ldots, r(\varepsilon))$ cover Σ. Assume on the contrary that there is an $\varepsilon_0>0$ for which the assertion does not hold. Let $\boldsymbol{\sigma}^{(1)} \in \Sigma$ be arbitrary. Since $E_{\varepsilon_0}(\boldsymbol{\sigma}^{(1)})$ does not cover Σ, there exists a vector $\boldsymbol{\sigma}^{(2)} \in \Sigma$ for which $\boldsymbol{\sigma}^{(2)} \in E_{\varepsilon_0}(\boldsymbol{\sigma}^{(1)})$. The sets $E_{\varepsilon_0}(\boldsymbol{\sigma}^{(1)})$ and $E_{\varepsilon_0}(\boldsymbol{\sigma}^{(2)})$ do not cover Σ either, thus there is a $\boldsymbol{\sigma}^{(3)} \in \Sigma$ such that $\boldsymbol{\sigma}^{(3)} \notin E_{\varepsilon_0}(\boldsymbol{\sigma}^{(1)}) \cup E_{\varepsilon_0}(\boldsymbol{\sigma}^{(2)})$. Proceeding in this fashion we get an infinite sequence $\boldsymbol{\sigma}^{(1)}, \boldsymbol{\sigma}^{(2)}, \ldots$ with the property

$$\boldsymbol{\sigma}^{(\nu)} \in \Sigma,\ \boldsymbol{\sigma}^{(\nu)} \in \bigcup_{i=1}^{\nu-1} E_{\varepsilon_0}(\boldsymbol{\sigma}^{(i)}), \quad (\nu=2, 3, \ldots).$$

Since the strategy sets Σ_k are weakly compact there exists a subsequence $\boldsymbol{\sigma}^{(\nu_1)}, \boldsymbol{\sigma}^{(\nu_2)}, \ldots$ of $\{\boldsymbol{\sigma}^{(\nu)}\}$ and a vector $\bar{\boldsymbol{\sigma}}^{(0)}=(\bar{\sigma}_1^{(0)}, \ldots, \bar{\sigma}_n^{(0)})$ to satisfy

$$\lim_{j\to\infty} K_l(\sigma_1, \ldots, \sigma_{k-1}, \sigma_k^{(\nu_j)}, \sigma_{k+1}, \ldots, \sigma_n)=$$
$$=K_l(\sigma_1, \ldots, \sigma_{k-1}, \bar{\sigma}_k^{(0)}, \sigma_{k+1}, \ldots, \sigma_n), \quad (l=1, \ldots, n)$$

uniformly on $\mathop{\times}\limits_{\substack{t=1\\ t\neq k}}^{n} \Sigma_i$ where $\sigma_k^{(\nu_j)}$ denotes the kth component of $\boldsymbol{\sigma}^{(\nu_j)}$.

By the continuity of the pay-off functions

$$\boldsymbol{\sigma}^{(\nu_j)} \in E_{\frac{\varepsilon_0}{2}}(\bar{\boldsymbol{\sigma}}^{(0)}) \quad \text{and} \quad \boldsymbol{\sigma}^{(\nu_{j+1})} \in E_{\frac{\varepsilon_0}{2}}(\bar{\boldsymbol{\sigma}}^{(0)})$$

for sufficiently large j. By the "triangle inequality" we get $\boldsymbol{\sigma}^{(\nu_{j+1})} \in E_{\varepsilon_0}(\boldsymbol{\sigma}^{(\nu_j)})$ contradicting the construction of the sequence $\{\boldsymbol{\sigma}^{(\nu)}\}$.

(b) We now prove the existence of an $\varepsilon_0>0$ such that for any $\boldsymbol{\sigma}^{(0)} \in \Sigma$ the set $E_{\varepsilon_0}(\boldsymbol{\sigma}^{(0)})$ can be covered by finitely many of the sets H_τ.

Assume on the contrary that for any $\varepsilon>0$ there exists a $\boldsymbol{\sigma}_\varepsilon \in \Sigma$ such that $E_\varepsilon(\boldsymbol{\sigma}_\varepsilon)$ cannot be covered by finitely many of the sets H_τ. For notational simplicity let $\boldsymbol{\sigma}^{(m)}=\boldsymbol{\sigma}_{\frac{1}{m}}$, $(m=1, 2, \ldots)$. Since the sets Σ_k are weakly compact and the pay-off functions $K_1, \ldots, K_n$ are continuous there exists a subsequence

$\boldsymbol{\sigma}^{(m_1)}$, $\boldsymbol{\sigma}^{(m_2)}$, ... of the sequence $\{\boldsymbol{\sigma}^{(m)}\}$ and a vector $\boldsymbol{\sigma}^{(0)}=(\sigma_1^{(0)}, \ldots, \sigma_n^{(0)})$ such that all elements of the sequence $\{\boldsymbol{\sigma}^{(m_i)}\}$ – except for finitely many – belong to $E_\varepsilon(\boldsymbol{\sigma}^{(0)})$. Since Σ can be covered by the family of sets $H_{\boldsymbol{\tau}}$, there exists a $\boldsymbol{\tau}^* \in \Sigma$ for which $\boldsymbol{\sigma}^{(0)} \in H_{\boldsymbol{\tau}^*}$ and $\Phi(\boldsymbol{\sigma}^{(0)}, \boldsymbol{\tau}^*) - \Phi(\boldsymbol{\sigma}^{(0)}, \boldsymbol{\sigma}^0) = \delta > 0$. Let $\varepsilon_0 = \dfrac{\delta}{2n^2}$ and $\boldsymbol{\sigma} = (\sigma_1, \ldots, \sigma_n) \in E_{\varepsilon_0}(\boldsymbol{\sigma}^{(0)})$ be arbitrary. Then for any $k = 1, \ldots, n$ we get

$$|K_k(\sigma_1, \ldots, \sigma_n) - K_k(\sigma_1^{(0)}, \ldots, \sigma_n^{(0)})| \leqq$$
$$\leqq \sum_{i=1}^{n} |K_k(\sigma_1^{(0)}, \ldots, \sigma_{i-1}^{(0)}, \sigma_i, \ldots, \sigma_n) -$$
$$- K_k(\sigma_1^{(0)}, \ldots, \sigma_i^{(0)}, \sigma_{i+1}, \ldots, \sigma_n)| < n\varepsilon_0 = \frac{\delta}{2n}.$$

Similarly

$$|K_k(\sigma_1, \ldots, \sigma_{k-1}, \tau_k^*, \sigma_{k+1}, \ldots, \sigma_n) -$$
$$- K_k(\sigma_1^{(0)}, \ldots, \sigma_{k-1}^{(0)}, \tau_k^*, \sigma_{k+1}^{(0)}, \ldots, \sigma_n^{(0)}| < \frac{\delta}{2n}.$$

By the definition of the function Φ we have

$$|\Phi(\boldsymbol{\sigma}, \boldsymbol{\sigma}) - \Phi(\boldsymbol{\sigma}^{(0)}, \boldsymbol{\sigma}^{(0)})| < \frac{\delta}{2}$$
$$|\Phi(\boldsymbol{\sigma}, \boldsymbol{\tau}^*) - \Phi(\boldsymbol{\sigma}^{(0)}, \boldsymbol{\tau}^*)| < \frac{\delta}{2}.$$

Thus

$$\Phi(\boldsymbol{\sigma}, \boldsymbol{\tau}^*) - \Phi(\boldsymbol{\sigma}, \boldsymbol{\sigma}) = \underbrace{\Phi(\boldsymbol{\sigma}, \boldsymbol{\tau}^*) - \Phi(\boldsymbol{\sigma}^{(0)}, \boldsymbol{\tau}^*) +}_{> -\frac{\delta}{2}}$$
$$\underbrace{+ \Phi(\boldsymbol{\sigma}^{(0)}, \boldsymbol{\sigma}^{(0)}) - \Phi(\boldsymbol{\sigma}, \boldsymbol{\sigma})}_{> -\frac{\delta}{2}} + \underbrace{\Phi(\boldsymbol{\sigma}^0, \boldsymbol{\tau}^*) - \Phi(\boldsymbol{\sigma}^{(0)}, \boldsymbol{\sigma}^{(0)})}_{= \delta} > 0,$$

implying $E_{\varepsilon_0}(\boldsymbol{\sigma}^{(0)}) \subseteqq H_{\boldsymbol{\tau}^*}$. For sufficiently large j we have

$$\boldsymbol{\sigma}^{(m_j)} \in E_{\frac{\varepsilon_0}{2}}(\boldsymbol{\sigma}^{(0)}) \quad \text{and} \quad \frac{1}{m_j} < \frac{\varepsilon_0}{2}.$$

Then

$$E_{\frac{1}{m_j}}(\boldsymbol{\sigma}^{(m_j)}) \subseteqq E_{\frac{\varepsilon_0}{2}}(\boldsymbol{\sigma}^{(m_j)}) \subseteqq E_{\varepsilon_0}(\boldsymbol{\sigma}^{(0)}) \subseteqq H_{\boldsymbol{\tau}^*}$$

which means that $E_{\frac{1}{m_j}}(\boldsymbol{\sigma}^{(m_j)})$ can be covered by the set $H_{\boldsymbol{\tau}^*}$ contradicting the way the vectors $\boldsymbol{\sigma}^{(m_j)}$ have been selected.

Parts (a) and (b) together obviously imply that Σ can be covered by finitely many of the sets $H_{\boldsymbol{\tau}}$.

Now let $H_{\boldsymbol{\tau}_1}, \ldots, H_{\boldsymbol{\tau}_l}$ be a finite collection of sets covering Σ, and define nonnegative functions

$$g_j(\boldsymbol{\sigma}) = \max\{\Phi(\boldsymbol{\sigma}, \boldsymbol{\tau}_j) - \Phi(\boldsymbol{\sigma}, \boldsymbol{\sigma}), 0\}, \quad (j = 1, \ldots, l)$$

Then obviously

$$g(\boldsymbol{\sigma}) = \sum_{j=1}^{l} g_j(\boldsymbol{\sigma}) > 0 \quad \text{for any} \quad \boldsymbol{\sigma} \in \Sigma.$$

Choose now a simplex S in $\mathbb{R}^{l-1}$ with vertices $\mathbf{p}_1, \ldots, \mathbf{p}_l$. It is well known that any point $\mathbf{p} \in S$ can be expressed uniquely as a convex linear combination of the vertices

$$\mathbf{p} = \sum_{j=1}^{l} \lambda_j \mathbf{p}_j, \quad (\lambda_j \geqq 0;\ j = 1, \ldots, l;\ \sum_{j=1}^{l} \lambda_j = 1)$$

and the barycentric coordinates $\lambda_j(\mathbf{p})$ are continuous functions of $\mathbf{p}$, $(j = 1, \ldots, l)$. The continuous mapping $\mathbf{p} \to \Psi(\mathbf{p})$ sends every point of S to Σ

$$\Psi(\mathbf{p}) = \sum_{j=1}^{l} \lambda_j(\mathbf{p}) \boldsymbol{\tau}_j$$

since Σ is convex.

The mapping

$$\mathbf{p} \to \sum_{j=1}^{l} \frac{g_j[\Psi(\mathbf{p})]}{g[\Psi(\mathbf{p})]} \mathbf{p}_j$$

is a continuous mapping of S into itself. Then by Brouwer's fixed point theorem there exists a vector $\mathbf{p}^*$ for which

$$\mathbf{p}^* = \sum_{j=1}^{l} \frac{g_j[\Psi(\mathbf{p}^*)]}{g[\Psi(\mathbf{p}^*)]} \mathbf{p}_j$$

By the concavity of Φ we obtain the inequality

$$\Phi(\Psi(\mathbf{p}^*), \Psi(\mathbf{p}^*)) = \Phi\left(\Psi(\mathbf{p}^*), \sum_{j=1}^{l} \frac{g_j[\Psi(\mathbf{p}^*)]}{g[\Psi(\mathbf{p}^*)]} \right) \geqq \tag{3}$$

$$\geqq \sum_{j=1}^{l} \frac{g_j[\Psi(\mathbf{p}^*)]}{g[\Psi(\mathbf{p}^*)]} \Phi(\Psi(\mathbf{p}^*), \tau_j).$$

On the other hand by the definition of the functions g_j there exists an index i for which $g_i[\Psi(\mathbf{p}^*)] > 0$ and for all these indices $\Phi[\Psi(\mathbf{p}^*), \Psi(\mathbf{p}^*)] < \Phi[\Psi(\mathbf{p}^*), \tau_i)]$ implying the following relations

$$\sum_{j=1}^{l} \frac{g_j[\Psi(\mathbf{p}^*)]}{g[\Psi(\mathbf{p}^*)]} \Phi(\Psi(\mathbf{p}^*), \tau_j) =$$

$$= \sum_{g_i[\Psi(\mathbf{p}^*)]>0} \frac{g_i[\Psi(\mathbf{p}^*)]}{g[\Psi(\mathbf{p}^*)]} \Phi[\Psi(\mathbf{p}^*), \tau_i] >$$

$$- \sum_{g_i[\Psi(\mathbf{p}^*)]>0} \frac{g_i[\Psi(\mathbf{p}^*)]}{g[\Psi(\mathbf{p}^*)]} \Phi(\Psi(\mathbf{p}^*), \Psi(\mathbf{p}^*)) = \Phi(\Psi(\mathbf{p}^*), \Psi(\mathbf{p}^*))$$

which contradicts (3). Thus our proof has been completed. ∎

REMARK. If the strategy sets $\Sigma_1, \ldots, \Sigma_n$ of the above game Γ are compact, then they can easily be shown to be weakly compact with respect to the continuous pay-off functions $K_1, \ldots, K_n$. Thus Theorem 5 can be regarded as a generalization of the Nikaido–Isoda theorem (see the remark after Theorem 2) to metric spaces.

Compactness of the strategy sets is of no use if we are dealing with mixed extensions of continuous games, since with the usual supremum metric the space of distribution functions defined on a finite interval is not compact but it is weakly compact with respect to continuous pay-off functions as we shall show it right away. Let $\Gamma=\{\bar{\Sigma}_1, \ldots, \bar{\Sigma}_n, \bar{K}_1, \ldots, \bar{K}_n\}$ be the mixed extension of a continuous game as defined in Chapter 2.

THEOREM 6. [31], [179] Game Γ possesses at least one equilibrium point.

Proof. We shall show that game Γ meets all the conditions of Theorem 5. $\bar{\Sigma}_j$ is a convex subset of the linear function space L^{∞}, (supplied with the usual supremum metric). The pay-off functions

$$\bar{K}_k(F_1, \ldots, F_n)=$$

$$= \int_{a_1}^{b_1} \ldots \int_{a_n}^{b_n} K_k(\sigma_1, \ldots, \sigma_n)\,\mathrm{d}F_1(\sigma_1), \ldots \mathrm{d}F_n(\sigma_n)$$

$$(k=1, \ldots, n)$$

depend linearly on the distribution functions F_j if the remaining ones $F_l(l \neq j, l=1, \ldots, n)$ are held fixed. Thus weak compactness of the strategy sets $\bar{\Sigma}_1, \ldots, \bar{\Sigma}_n$ with respect to the functions $\bar{K}_1, \ldots, \bar{K}_n$ is only to be shown in order to be able to apply Theorem 5. Our proof consists of three steps.

(a) Let $\{F_k^{(i)}\}_{i=1}^{\infty}$ be a sequence of elements from $\bar{\Sigma}_k$. We first show that there exists a subsequence $\{F_k^{(i_j)}\}$ and a distribution function $F_k^{(0)} \in \bar{\Sigma}_k$ such that at every point of continuity of $F_k^{(0)}$, $F^{(i_j)}(\sigma) \to F_k^{(0)}(\sigma)$.

Denote $r_1, r_2, \ldots$ the sequence of rational numbers in the interval $[a_k, b_k]$. Since for any $l=1, 2, \ldots$ the sequence $\{F_k^{(i)}(r_l)\}$ is bounded, there exists a convergent subsequence $\{F_k^{(i_j^{(l)})}(r_l)\}$ for each l. Furthermore there is a subsequence $\{F_k^{(i_j^{(j)})}\}$ converging for any rational number r_l. Denote for the sake of simplicity $\{F_k^{(i_j^{(j)})}\}$

by $\{F_k^{(i_j)}\}$ and

$$\lim_{j\to\infty} F_k^{(i_j)}(r_l)=c_l\,.$$

Define for any real $\sigma\in(-\infty,\infty)$ the functions

$$(4)\qquad F_k^{(0)}(\sigma)=\begin{cases}0, & \text{if}\quad \sigma\leqq a_k\\ \sup\limits_{\sigma\geqq r_l} c_l & \text{if}\quad a_k<\sigma\leqq b_k\\ 1 & \text{if}\quad \sigma>b_k\,.\end{cases}$$

Obviously $0\leqq F_k^{(0)}(\sigma)\leqq 1$, since $0\leqq c_l\leqq 1$ and for $a_k<\sigma<\sigma'\leqq b_k$

$$F_k^{(0)}(\sigma)=\sup_{\sigma\geqq r_l} c_l\leqq \sup_{\sigma'\geqq r} c_l=F_k^{(0)}(\sigma'),$$

thus $F_k^{(0)}$ is monotone nondecreasing. Now we shall show that $F_k^{(0)}$ is continuous from the left. At points being outside of the interval $(a_k, b_k]$ function $F_k^{(0)}$ is continuous from the left by Definition (4). Let now $a_k<\sigma\leqq b_k$ and $\varepsilon>0$ be fixed. Then there exists a rational number $r_t<x$, such that $c_t>F_k^{(0)}(\sigma)-\varepsilon$. Thus for any σ' satisfying $r_t<\sigma'<\sigma$ we have

$$F_k^{(0)}(\sigma')=\sup_{\sigma'\geqq r} c_l\geqq c_t>F_k^{(0)}(\sigma)-\varepsilon\,,$$

which means that $F_k^{(0)}$ is continuous from the left at σ therefore it is a distribution function on $[a_k, b_k]$.

Now let σ be a point of continuity of F_k^0. We are going to show that

$$(5)\qquad \lim_{j\to\infty} F_k^{(i_j)}(\sigma)=F_k^{(0)}(\sigma)\,.$$

If $\sigma=a_k$ or $\sigma=b_k$, then (5) obviously holds. Let $a_k<\sigma<b_k$ and $\varepsilon>0$ be fixed. Then by (4) there exists a rational number r_l, such that $r_l<\sigma$, $c_l>F_k^{(0)}(\sigma)-\dfrac{\varepsilon}{2}$ and for sufficiently large j

$$|F_k^{(i_j)}(r_l)-c_l|<\frac{\varepsilon}{2}\,.$$

Thus

$$(6) \qquad F_k^{(i_j)}(\sigma) - F_k^{(0)}(\sigma) \geqq F_k^{(i_j)}(r_l) - F_k^{(0)}(\sigma) =$$
$$= [F_k^{(i_j)}(r_l) - c_l] + [c_l - F_k^{(0)}(\sigma)] > -\varepsilon .$$

Since $F_k^{(0)}$ is continuous at σ there is a $\sigma' > \sigma$ for which $|F_k^{(0)}(\sigma') - F_k^{(0)}(\sigma)| < \frac{\varepsilon}{2}$. Let r_t be a rational number satisfying $\sigma' > r_t > \sigma$. Then by (4) $F_k^{(0)}(\sigma') \geqq c_t$ and $F_k^{(0)}(\sigma) \geqq c_t - \frac{\varepsilon}{2}$. For sufficiently large j

$$|F_k^{(i_j)}(r_t) - c_t| < \frac{\varepsilon}{2},$$

hence

$$F_k^{(i_j)}(\sigma) - F_k^{(0)}(\sigma) \leqq F_k^{(i_j)}(r_t) - F_k^{(0)}(\sigma) =$$
$$= [F_k^{(i_j)}(r_t) - c_t] + [c_t - F_k^{(0)}(\sigma)] < \varepsilon ,$$

which together with (6) gives us

$$|F_k^{(i_j)}(\sigma) - F_k^{(0)}(\sigma)| < \varepsilon$$

if j is large enough. Thereby we have proved the assertion of paragraph (a).

(b) Let B be a family of equicontinuous and uniformly bounded functions defined on the interval $[a_k, b_k]$. We are going to show that for the sequence $\{F_k^{(i_j)}\}$ defined in paragraph (a)

$$\lim_{j \to \infty} \int_{a_k}^{b_k} f(\sigma) \, dF_k^{(i_j)}(\sigma) = \int_{a_k}^{b_k} f(\sigma) \, dF_k^{(0)}(\sigma)$$

uniformly on B.

Choose an arbitrary $\varepsilon > 0$. By the equicontinuity there is a $\delta > 0$ such that for $\sigma, \sigma' \in [a_k, b_k]$, $|\sigma - \sigma'| < \delta$ the inequality

$$|f(\sigma) - f(\sigma')| < \frac{\varepsilon}{3}$$

holds for any $f \in B$. Subdivide the interval $[a_k, b_k]$ into $N+1$

subintervals the length of each being less than $\frac{\delta}{2}$. Since $F_k^{(0)}$ is monotone nondecreasing the set of its points of discontinuity is denumerable. Thus in each subinterval of $[a_k, b_k]$ there is at least one point at which $F_k^{(0)}$ is continuous. Let these points be $\sigma_0 = a_k$, $\sigma_1, \ldots, \sigma_{N-1}$, $\sigma_N = b_k$. Obviously $|\sigma_{i+1} - \sigma_i| < \delta$, $(i = 0, 1, \ldots, N-1)$. Thus for any $f \in B$ and $\nu = 0, i_1, i_2, \ldots$

$$\int_{a_k}^{b_k} f(\sigma)\, dF_k^{(\nu)}(\sigma) = \int_{[\sigma_0, \sigma_1]} f(\sigma)\, dF_k^{(\nu)}(\sigma) +$$

$$+ \sum_{i=1}^{N-1} \int_{(\sigma_i, \sigma_{i+1})} f(\sigma)\, dF_k^{(\nu)}(\sigma) = \int_{[\sigma_0, \sigma_1]} f(\sigma_0)\, dF_k^{(\nu)}(\sigma) +$$

$$+ \sum_{i=1}^{N-1} \int_{[\sigma_i, \sigma_{i+1}]} f(\sigma_i)\, dF_k^{(\nu)}(\sigma) +$$

$$+ \int_{[\sigma_0, \sigma_1]} (f(\sigma) - f(\sigma_0))\, dF_k^{(\nu)}(\sigma) + \sum_{i=1}^{N-1} \int_{[\sigma_i, \sigma_{i+1}]} (f(\sigma) - f(\sigma_i))\, dF_k^{(\nu)}(\sigma).$$

The absolute value of the sum of the third and fourth term in the right-hand side of the above expression is less than $\frac{\varepsilon}{3}$, hence for sufficiently large ν we have

$$\left| \int_{a_k}^{b_k} f(\sigma)\, dF_k^{(\nu)}(\sigma) - \int_{a_k}^{b_k} f(\sigma)\, dF_k^{(0)}(\sigma) \right| < \left| f(\sigma_0) F_k^{(\nu)}(\sigma_i) + \right.$$

$$+ \sum_{i=1}^{N-1} f(\sigma_i)\, [F_k^{(\nu)}(\sigma_{i+1}) - F_k^{(\nu)}(\sigma_i)] - f(\sigma_0) F_k^{(0)}(\sigma_1) -$$

$$\left. - \sum_{i=1}^{N-1} f(\sigma_i)\, [F_k^{(0)}(\sigma_{i+1}) - F_k^{(0)}(\sigma_i)] \right| < \frac{2\varepsilon}{3} < \varepsilon$$

which proves the assertion of paragraph (b).

(c) The pay-off functions K_k are uniformly continuous on the Cartesian product $\times_{l=1}^{n} [a_l, b_l]$, thus the Stieltjes integrals

$$\int_{a_1}^{b_1} \cdots \int_{a_{k-1}}^{b_{k-1}} \int_{a_{k+1}}^{b_{k+1}} \cdots \int_{a_n}^{b_n} K_k(\sigma_1, \ldots, \sigma_n) \, \mathrm{d}F_1(\sigma_1) \ldots$$

$$\ldots \mathrm{d}F_{k-1}(\sigma_{k-1}) dF_{k+1}(\sigma_{k+1}) \ldots \mathrm{d}F_n(\sigma_n)$$

form an equicontinuous and uniformly bounded family of functions defined on $[a_k, b_k]$.

Paragraphs (a), (b) and (c) together imply that the strategy sets $\bar{\Sigma}_1, \ldots, \bar{\Sigma}_n$ are weakly compact with respect to the pay-off functions $\bar{K}_1, \ldots, \bar{K}_n$. Thus game $\bar{\Gamma} = \{\bar{\Sigma}_1, \ldots, \bar{\Sigma}_n; \bar{K}_1, \ldots, \bar{K}_n\}$ has at least one equilibrium point by Theorem 5. ∎

THEOREM 7. [179] The mixed extension of a continuous game has a group-equilibrium point for any disjoint partition of the players.

Proof. By the techniques applied in the proof of Theorem 6 together with Theorem 5 we can prove Theorem 7 in a straightforward way. We do not include the details of the proof here, it is left for the reader as an exercise. ∎

4. Special n-person games and methods to solve them

Unfortunately there is no general method for the solution of an arbitrary n-person game given in normal form. By solution we mean finding all equilibrium points of the game. The situation is not any better if we want to tackle the much less difficult problem of determining only one equilibrium point of a general n-person game. However this latter task is more amenable to analysis in case of special n-person games. Mathematical programming methods, special iterative processes and combinatorial methods have proved to be most successful for the solution of special n-person games including those such as concave games, polyhedral games, finite n-person games and the oligopoly game. Intensive research work is being done in this field and emergence of new, more efficient methods, as well as the extension of classes of games solvable by them is expected in the near future.

4.1. MATHEMATICAL PROGRAMMING METHODS FOR THE SOLUTION OF n-PERSON CONCAVE GAMES

By an n-person concave game given in normal form we mean

$$\Gamma = \{\Sigma_1, \ldots, \Sigma_n;\ K_1, \ldots, K_n\}$$

where $\Sigma_1, \ldots, \Sigma_n$ are compact, convex subsets of finite dimensional Euclidean spaces; and the pay-off functions $K_j(\boldsymbol{\sigma}_1, \ldots, \boldsymbol{\sigma}_{j-1}, \boldsymbol{\sigma}_j, \boldsymbol{\sigma}_{j+1}, \ldots, \boldsymbol{\sigma}_n)$ are concave on Σ_j for any fixed $\boldsymbol{\sigma}_1, \ldots, \boldsymbol{\sigma}_{j-1}, \boldsymbol{\sigma}_{j+1}, \ldots, \boldsymbol{\sigma}_n$. We already know that Γ has at least one equilibrium point (see Theorem 2 of Chapter 3).

In order to be able to appeal to mathematical programming methods we have to assume that the strategy sets $\Sigma_1, \ldots, \Sigma_n$ are defined by a finite number of inequalities, i.e.,

$$\Sigma_k = \{\mathbf{x}_k \mid \mathbf{x}_k \in \mathbb{R}^{m_k},\ \mathbf{g}_k(\mathbf{x}_k) \geqq \mathbf{0}\}, \quad (k=1, \ldots, n)$$

where each component of $\mathbf{g}_k$ is a continuously differentiable concave function ($k=1, \ldots, n$). To conform to mathematical programming notations we denote the strategies by $\mathbf{x}$ instead of $\boldsymbol{\sigma}$.

We suppose furthermore that $\mathbf{g}_k$ satisfies a "constraint qualification" (see [107]) which is a mild regularity assumption. (A simple constraint qualification is Slater's: the inequality system $\mathbf{g}_k(\mathbf{x}_k) \geqq 0$ is assumed to be superconsistent, i.e., there exists a $\bar{\mathbf{x}}_k$ for which $\mathbf{g}_k(\bar{\mathbf{x}}_k) > 0$.) The pay-off functions $K_1, \ldots, K_n$ are also assumed to be continuously differentiable.

Let $\mathbf{x}^* = (\mathbf{x}_1^*, \ldots, \mathbf{x}_n^*)$ be an equilibrium point of Γ. Then the pay-off function K_k has a maximum point at $\mathbf{x}_k^*$ for fixed $\mathbf{x}_1^*, \ldots, \mathbf{x}_{k-1}^*, \mathbf{x}_{k+1}^*, \ldots, \mathbf{x}_n^*$. Thus by the Kuhn–Tucker theorem (see e.g. [107]) there exists a nonnegative vector $\mathbf{u}_k^*$ such that $\mathbf{x}^*, \mathbf{u}_k^*$ satisfy the so-called Kuhn–Tucker conditions

$$(1) \qquad \begin{aligned} \nabla_k K_k(\mathbf{x}^*) + \mathbf{u}_k^* \nabla_k \mathbf{g}_k(\mathbf{x}_k^*) &= \mathbf{0} \\ \mathbf{u}_k^* \mathbf{g}_k(\mathbf{x}_k^*) &= 0 \end{aligned}$$

where $\nabla_k K_k$ denotes the gradient of K_k with respect to $\mathbf{x}_k$ and $\nabla_k \mathbf{g}_k$ stands for the Jacobi-matrix of $\mathbf{g}_k$.

Denote briefly the left-hand side of the first equality in (1) by $\Psi_k(\mathbf{x}^*, \mathbf{u}_k)$ and consider the following mathematical programming problem:

$$(2) \qquad \begin{aligned} \sum_{k=1}^{n} \mathbf{u}_k \mathbf{g}_k(\mathbf{x}_k) &\to \min \\ \mathbf{u}_k &\geqq \mathbf{0} \\ \mathbf{g}_k(\mathbf{x}_k) &\geqq \mathbf{0} \\ \Psi_k(\mathbf{x}, \mathbf{u}_k) &= 0 \quad (k=1, \ldots, n) \end{aligned}$$

THEOREM 1. [179] An n-tuple of strategies $\mathbf{x}^* = (\mathbf{x}_1^*, \ldots, \mathbf{x}_n^*)$ is an equilibrium point of the concave game Γ if and only if there exist vectors $\mathbf{u}_1^*, \ldots, \mathbf{u}_n^*$ such that $\mathbf{x}^*, \mathbf{u}_1^*, \ldots, \mathbf{u}_n^*$ are optimal solutions of the programming problem (2), the optimal objective function's value of which is 0.

Proof. For any feasible solution of (2) the objective function's value is nonnegative. If $\mathbf{x}^*$ is an equilibrium point, then there are vectors $\mathbf{u}_1^*, \ldots, \mathbf{u}_n^*$ such that $\mathbf{x}^*, \mathbf{u}_1^*, \ldots, \mathbf{u}_n^*$ satisfy the Kuhn–Tucker conditions (1). At this point the objective function's value of (2) is 0, which means optimality by the nonnegativity of $\sum_{k=1}^{n} \mathbf{u}_k \mathbf{g}_k(\mathbf{x}_k)$. On the other hand, since the constraint functions are concave the Kuhn–Tucker constraints are sufficient for $\mathbf{x}_k^*$ to be a maximum point of $K_k(\mathbf{x}_1^*, \ldots, \mathbf{x}_{k-1}^*, \mathbf{x}_k, \mathbf{x}_{k+1}^*, \ldots, \mathbf{x}_n^*)$ which means that $\mathbf{x}^*$ is an equilibrium point. ∎

To solve programming problem (2) is not an easy job. Its feasible set is not convex thus gradient-type methods are of no use. However this difficulty can easily be overcome by a simple transformation. Consider the problem:

$$(3) \qquad \sum_{k=1}^{n} \mathbf{u}_k \mathbf{g}_k(\mathbf{x}_k) + \sum_{k=1}^{n} \Psi_k^2(\mathbf{x}, \mathbf{u}_k) \to \min$$

$$\mathbf{u}_k \geqq \mathbf{0}$$

$$\mathbf{g}_k(\mathbf{x}_k) \geqq \mathbf{0}$$

It is obvious that the optimal solutions of (2) and (3) coincide and the objective function's value of each is 0. The feasible set of (3) is convex, its objective function is continuously differentiable, thus gradient-type methods can be applied to locate stationary points. Unfortunately (3) may have local but not global minima. However identifying a global minimum is easy since all global minima have an objective function value equal to 0.

More efficient methods are available if further restrictions are imposed on the functions $\mathbf{g}_1, \ldots, \mathbf{g}_n, K_1, \ldots, K_n$. We shall in turn discuss a few important special cases.

4.2. GENERALIZED POLYHEDRAL GAMES

If the stationary sets $\Sigma_1, \ldots, \Sigma_n$ of an n-person concave game Γ are polyhedrons and the pay-off functions $K_1, \ldots, K_n$ are multilinear functions of the strategy vectors $\mathbf{x}_1, \ldots, \mathbf{x}_n$, then Γ is called a *(generalized) polyhedral game.* (Sometimes the name polyhedral game is reserved for the special case $n=2$.) More specifically

$$\Sigma_k = \{\mathbf{x}_k \mid \mathbf{b}_k - \mathbf{A}_k\mathbf{x}_k \geqq \mathbf{0}\}$$

$$K_k(\mathbf{x}) = \sum_{i_1=1}^{m_1} \cdots \sum_{i_n=1}^{m_n} c^{(k)}_{i_1 \ldots i_n} x^{(1)}_{i_1} \ldots x^{(n)}_{i_n} \quad (k=1, \ldots, n)$$

where $\mathbf{x}_l = (x^{(l)}_1, \ldots, x^{(l)}_{m_l})$, $\mathbf{b}_k$, $\mathbf{A}_k$, $c_{i_1}, \ldots {}_{i_n}$ are constant vectors, matrices and scalars resp.

Now programming problem (2) takes a simpler form.

Let

$$\mathbf{u}_k(\mathbf{x}) = (u^{(k)}_1(\mathbf{x}), \ldots, u^{(k)}_{m_k}(\mathbf{x}))$$

where

$$u^{(k)}_i(\mathbf{x}) = \sum_{i_1=1}^{m_1} \cdots \sum_{i_{k-1}=1}^{m_{k-1}} \sum_{i_{k+1}=1}^{m_{k+1}} \cdots$$

$$\cdots \sum_{i_n=1}^{m_n} c^{(k)}_{i_1 \ldots i_{k-1} i_{k+1} \ldots i_n} x^{(1)}_{i_1} \ldots x^{(k-1)}_{i_{k-1}} x^{(k+1)}_{i_{k+1}} \ldots x^{(n)}_{i_n}$$

for $i=1, \ldots, m_k$; $k=1, \ldots, n$. Notice that $\mathbf{u}_k(\mathbf{x})$ does not depend on $\mathbf{x}_k$ thus (2) can be written as

$$\sum_{k=1}^{n} \mathbf{u}_k(\mathbf{b}_k - \mathbf{A}_k\mathbf{x}_k) \to \min$$

$$\mathbf{u}_k \geqq \mathbf{0}$$
$$\mathbf{b}_k - \mathbf{A}_k \mathbf{x}_k \geqq \mathbf{0}$$
$$\mathbf{u}_k(\mathbf{x}) - \mathbf{A}_k^T \mathbf{u}_k = \mathbf{0}, \quad (k=1, \ldots, n).$$

Taking into account that

$$K_k(\mathbf{x}) = \mathbf{u}_k(\mathbf{x}) \mathbf{x}_k,$$

we obtain the programming problem

$$(4) \qquad -\sum_{k=1}^{n} K_k(\mathbf{x}) + \sum_{k=1}^{n} \mathbf{u}_k \mathbf{b}_k \to \min$$
$$\mathbf{u}_k \geqq \mathbf{0}$$
$$\mathbf{b}_k - \mathbf{A}_k \mathbf{x}_k \geqq \mathbf{0}$$
$$\mathbf{u}_k(\mathbf{x}) - \mathbf{A}_k^T \mathbf{u}_k = \mathbf{0}, \quad (k=1, \ldots, n).$$

In the special case $n=2$ (two-person game) the pay-off functions are bilinear

$$K_1(\mathbf{x}_1, \mathbf{x}_2) = \mathbf{x}_1 \mathbf{C}_1 \mathbf{x}_2,$$
$$K_2(\mathbf{x}_1, \mathbf{x}_2) = \mathbf{x}_1 \mathbf{C}_2 \mathbf{x}_2;$$

and (4) reduces to a (nonconvex) quadratic programming problem:

$$(5) \qquad -\mathbf{x}_1 \mathbf{C}_1 \mathbf{x}_2 - \mathbf{x}_1 \mathbf{C}_2 \mathbf{x}_2 + \mathbf{u}_1 \mathbf{b}_1 + \mathbf{u}_2 \mathbf{b}_2 \to \min$$
$$\mathbf{u}_1, \mathbf{u}_2 \geqq \mathbf{0}$$
$$\mathbf{A}_1 \mathbf{x}_1 \leqq \mathbf{b}_1$$
$$\mathbf{A}_2 \mathbf{x}_2 \leqq \mathbf{b}_2$$
$$\mathbf{C}_1 \mathbf{x}_2 - \mathbf{A}_1^T \mathbf{u}_1 = \mathbf{0}$$
$$\mathbf{C}_2^T \mathbf{x}_1 - \mathbf{A}_2^T \mathbf{u}_2 = \mathbf{0}$$

If

$$\mathbf{A}_1 = \begin{bmatrix} -\mathbf{E} \\ \mathbf{1} \end{bmatrix}, \quad \mathbf{b}_1 = \begin{bmatrix} \mathbf{0} \\ 1 \end{bmatrix}, \quad \mathbf{A}_2 = \begin{bmatrix} -\mathbf{E} \\ \mathbf{1} \end{bmatrix}, \quad \mathbf{b}_2 = \begin{bmatrix} \mathbf{0} \\ 1 \end{bmatrix},$$

then our two-person polyhedral game is in fact a mixed extension of a two-person finite game (a bimatrix game) to be discussed later in Chapter 8. In this case, for finding an equilibrium point, there exist more powerful methods than the cumbersome minimization of a nonconvex quadratic function.

If our two-person polyhedral game is zero-sum, i.e., $\mathbf{C}_1 = -\mathbf{C}_2$, then the first two terms in the objective functions of (5) sum up to 0 and (5) is a linear programming problem.

4.3. SOLUTION OF n-PERSON ZERO-SUM CONCAVE-CONVEX GAMES

Let $\Gamma = \{\Sigma_1, \ldots, \Sigma_n; K_1, \ldots, K_n\}$ be an n-person zero-sum game where the strategy sets $\Sigma_1, \ldots, \Sigma_n$ are compact subsets of finite dimensional Euclidean spaces. Let us write the pay-off function K_k in the following form:

$$K_k(\mathbf{x}_k, \mathbf{y}), \quad (k = 1, \ldots, n),$$

where $\mathbf{y} = (\mathbf{x}_1, \ldots, \mathbf{x}_{k-1}, \mathbf{x}_{k+1}, \ldots, \mathbf{x}_n)$. The pay-off function $K_k(\mathbf{x}_k, \mathbf{y})$ is assumed to be continuous, concave in $\mathbf{x}_k$ for any fixed $\mathbf{y} \in \Sigma^{(k)}$ and convex in $\mathbf{y}$ for any fixed $\mathbf{x}_k \in \Sigma_k$, $(k = 1, \ldots, n)$ where $\Sigma^{(k)} = \Sigma_1 \times \ldots \times \Sigma_{k-1} \times \Sigma_{k+1} \times \ldots \times \Sigma_n$.

Our aim is to find an equilibrium point of Γ the existence of which is assured by the general existence theorem for concave games (see Theorem 2 of Chapter 3). In the following we need a special characterization of equilibrium points.

Consider the function

$$U(\mathbf{y}) = \sum_{k=1}^{n} K_k(\hat{\mathbf{x}}_k, \mathbf{y}), \quad \mathbf{y} \in \Sigma^{(k)}$$

where

$$K_k(\hat{\mathbf{x}}_k, \mathbf{y}) = \max_{\mathbf{z}_k \in \Sigma_k} K_k(\mathbf{z}_k, \mathbf{y}), \quad (k = 1, \ldots, n).$$

Clearly $U(\mathbf{y}) \geqq 0$, since $\sum_{k=1}^{n} K_k(\mathbf{x}_k, \mathbf{y}) = 0$.

THEOREM 2. [30] $\mathbf{x} \in \Sigma$ is an equilibrium point of the n-person zero-sum concave-convex game Γ if and only if $U(\mathbf{x})=0$.

Proof. If $\mathbf{x}^*=(\mathbf{x}_1^*, \ldots, \mathbf{x}_n^*)$ is an equilibrium point, then by definition $K_k(\mathbf{x}_k^*, \mathbf{x}^*)=K_k(\hat{\mathbf{x}}_k, \mathbf{x}^*)$, $(k=1, \ldots, n)$. Hence $U(\mathbf{x}^*)=0$.
On the other hand if $U(\mathbf{x})=0$ and $\mathbf{x}$ is not an equilibrium point, then $K_k(\hat{\mathbf{x}}_k, \mathbf{x}) \geqq K_k(\mathbf{x}_k, \mathbf{x})$ for any $k=1, \ldots, n$ and for at least one k we have $K_k(\hat{\mathbf{x}}_k, \mathbf{x})>K_k(\mathbf{x}_k, \mathbf{x})$ implying $U(\mathbf{x})>0$, which is a contradiction. ∎

The algorithm for the solution of Γ which we are considering consists of iterational steps. In addition to the assumptions we have made so far we have to assume that the pay-off functions $K_1, \ldots, K_n$ are continuously differentiable with respect to each variable. Suppose we are given a vector $\mathbf{x}^k=(\mathbf{x}_1^k, \ldots, \mathbf{x}_n^k)$. Then we solve n maximization problems

$$
(6) \qquad \max_{\mathbf{x}_1} \{K_1(\mathbf{x}_1, \mathbf{x}_2^k, \ldots, \mathbf{x}_n^k) | \mathbf{x}_1 \in \Sigma_1\} =
$$

$$
= K_1(\bar{\mathbf{x}}_1^k, \mathbf{x}_2^k, \ldots, \mathbf{x}_n^k) = \Psi_1(\mathbf{x}_2^k, \ldots, \mathbf{x}_n^k)
$$

$$
\cdots\cdots\cdots\cdots\cdots\cdots
$$

$$
\max_{\mathbf{x}_n} \{K_n(\mathbf{x}_1^k, \ldots, \mathbf{x}_{n-1}^k, \mathbf{x}_n) | \mathbf{x}_n \in \Sigma_n\} =
$$

$$
= K_n(\mathbf{x}_1^k, \ldots, \mathbf{x}_{n-1}^k, \bar{\mathbf{x}}_n^k) = \Psi_n(\mathbf{x}_1^k, \ldots, \mathbf{x}_{n-1}^k)
$$

which are concave mathematical programming problems. Consider now the convex set R_{n+1}^k defined by

$$
R_{n+1}^k = \{(\mathbf{x}_1, \ldots, \mathbf{x}_n, x_{n+1}) | \mathbf{x}_1 \in \Sigma_1, \ldots,
$$

$$
\ldots, \mathbf{x}_n \in \Sigma_n, K_1(\bar{\mathbf{x}}_1^i, \mathbf{x}_2^i, \ldots, \mathbf{x}_n^i) +
$$

$$
+ \ldots + K_n(\mathbf{x}_1^i, \ldots, \mathbf{x}_{n-1}^i, \bar{\mathbf{x}}_n^i) +
$$

$$
+ \nabla K_{1\mathbf{x}_2}(\bar{\mathbf{x}}_1^i, \mathbf{x}_2^i, \ldots, \mathbf{x}_n^i)(\mathbf{x}_2 - \mathbf{x}_2^i) +
$$

$$+\ldots+\nabla K_{1\mathbf{x}_n}(\bar{\mathbf{x}}_1^i, \mathbf{x}_2^i, \ldots, \mathbf{x}_n^i)(\mathbf{x}_n-\mathbf{x}_n^i)+$$
$$+\ldots+\nabla K_{n\mathbf{x}_1}(\mathbf{x}_1^i, \ldots, \mathbf{x}_{n-1}^i, \bar{\mathbf{x}}_n^i)(\mathbf{x}_1-\mathbf{x}_1^i)+$$
$$+\ldots+\nabla K_{n\mathbf{x}_{n-1}}(\mathbf{x}_1^i, \ldots, \mathbf{x}_{n-1}^i, \bar{\mathbf{x}}_n^i)(\mathbf{x}_{n-1}-\mathbf{x}_{n-1}^i)\leqq$$
$$\leqq x_{n+1}, \quad (i=1, \ldots, k)\}.$$

The next iterational vector $\mathbf{x}^{k+1}$ is determined by solving the programming problem

$$(7) \qquad x_{n+1}\to\min$$
$$(\mathbf{x}_1, \ldots, \mathbf{x}_n, x_{n+1})\in R_{n+1}^k$$

It is clear that at every iteration an additional constraint is adjoined to (7), thus the sequence of objective function's values $x_{n+1}^1, x_{n+1}^2, \ldots$ is monotone nondecreasing. The convergence of the above procedure is established in the following theorem:

THEOREM 3. [30] The iterative procedure described above converges to an equilibrium point of the game Γ, i.e., any cluster point of the sequence $\{\mathbf{x}^k\}$ is an equilibrium point.

Proof. Let $\mathbf{x}^k=(\mathbf{x}_1^k, \ldots, \mathbf{x}_n^k)$ be the vector we start out from at the kth step of the iteration. If $\mathbf{x}^k$ is not an equilibrium point, then

$$\max_{\mathbf{x}_i}\{K_i(\mathbf{x}_1^k, \ldots, \mathbf{x}_i, \ldots, \mathbf{x}_n^k)|\mathbf{x}_i\in\Sigma_i\}\geqq$$
$$\geqq K_i(\mathbf{x}_1^k, \ldots, \mathbf{x}_i^k, \ldots, \mathbf{x}_n^k)$$

for all $i=1, \ldots, n$ and there exists at least one t for which

$$\max_{\mathbf{x}_t}\{K_t(\mathbf{x}_1^k, \ldots, \mathbf{x}_t, \ldots, \mathbf{x}_n^k)|\mathbf{x}_t\in\Sigma_t\}>$$
$$>K_t(\mathbf{x}_1^k, \ldots, \mathbf{x}_t^k, \ldots, \mathbf{x}_n^k),$$

that is,

$$(8) \qquad \sum_{i=1}^{n} K_i(\mathbf{x}_1^k, \ldots, \bar{\mathbf{x}}_i^k, \ldots, \mathbf{x}_n^k) >$$

$$> \sum_{i=1}^{n} K_i(\mathbf{x}_1^k, \ldots, \mathbf{x}_i^k, \ldots, \mathbf{x}_n^k) = 0$$

Therefore the sequence

$$\left\{\sum_{i=1}^{n} K_i(\mathbf{x}_1^k, \ldots, \bar{\mathbf{x}}_i^k, \mathbf{x}_n^k)\right\}$$

is bounded from below by 0.

Moreover, from (6) we have

$$(9) \qquad K_1(\mathbf{x}_1^k, \mathbf{x}_2, \ldots, \mathbf{x}_n) \leqq \Psi_1(\mathbf{x}_2, \ldots, \mathbf{x}_n)$$
$$\ldots$$
$$K_n(\mathbf{x}_1, \ldots, \mathbf{x}_{n-1}, \mathbf{x}_n^k) \leqq \Psi_n(\mathbf{x}_1, \ldots, \mathbf{x}_{n-1}).$$

Define

$$R_{n+1}^{\Psi} = \{(\mathbf{x}_1, \ldots, \mathbf{x}_n, x_{n+1}) \mid \Psi_1(\mathbf{x}_2, \ldots, \mathbf{x}_n) + \ldots$$
$$\ldots + \Psi_n(\mathbf{x}_1, \ldots, \mathbf{x}_{n-1}) \leqq x_{n+1}\}.$$

and

$$R_{n+1}^{(0)} = \{(\mathbf{x}_1, \ldots, \mathbf{x}_n, x_{n+1}) \mid \sum_{i=1}^{n} K_i(\mathbf{x}_1, \ldots, \bar{\mathbf{x}}_i^k, \ldots, \mathbf{x}_n) \leqq x_{n+1}\}$$

By (9) $R_{n+1}^{\Psi} \subset R_{n+1}^{(0)}$ and by the convexity of the functions $K_i(\mathbf{x}_1, \ldots, \bar{x}_i^k, \ldots, \mathbf{x}_n)$ we have $R_{n+1}^{(0)} \subset R_{n+1}^k$, thus $R_{n+1}^{\Psi} \subset R_{n+1}^k$. By Theorem 2 of Chapter 3, R_{n+1}^{Ψ} is nonempty. In particular, $(\mathbf{x}_1^*, \ldots, \mathbf{x}_n^*, 0) \in R_{n+1}^{\Psi}$ if $(\mathbf{x}_1^*, \ldots, \mathbf{x}_n^*)$ is an arbitrary equilibrium point. Therefore 0 is an upper bound for the sequence $\{x_{n+1}^k\}$. Thus the sequence $\{(\mathbf{x}_1^k, \ldots, \mathbf{x}_n^k, x_{n+1}^k)\}$ is bounded and has at least one cluster point, say $(\mathbf{x}_1^0, \ldots, \mathbf{x}_n^0, x_{n+1}^0)$. Let a subsequence converging to $(\mathbf{x}_1^0, \ldots, \mathbf{x}_n^0, x_{n+1}^0)$ be $\{\mathbf{x}_1^t, \ldots, \mathbf{x}_n^t, x_{n+1}^t\}$. Denote C as an upper bound for the absolute value of the gradients $\nabla K_{j\mathbf{x}_i}$ $(i, j = 1, \ldots, n)$. By rearranging the terms in the inequalities defining R^t_{n+1} and taking absolute values of both sides we get the inequality for any $s>t$

$$nC\|\mathbf{x}^s-\mathbf{x}^t\| \geqq \left\| x^t_{n+1} - \sum_{i=1}^{n} K_i(\mathbf{x}^t_1, \ldots, \bar{\mathbf{x}}^t_i, \ldots, \mathbf{x}^t_n) \right\|.$$

Taking into account that $x^t_{n+1} \leqq 0$ and

$$-\sum_{i=1}^{n} K_i(\mathbf{x}^t_1, \ldots, \bar{\mathbf{x}}^t_i, \ldots \mathbf{x}^t_n) \leqq 0$$

(see (8)) we conclude that

$$x^t_{n+1} \to 0$$

$$\sum_{i=1}^{n} K_i(\mathbf{x}^t_1, \ldots, \bar{\mathbf{x}}^t_i, \ldots, \mathbf{x}^t_n) \to 0$$

as $t \to \infty$ because $\|\mathbf{x}^s - \mathbf{x}^t\| \to 0$. This means by virtue of Theorem 2 that $(\mathbf{x}^0_1, \ldots, \mathbf{x}^0_n)$ is an equilibrium point. ∎

It is worth mentioning that if Γ is a polyhedral game, then problem (7) is a linear programming problem and at each iteration a new row (cut) is added to the problem of the previous step. Thus the computational advantages of reoptimization techniques can speed up the iterative procedure.

4.4. CONCAVE GAMES WITH UNIQUE EQUILIBRIUM POINTS

Let $\Gamma = \{\Sigma_1, \ldots, \Sigma_n, K_1, \ldots, K_n\}$ be a concave n-person game as defined in Section 4.1. Since in case of most n-person games the existence of multiple equilibrium points may cause conceptional and interpretational problems, games with unique equilibrium points deserve special attention.

Knowing that a strictly concave function has a unique maximumpoint one might suspect that strict concavity of the pay-off functions in a concave game also assures the uniqueness of the equilibrium point. To illustrate that it is not so let us consider the following concave two-person game

$$\Gamma' = \{\Sigma_1, \Sigma_2, K_1, K_2\}.$$

$$\Sigma_1 = [0; 1],$$

$$\Sigma_2 = [0; 1].$$

$$K_1(x_1, x_2) = x_1 f(x_1 + x_2) - g_1(x_1),$$

$$K_2(x_1, x_2) = x_2 f(x_1 + x_2) - g_2(x_2),$$

where $g_1(x_1) = 0.5x_1$; $g_2(x_2) = 0.5x_2$ and

$$f(x) = \begin{cases} 1.75 - 0.5x & \text{if} \quad 0 \leqq x \leqq 1.5 \\ 2.5 - x, & \text{if} \quad 1.5 \leqq x \leqq 2 \end{cases}.$$

It can easily be seen that Γ' is a concave game and $K_1(K_2)$ is a strictly concave function of $x_1(x_2)$ for fixed values of $x_2(x_1)$.

It is an elementary exercise for the reader to show that any point of the set

$$X^* = \{(x_1, x_2) | 0.5 \leqq x_1 \leqq 1, 0.5 \leqq x_2 \leqq 1, x_1 + x_2 = 1.5\}$$

is an equilibrium point of Γ'.

Let us now return to our original concave n-person game Γ and suppose that the pay-off functions K_i are twice continuously differentiable. Consider an arbitrary nonnegative vector $\mathbf{r} \in \mathbb{R}^n$ and define the function $\mathbf{h}: \mathbb{R}^N \to \mathbb{R}^N$ by

$$(10) \qquad \mathbf{h}(\mathbf{x}, \mathbf{r}) = \begin{bmatrix} r_1 \nabla_1 K_1(\mathbf{x}) \\ \cdots\cdots \\ r_n \nabla_n K_n(\mathbf{x}) \end{bmatrix}$$

where $\nabla_i K_i(\mathbf{x})$ denotes the gradient of $K_i(\mathbf{x})$ with respect to $\mathbf{x}_i$, $(i = 1, \ldots, n)$ and N is the number of components in the vector $\mathbf{x} = (\mathbf{x}_1, \ldots, \mathbf{x}_n)$.

The concave game Γ is said to be *diagonally strictly concave* on Σ if for any $\mathbf{x}^{(0)} \neq \mathbf{x}^{(1)} \in \Sigma$ and for some $\mathbf{r} \geqq \mathbf{0}$ the inequality

$$(11) \qquad (\mathbf{x}^{(1)}-\mathbf{x}^{(0)})\mathbf{h}(\mathbf{x}^{(0)},\mathbf{r})+(\mathbf{x}^{(0)}-\mathbf{x}^{(1)})\mathbf{h}(\mathbf{x}^{(1)},\mathbf{r})>0$$

holds. ($\Sigma = \Sigma_1 \times \ldots \times \Sigma_n$.)

We first give a sufficient condition for a concave game Γ to be diagonally strictly concave. Let $\mathbf{H}$ denote the Jacobian of $\mathbf{h}$.

THEOREM 4. [148] If the matrix $\mathbf{H}(\mathbf{x},\mathbf{r})+\mathbf{H}^T(\mathbf{x},\mathbf{r})$ is negative definite for a fixed $\mathbf{r} \geqq \mathbf{0}$ and for arbitrary $\mathbf{x} \in \Sigma$, then the game Γ is diagonally strictly concave.

Proof. Let $\mathbf{x}^{(0)}, \mathbf{x}^{(1)} \in \Sigma$. Then by convexity of Σ we have

$$\mathbf{x}(\lambda)=\lambda\mathbf{x}^{(1)}+(1-\lambda)\mathbf{x}^{(0)} \in \Sigma, \quad (0 \leqq \lambda \leqq 1)$$

and

$$\frac{d\mathbf{h}(\mathbf{x}(\lambda),\mathbf{r})}{d\lambda}=\mathbf{H}(\mathbf{x}(\lambda),\mathbf{r})\frac{d\mathbf{x}(\lambda)}{d\lambda}=\mathbf{H}(\mathbf{x}(\lambda),\mathbf{r})\,(\mathbf{x}^{(1)}-\mathbf{x}^{(0)}).$$

Integrating both sides on the interval [0, 1] we immediately get

$$\mathbf{h}(\mathbf{x}^{(1)},\mathbf{r})-\mathbf{h}(\mathbf{x}^{(0)},\mathbf{r})=\int_0^1 \mathbf{H}(\mathbf{x}(\lambda),\mathbf{r})\,(\mathbf{x}^{(1)}-\mathbf{x}^{(0)})d\lambda.$$

Premultiplying both sides by $(\mathbf{x}^{(0)}-\mathbf{x}^{(1)})$ we obtain

$$(\mathbf{x}^{(0)}-\mathbf{x}^{(1)})\mathbf{h}(\mathbf{x}^{(1)},\mathbf{r})+(\mathbf{x}^{(1)}-\mathbf{x}^{(0)})\mathbf{h}(\mathbf{x}^{(0)},\mathbf{r})=$$

$$=-\frac{1}{2}\int_0^1 (\mathbf{x}^{(1)}-\mathbf{x}^{(0)})\,[\mathbf{H}(\mathbf{x}(\lambda),\mathbf{r})+\mathbf{H}^{\mathrm{T}}(\mathbf{x}(\lambda),\mathbf{r}]\,(\mathbf{x}^{(1)}-\mathbf{x}^{(0)})d\lambda.$$

The integrand is continuous and negative by the assumption of our theorem, therefore the left hand side is positive which is exactly the definition of diagonally strict concavity. ∎

Now we can state Rosen's uniqueness theorem.

THEOREM 5. [148] Each diagonally strictly concave game has precisely one equilibrium point.

Proof. Let us suppose that there are two distinct equilibrium points $\mathbf{x}^{(0)}=(\mathbf{x}_1^0, \ldots, \mathbf{x}_n^0)$ and $\mathbf{x}^{(1)}=(\mathbf{x}_1^{(1)}, \ldots, \mathbf{x}_n^{(1)})$. Then, by Theorem 1. there exist nonnegative vectors $\mathbf{u}_k^{(l)}$ to satisfy the Kuhn–Tucker conditions

$$(12) \qquad \mathbf{u}_k^{(l)}\mathbf{g}_k(\mathbf{x}_k^{(l)})=0\,, \quad (l=1,2)$$

$$(13) \qquad \nabla_k K_k(\mathbf{x}^{(l)})+\mathbf{u}_k^{(l)}\nabla_k \mathbf{g}_k(\mathbf{x}_k^{(l)})=0, \quad (k=1,\ldots,n)\,.$$

Supposing that the number of inequalities defining Σ_k is p_k, (13) can be written in a detailed form

$$(14) \qquad \nabla_k K_k(\mathbf{x}^{(l)})+\sum_{j=1}^{p_k} u_{kj}^{(l)}\nabla_k g_{kj}(\mathbf{x}_k^{(l)})=0,\ (l=1,2), \quad (k=1,\ldots,n)$$

where $u_{kj}^{(l)}$ and g_{kj} denote the k-th component of $\mathbf{u}_k^{(l)}$ and $\mathbf{g}_k$ resp. Multiplying (14) by $r_k(\mathbf{x}_k^{(1)}-\mathbf{x}_k^{(0)})$ for $l=0$ and by $r_k(\mathbf{x}_k^{(0)}-\mathbf{x}_k^{(1)})$ for $l=1$ and summing up for $k=1,\ldots,n$ we get

$$0=\{(\mathbf{x}^{(1)}-\mathbf{x}^{(0)})\mathbf{h}(\mathbf{x}^0,\mathbf{r})+(\mathbf{x}^{(0)}-\mathbf{x}^{(1)})\mathbf{h}(\mathbf{x}^{(1)},\mathbf{r})\}+$$
$$+\sum_{k=1}^{n}\left\{\sum_{j=1}^{p_k} r_k[u_{kj}^{(0)}(\mathbf{x}_k^{(1)}-\mathbf{x}_k^{(0)})\nabla_k g_{kj}(\mathbf{x}_k^{(0)})+\right.$$
$$\left.+u_{kj}^{(1)}(\mathbf{x}_k^{(0)}-\mathbf{x}_k^{(1)})\nabla_k g_{kj}(\mathbf{x}_k^{(1)})]\right\}.$$

The first term of the right-hand side is positive by the diagonally strict concavity of the game Γ. Thus the second term is negative. By the concavity of the functions $\mathbf{g}_k$, $(k=1,\ldots,n)$ we get

$$0>\sum_{k=1}^{n}\left\{\sum_{j=1}^{p_k} r_k[u_{kj}^{(0)}(g_{kj}(\mathbf{x}_k^{(1)})-g_{kj}(\mathbf{x}_k^{(0)}))+\right.$$
$$\left.+u_{kj}^{(1)}(g_{kj}(\mathbf{x}_k^{(0)})-g_{kj}(\mathbf{x}_k^{(1)})]\right\}.$$

Using (12) we obtain

$$0 > \sum_{k=1}^{n} \left\{ \sum_{j=1}^{p_k} r_k [u_{kj}^{(0)'} g_{kj}(\mathbf{x}_k^{(1)}) + u_{kj}^{(1)'} g_{kj}(\mathbf{x}_k^{(0)})] \right\} \geqq 0,$$

a contradiction. ∎

As an example let us consider an n-person game Γ where the strategy set of the ith player $\Sigma_i = \{\mathbf{x}_i | \mathbf{x}_i \geqq 0, \mathbf{1}\mathbf{x}_i = 1\}, (i = 1, \ldots, n)$ is a simplex of $\mathbb{R}^{m_i}$ and the pay-off function is quadratic

$$K_i(\mathbf{x}) = \sum_{j=1}^{n} [\mathbf{c}_{ij} + \mathbf{x}_i \mathbf{C}_{ij}] \mathbf{x}_j, \quad (i = 1, \ldots, n).$$

By simple calculation

$$\mathbf{H}(\mathbf{x}, \mathbf{r}) = \mathbf{D}\mathbf{C}$$

where

$$\mathbf{C} = \begin{bmatrix} 2\mathbf{C}_{11} & \mathbf{C}_{12} & \ldots & \mathbf{C}_{1n} \\ \mathbf{C}_{21} & 2\mathbf{C}_{22} & \ldots & \mathbf{C}_{2n} \\ \ldots & \ldots & \ldots & \ldots \\ \mathbf{C}_{n1} & \mathbf{C}_{n2} & \ldots & 2\mathbf{C}_{nn} \end{bmatrix}$$

and

$$\mathbf{D} = \begin{bmatrix} r_1 \mathbf{E}_{m_1} & \mathbf{0} & \ldots & \mathbf{0} \\ \mathbf{0} & r_2 \mathbf{E}_{m_2} & \ldots & \mathbf{0} \\ \vdots & \vdots & & \\ \mathbf{0} & \mathbf{0} & \ldots & r_n \mathbf{E}_{m_n} \end{bmatrix}$$

($\mathbf{E}_{m_k}$ denotes the identity matrix of order m_k, $(k = 1, \ldots, n)$). By Theorem 5 if there exists a vector $\mathbf{r} \geqq \mathbf{0}$ such that the matrix $\mathbf{D}\mathbf{C} + \mathbf{C}^T\mathbf{D}$ is negative definite, then the game defined above has a unique equilibrium point.

As we have seen for zero-sum convex-concave games (see Section 4.3), the iterative solution of mathematical programming problems could produce an equilibrium point. A similar procedure can be devised for diagonally strict concave games without assuming the game to be either zero-sum or convex-concave.

Let $\Gamma = \{\Sigma_1, \ldots, \Sigma_n; K_1, \ldots, K_n\}$ be a diagonally strict concave game where $\Sigma_i \subset \mathbb{R}^{m_i}$, $(i=1, \ldots, n)$ and (11) holds for some $\mathbf{r} > \mathbf{0}$. Consider the function

$$\Phi(\mathbf{x}, \mathbf{y}) = \sum_{k=1}^{n} r_k K_k(\mathbf{x}_1, \ldots, \mathbf{x}_{k-1}, \mathbf{y}_k, \mathbf{x}_{k+1}, \ldots, \mathbf{x}_n)$$

where $\mathbf{r} = (r_1, \ldots, r_n) > \mathbf{0}$. It can easily be seen that any vector $\mathbf{x}^* \in \Sigma$ satisfying

$$\text{(15)} \qquad \Phi(\mathbf{x}^*, \mathbf{x}^*) \geqq \Phi(\mathbf{x}^*, \mathbf{y}), \quad (\mathbf{y} \in \Sigma)$$

is an equilibrium point. To see this all we have to do is substitute $\mathbf{y} = (\mathbf{x}_1^*, \ldots, \mathbf{x}_{k-1}^*, \mathbf{y}_k, \mathbf{x}_{k+1}^*, \ldots, \mathbf{x}_n^*)$ into (15) for fixed k and arbitrary $\mathbf{y}_k \in \Sigma_k$. All terms but the k-th one vanish and we get

$$r_k K_k(\mathbf{x}^*) \geqq r_k K_k(\mathbf{y})$$

which is the definition of an equilibrium point if k goes from 1 to n. (Notice that we did not assert that any equilibrium point satisfies (15).)

In the following our aim is to find a point $\mathbf{x}^*$ to satisfy (15). The next two lemmas provide useful properties of equilibrium points satisfying (15).

LEMMA 1. A vector $\mathbf{x}^* \in \Sigma$ satisfies inequality (15) if and only if

$$\text{(16)} \qquad \max\{\mathbf{h}(\mathbf{x}^*, \mathbf{r})(\mathbf{x} - \mathbf{x}^*) \mid \mathbf{x} \in \Sigma\} = 0.$$

($\mathbf{h}$ is the function defined in (10).)

Proof. Let $\mathbf{x}^*$ satisfy (15). Since $\Phi(\mathbf{x}^*, \mathbf{y})$ attains its maximum at $\mathbf{x}^*$ on Σ, by the differentiability of Φ with respect to $\mathbf{y}$ we have

$$\text{(17)} \qquad \nabla_{\mathbf{y}} \Phi(\mathbf{x}^*, \mathbf{x}^*)(\mathbf{y} - \mathbf{x}^*) \leqq 0, \quad \text{for all} \quad \mathbf{y} \in \Sigma.$$

Computing $\nabla_{\mathbf{y}}\Phi(\mathbf{x}^*, \mathbf{x}^*)$ we get $\mathbf{h}(\mathbf{x}^*, \mathbf{r})$ and (17) becomes

$$\mathbf{h}(\mathbf{x}^*, \mathbf{r})(\mathbf{y}-\mathbf{x}^*)\leqq 0, \quad \text{for all} \quad \mathbf{y}\in\Sigma$$

which is exactly (16).

On the other hand, assuming $\mathbf{h}(\mathbf{x}^*, \mathbf{r})(\mathbf{y}-\mathbf{x}^*)\leqq 0$, $(\mathbf{y}\in\Sigma)$ we derive the inequalities

$$\Phi(\mathbf{x}^*, \mathbf{x}^*)-\Phi(\mathbf{x}^*, \mathbf{y})\geqq \mathbf{h}(\mathbf{y}, \mathbf{r})(\mathbf{x}^*-\mathbf{y})\geqq$$

$$\geqq \mathbf{h}(\mathbf{y}, \mathbf{r})(\mathbf{x}^*-\mathbf{y})+\mathbf{h}(\mathbf{x}^*, \mathbf{r})(\mathbf{y}-\mathbf{x}^*)>0$$

which means that $\mathbf{x}^*$ satisfies (15). (Here we used the concavity of $\Phi(\mathbf{x}^*, \mathbf{y})$ and the diagonally strict concavity of the game Γ). ▮

LEMMA 2. A vector $\mathbf{x}^*\in\Sigma$ satisfies inequality (15) if and only if $(\mathbf{x}^*, \mathbf{x}^*)$ is an equilibrium point of the two-person zero-sum game $G=\{\Sigma, \Sigma, f\}$, where

$$f(\mathbf{x}, \mathbf{y})=\mathbf{h}(\mathbf{y}, \mathbf{r})(\mathbf{x}-\mathbf{y}).$$

Proof. (a) Suppose first that $\mathbf{x}^*$ satisfies (15). Then by Lemma 1 we get

$$\mathbf{h}(\mathbf{x}^*, \mathbf{r})(\mathbf{x}-\mathbf{x}^*)\leqq 0$$

for all $\mathbf{x}\in\Sigma$ that is

$$f(\mathbf{x}, \mathbf{x}^*)\leqq 0=f(\mathbf{x}^*, \mathbf{x}^*) \tag{18}$$

for any $\mathbf{x}\in\Sigma$. Thus we only need to show that

$$f(\mathbf{x}^*, \mathbf{x})\geqq 0, \quad (\mathbf{x}\in\Sigma).$$

Assume on the contrary that there exists a $\mathbf{y}\in\Sigma$ such that $f(\mathbf{x}^*, \mathbf{y})<0$. Then by (11) we have

$$0>f(\mathbf{x}^*, \mathbf{y})=\mathbf{h}(\mathbf{y}, \mathbf{r})(\mathbf{x}^*-\mathbf{y})>\mathbf{h}(\mathbf{x}^*, \mathbf{r})(\mathbf{x}^*-\mathbf{y})$$

or

$$\mathbf{h}(\mathbf{x}^*, \mathbf{r})(\mathbf{y}-\mathbf{x}^*)>0$$

contradicting (18).

(b) Let now $(\mathbf{x}^*, \mathbf{x}^*)$ be an equilibrium point of G. Then for any pair $\mathbf{x}, \mathbf{y} \in \Sigma$ the inequalities

$$f(\mathbf{x}, \mathbf{x}^*) \leqq f(\mathbf{x}^*, \mathbf{x}^*) \leqq f(\mathbf{x}^*, \mathbf{y})$$

hold. The first inequality

$$\mathbf{h}(\mathbf{x}^*, \mathbf{r})(\mathbf{x} - \mathbf{x}^*) \leqq 0$$

and Lemma 1 imply that $\mathbf{x}^*$ satisfies (15). ∎

As a consequence of Lemma 1, to determine an equilibrium point of Γ it is sufficient to find an $\mathbf{x}^* \in \Sigma$ which maximizes the function $\min_{\mathbf{y} \in \Sigma} f(\mathbf{x}, \mathbf{y})$. The following algorithm does this in an iterative manner.

Let $\mathbf{x}^{(1)} \in \Sigma$ be an arbitrary initial vector and solve the programming problem

$$(19) \quad f(\mathbf{x}, \mathbf{x}^{(1)}) \to \max$$
$$\mathbf{x} \in \Sigma .$$

Denote $\mathbf{x}^{(2)}$ an optimal solution and let $\mu_1 = f(\mathbf{x}^{(2)}, \mathbf{x}^{(1)})$. If $\mu_1 = 0$, then $\mathbf{x}^{(1)}$ is an equilibrium point by Lemma 2. Since $f(\mathbf{x}^{(1)}, \mathbf{x}^{(1)}) = 0$ we may assume that $\mu_1 > 0$. At the kth step of the algorithm we have k vectors $\mathbf{x}^{(1)}, \mathbf{x}^{(2)}, \ldots, \mathbf{x}^{(k)}$ and $k-1$ scalars $\mu_1, \mu_2, \ldots, \mu_{k-1} > 0$. Then $\mathbf{x}^{(k+1)}$ and μ_k are defined as solutions of the programming problem

$$(20) \quad \mu \to \max$$
$$f(\mathbf{x}, \mathbf{x}^{(i)}) \geqq \mu, \quad (i = 1, \ldots, k)$$
$$\mathbf{x} \in \Sigma .$$

Since

$$f(\mathbf{x}^{(k)}, \mathbf{x}^{(i)}) \geqq \mu_{k-1} \geqq 0, \quad (i = 1, \ldots, k-1)$$
$$f(\mathbf{x}^{(k)}, \mathbf{x}^{(k)}) = 0$$

therefore $\mu_k \geqq 0$.

The following theorem establishes the convergence of the above algorithm.

THEOREM 6. [213] There exists a subsequence $\{\mathbf{x}^{(k_i)}\}$ of $\{\mathbf{x}^{(k)}\}$ converging to an equilibrium point of Γ.

Proof. First we show that $\mu_k \to 0$. Since at each iteration an additional constraint is adjoined to (20), the sequence $\{\mu_k\}$ is nonincreasing and nonnegative, therefore convergent. Let $\{\mathbf{x}^{(k_i)}\}$ be a convergent subsequence of $\{\mathbf{x}^{(k)}\}$ which must exist because Σ is closed and bounded. Then by the definition of μ_{k_i-1} we have

$$0 \leqq \mu_{k_i-1} = \max\left\{\min_{1\leqq k\leqq k_i-1} \mathbf{h}(\mathbf{x}^{(k)}, \mathbf{r})(\mathbf{x}-\mathbf{x}^{(k)}) \mid \mathbf{x}\in\Sigma\right\} =$$

$$= \min_{1\leqq k\leqq k_i-1} \mathbf{h}(\mathbf{x}^{(k)}, \mathbf{r})(\mathbf{x}^{(k_i)}-\mathbf{x}^{(k)}) \leqq$$

$$\leqq \mathbf{h}(\mathbf{x}^{(k_i-1)}, \mathbf{r})(\mathbf{x}^{(k_i)}-\mathbf{x}^{(k_i-1)})$$

implying that $\mu_{k_i-1}\to 0$ as $k_i\to\infty$. Since $\{\mu_k\}$ is monotone nonincreasing therefore $\mu_k\to 0$ as $k\to\infty$.

Let now $\mathbf{x}^*$ be an equilibrium point of Γ and define

$$\text{(21)}\qquad \delta(t) = \min\{[\mathbf{h}(\mathbf{x}, \mathbf{r}) - \mathbf{h}(\mathbf{y}, \mathbf{r})](\mathbf{y}-\mathbf{x}) \mid \|\mathbf{x}-\mathbf{y}\| \geqq$$

$$\geqq t, \mathbf{x}, \mathbf{y}\in\Sigma\}.$$

By (11) $\delta(t)\geqq 0$. Define now the indices k_i according to

$$\delta(\|\mathbf{x}^{(k_i)}-\mathbf{x}^*\|) = \min_{1\leqq k\leqq i} \delta(\|\mathbf{x}^{(k)}-\mathbf{x}^*\|), \quad (i=1, 2, \ldots).$$

Then, for $k=1, 2, \ldots, i$ we obtain

$$\delta(\|\mathbf{x}^{(k_i)}-\mathbf{x}^*\|) \leqq [\mathbf{h}(\mathbf{x}^{(k)}, \mathbf{r}) - \mathbf{h}(\mathbf{x}^*, \mathbf{r})](\mathbf{x}^*-\mathbf{x}^{(k)}) =$$

$$= \mathbf{h}(\mathbf{x}^{(k)}, \mathbf{r})(\mathbf{x}^*-\mathbf{x}^{(k)}) + \mathbf{h}(\mathbf{x}^*, \mathbf{r})(\mathbf{x}^*-\mathbf{x}^{(k)}) \leqq$$

$$\leqq \mathbf{h}(\mathbf{x}^{(k)}, \mathbf{r})(\mathbf{x}^*-\mathbf{x}^{(k)})$$

since $\mathbf{h}(\mathbf{x}^*, \mathbf{r})(\mathbf{x}^*-\mathbf{x}^{(k)})\leqq 0$ by Lemma 2. Thus

$$\delta(\|\mathbf{x}^{(k_i)}-\mathbf{x}^*\|)\leqq \min_{1\leqq k\leqq i} \mathbf{h}(\mathbf{x}^{(k)},\mathbf{r})(\mathbf{x}^*-\mathbf{x}^{(k)})\leqq \tag{22}$$

$$=\max_{\mathbf{x}\in\Sigma}\min_{1\leqq k\leqq i} \mathbf{h}(\mathbf{x}^{(k)},\mathbf{r})(\mathbf{x}-\mathbf{x}^{(k)})=$$

$$=\min_{1\leqq k\leqq i} \mathbf{h}(\mathbf{x}^{(k)},\mathbf{r})(\mathbf{x}^{(i+1)}-\mathbf{x}^{(k)})=\mu_i\,.$$

Therefore $\delta(\|\mathbf{x}^{(k_i)}-\mathbf{x}^*\|)\to 0$ as $i\to\infty$. Because of the following properties of the function $\delta(t)$ $\delta(t_i)\to 0$ implies $t_i\to 0$ if $\{t_i\}$ is convergent:

(i) $\delta(t)$ is a continuous function of t,

(ii) $\delta(t)>0$ if $t>0$ by the diagonally strict concavity of Γ.

Hence $\|\mathbf{x}^{(k_i)}-\mathbf{x}^*\|\to 0$ should hold which is exactly what our theorem asserts. ∎

It should be noted that we have not utilized here the assumption that the strategy sets Σ_i are defined by a finite number of constraints. Therefore the uniqueness of the equilibrium point for the game Γ is not assured. Of course, under the assumptions of Theorem 5, the sequence $\{\mathbf{x}_k\}$ generated by the programming problems (20) converges to the unique equilibrium point.

The algorithm of this section and Section 4.3 are very similar as far as the "cutting-plane like" nature of the iterations is concerned but the conditions under which they can be applied are somewhat different. The algorithm of Section 4.3 can be applied for games which may have multiple equilibria whereas the algorithm of this Section is of no use for polyhedral games because they are not diagonally strict concave games. (The reader himself can readily verify this statement.)

5. The Scarf–Hansen algorithm for approximating an equilibrium point of a finite n-person game

Let $\Gamma=\{\Sigma_1, \ldots, \Sigma_N, K_1, \ldots, K_N\}$ be the mixed extension of a finite n-person game (not necessarily zero-sum!). Based on the notion of primitive sets first defined and studied by Scarf [153], an interesting combinatorial algorithm was developed by Hansen for approximating an equilibrium point of Γ [65].

First we will briefly study primitive sets. Let $X=\{\mathbf{x}^1, \ldots, \mathbf{x}^h\}$ be a collection of vectors $\mathbf{x}^j \in \mathbb{R}^n$

$$\text{(1)} \qquad \mathbf{x}^j=\left(\frac{m_1^j}{D}, \ldots \frac{m_n^j}{D}\right), \quad (j=1, \ldots, h)$$

where $m_i^j \geqq -1$, $(i=1, \ldots, n)$, $D>0$ are integers and

$$\sum_{i=1}^{n} m_i^j = D .$$

Let $\mathbf{M}$ be a quadratic matrix of order n, the columns of which belong to X, and $\mathbf{M}'=D\,\mathbf{M}$. The columns of $\mathbf{M}$ form a *primitive set* in X if after proper rearrangement of the columns of $\mathbf{M}$ (and $\mathbf{M}'$), there exists a permutation $I(l)$ of the numbers $1, \ldots, n$ such that

(i) the lth and $(l-1)$st columns of $\mathbf{M}$ (and $\mathbf{M}'$) are identical except for the $(I(l)-1)$st and $I(l)$th components,

$$\text{(ii)} \qquad m_{kl}=m_{k,l-1}+1, \quad \text{if} \quad k=I(l)-1,$$

$$m_{kl}=m_{k,l-1}-1, \quad \text{if} \quad k=I(l),$$

if $l=1$ and $I(l)=1$, then by definition $l-1=n$ and $I(l)-1=n$ resp.

The columns of the following matrix **M** form a primitive set in $X(n=4, D=80)$

$$\text{(2)} \qquad \mathbf{M}=\frac{1}{80}\begin{bmatrix} 20 & 20 & 20 & 21 \\ 30 & 31 & 31 & 30 \\ 16 & 15 & 16 & 16 \\ 14 & 14 & 13 & 13 \end{bmatrix}.$$

$$I(l)=(1, 3, 4, 2).$$

The following properties are direct consequences of the definition of primitive sets:

1. In any row of $\mathbf{M}'$ the difference between the largest and smallest elements is exactly 1. Therefore if D is large, then the columns of **M** are "almost" identical.

2. Adding up q consecutive rows of $\mathbf{M}'$, $(q<n)$ we get a vector whose largest and smallest component differ exactly by 1.

3. Let $\mathbf{m}_l$ and $\mathbf{m}_{l-1}$ be columns of $\mathbf{M}'$. Then

$$\mathbf{m}_l-\mathbf{m}_{l-1}=(0, \ldots, 0, 1, -1, 0, \ldots 0)$$

where -1 is on the $I(l)$th place. (If $l=1$, then $l-1=n$ by definition.) The replacement of an element of a primitive set by a vector from X in such a way that the set obtained remains a primitive set is called an *elementary transformation.* If $\mathbf{m}_\alpha$ is the vector leaving the primitive set, then the entering vector is defined by

$$\text{(3)} \qquad \mathbf{m}_\alpha^*=\mathbf{m}_{\alpha-1}+\mathbf{m}_{\alpha+1}-\mathbf{m}_\alpha .$$

(If $\alpha=1$, then $\alpha-1=n$ and if $\alpha=n$, then $\alpha+1=1$ by definition.)

The new matrix

$$\mathbf{M}^*=(\mathbf{m}_1, \ldots, \mathbf{m}_{\alpha-1}, \mathbf{m}_\alpha^*, \mathbf{m}_{\alpha+1}, \ldots, \mathbf{m}_n)$$

represents a primitive set, since

$$\mathbf{m}_\alpha^*-\mathbf{m}_{\alpha-1}=\mathbf{m}_{\alpha+1}-\mathbf{m}_\alpha$$

$$\mathbf{m}_{\alpha+1}-\mathbf{m}_\alpha^*=\mathbf{m}_\alpha-\mathbf{m}_{\alpha-1}$$

and thus $\mathbf{m}_\alpha^*$ differs from its neighbours in two components. The new permutation $I^*(l)$ belonging to $\mathbf{M}^*$ is given by

$$I^*(l) = I(l)$$

$$I^*(\alpha) = I(\alpha+1), \quad (l = 1, \ldots, n,\ l \neq \alpha,\ \alpha+1)$$

$$I^*(\alpha+1) = I(\alpha).$$

As an illustration let us replace the second column in the matrix $\mathbf{M}$ of (2). The transformed matrix $\mathbf{M}^*$ and permutation I^* is as follows

$$\mathbf{M}^* = \begin{bmatrix} 20 & 20 & 20 & 21 \\ 30 & 30 & 31 & 30 \\ 16 & 17 & 16 & 16 \\ 14 & 13 & 13 & 13 \end{bmatrix}$$

$$I^*(l) = (1, 4, 3, 2).$$

The elementary transformation is said to be *admissible* if each component of the new column satisfies

$$(4) \qquad -1 \leqq m_{k\alpha}^* \leqq D+1, \quad (k = 1, \ldots, n).$$

Assign to each vector $\mathbf{x}^j \in X$ one of the numbers $1, \ldots, n$ and call it an *indicator*. If $\mathbf{x}^j \geqq \mathbf{0}$, then its indicator is arbitrary. Otherwise the indicator of $\mathbf{x}^j$ is defined to be k if its first negative component is x_k^j.

The Scarf–Hansen algorithm for finding an equilibrium point to the mixed-extension of a finite game is based on the following theorem of Scarf.

THEOREM 1. [153], [65] There exists at least one primitive set with no two vectors having the same indicator.

Proof. Our proof is constructive. We are going to give an algorithm producing a primitive set with the desired property in a finite number of steps.

Let us start from the primitive set represented by the matrix $\mathbf{M}'$

$$\text{(5)} \qquad \mathbf{M}' = \begin{bmatrix} D & D+1 & D+1 & \dots & D+1 \\ 0 & -1 & 0 & \dots & 0 \\ 0 & 0 & -1 & \dots & 0 \\ \dots & \dots & \dots & \dots & \dots \\ 0 & 0 & 0 & \dots & -1 \end{bmatrix}$$

$$I(l) = (1, \dots, n).$$

By definition the indicator of the lth column of $\mathbf{M}'$ is l if $l \geqq 2$. If the first column has 1 as its indicator, then $\mathbf{M}'$ has the desired property. Let us assume therefore that the indicator of the first column is α $(2 \leqq \alpha \leqq n)$. Then we perform an elementary transformation on $\mathbf{M}'$ and replace $\mathbf{m}_\alpha$ by $\mathbf{m}_\alpha^*$ according to (3). If the indicator of $\mathbf{m}_\alpha^*$ is 1, then we have finished, otherwise the transformed matrix $\mathbf{M}^*$ has a column having the same indicator as $\mathbf{m}_\alpha^*$. In the next step this particular vector is to be replaced by an elementary transformation. Proceeding in this fashion – performing an elementary transformation on the column having the same indicator as the vector which has just entered the primitive set – we have at each iteration a primitive set with the following properties:

1. No vector of the primitive set has indicator 1.
2. Each vector except for exactly two has different indicators.
3. One of the vectors with identical indicators has just entered the primitive set.

We have to prove that after a finite number of steps a vector with indicator 1 enters the primitive set.

First we shall prove that any elementary transformation of the kind described above is admissible, i.e., satisfies (4). Assume on the contrary that (4) does not hold. If $m_{v\alpha}^* < -1$ for some row index v, then

$$m_{v,\alpha-1} = m_{v,\alpha+1} = -1 \quad \text{and} \quad m_{v,\alpha} = 0.$$

Thus

$$m_{v,\alpha-1} < m_{v,\alpha} > m_{v,\alpha+1}$$

which means that the vth component of any vector of $\mathbf{M}'$ except for the one to be replaced is -1. Each of them has different indicators and none of the indicators is 1. Therefore $v=n$, otherwise there would be no vector with indicator n. By the definition of indicators the lth component of the vector with indicator l is negative ($l \neq \alpha$, $l=2, \ldots, n$). Without loss of generality we may assume that the first column of $\mathbf{M}'$ has indicator 2. By Property (i)

$$m_{21} \leqq -1$$

$$m_{l1} \leqq 0 \qquad (l=3, \ldots, n-1)$$

$$m_{n1} = -1 ,$$

implying $m_{11} \geqq D+2$ which is a contradiction. Therefore $m^*_{k\alpha} \geqq -1$.

Suppose now that $m^*_{k\alpha} > D+1$ for some k. By (3)

$$m_{k\alpha} = m_{k,\alpha+1} = D+1 , \quad m_{k\alpha} = D .$$

Thus

$$m_{k,\alpha-1} > m_{k\alpha} < m_{k\alpha+1}$$

implying that

$$m_{kl} = D+1 , \quad (l \neq \alpha) .$$

Therefore any vector of $\mathbf{M}'$, except for the one to be replaced, has a negative component. Since any of these has different indicators, and none of the indicators is 1 we have $k=1$, and each column of $\mathbf{M}'$ except for the first has a negative component in different rows. This means that the primitive set under consideration is precisely the initial one given by (5). But at the first step the vector to replaced is not the first one (remember the definition of the algorithm!) and the initial primitive set can never return. Should this situation ever occur, the primitive set of the previous step could be obtained by replacing the first column, which is the only column not replaceable in $\mathbf{M}'$. (Note that the operation of an elementary transformation is reversible. According to (3) replacing $\mathbf{m}_\alpha^*$ we get back $\mathbf{m}_\alpha$).

Thus our algorithm provides a primitive set which consists of vectors, with distinct indicators, in a finite number of steps. The finiteness follows from the fact that the number of all possible primitive sets is finite and no cycle can occur, since from the initial primitive set there is only one way to go. In all other cases there are two directions: One of these is the primitive set we have just performed an elementary transformation on and the other is the new primitive set after the transformation. The existence of any cycle would imply the existence of a point along the "path" of primitive sets with at least three possible directions. ∎

Let us consider now the mixed extension of an N-person finite game given in normal form. Denote s_i the number of pure strategies of the ith player and p_{ki} the probability that player i applies his kth pure strategy. Of course

$$\sum_k p_{ki} = 1 \qquad \text{(for all } i, k)$$
$$p_{ki} \geqq 0 .$$

Let K_v be the pay-off for player v; i.e.,

$$K_v = K_v(\mathbf{p}_1, \ldots, \mathbf{p}_N)$$

is the expected pay-off of player v if player i plays his kth strategy with probability p_{ik}, $(k=1, \ldots, s_i;\ i=1, \ldots, N)$ and $\mathbf{p}_i = (p_{1i}, \ldots, p_{s_i i})$.

As we already know the N-tuple $\bar{\mathbf{p}} = (\bar{\mathbf{p}}_1, \ldots, \bar{\mathbf{p}}_N)$ is said to be an equilibrium point if for any probability vector $\mathbf{p}_v$

$$K_v(\bar{\mathbf{p}}_1, \ldots, \mathbf{p}_{v-1}, \mathbf{p}_v, \bar{\mathbf{p}}_{v+1}, \ldots, \bar{\mathbf{p}}_N) \leqq K_v(\bar{\mathbf{p}}_1, \ldots, \bar{\mathbf{p}}_N), \qquad (v=1, \ldots, N).$$

Consider now the set X defined by (1) assuming that D is divisible by N. Let

$$n=\sum_{i=1}^{N} s_i-N+1$$

and $\mathbf{x}^j$ be an arbitrary nonnegative element of X. To any such $\mathbf{x}^j$ we assign an n-tuple $(\mathbf{p}_1^j, \ldots, \mathbf{p}_N^j)$ in the following way:

Let

$$\text{(6)} \qquad t_0=0 \quad \text{and} \quad t_i=s_i-1, \quad (i=1, \ldots, N),$$

$$r_{ki}=1+\sum_{v=0}^{i-1} t_v+k, \quad (k=1, \ldots, t_i;\, i=1, \ldots, N),$$

$$p_{ki}^j=N\, x_{r_{ki}}^j, \quad (k=1, \ldots, t_i;\, i=1, \ldots, N),$$

$$p_{s_i i}^j=1-\sum_{k=1}^{t_i} p_{ki}^j .$$

Note that $p_{ki}^j \geqq 0$, $(k=1, \ldots, t_i;\, i=1, \ldots, N)$ while $p_{s_i i}^j$ can be negative but

$$\sum_k p_{ki}^j=1 .$$

The following example helps to understand the rather cumbersome assignment rule. Suppose $N=4$, $D=100$, $s_1=s_2=2$, $s_3=s_4=3$, $n=7$.

Let

$$\mathbf{x}^j=(0.10,\ 0.28,\ 0.09,\ 0.15,\ 0.16,\ 0.22,\ 0) .$$

Then

$$\mathbf{p}_1^j=(1.12,\ -0.12),$$

$$\mathbf{p}_2^j=(0.36,\ 0.64),$$

$$\mathbf{p}_3^j=(0.60,\ 0.64,\ -0.24),$$

$$\mathbf{p}_4^j=(0.88,\ 0,\ 0.12) .$$

To the elements of X we assign indicators in the following manner.

Let $\mathbf{x}^j \in X$ be a nonnegative vector. By the assignment rule (6) calculate the vectors $\mathbf{p}_1^j, \ldots, \mathbf{p}_N^j$.

CASE 1. If $p_{s_i i}^j < 0$ for some i, then the indicator of $\mathbf{x}^j$ is 1.

CASE 2. $p_{s_i i}^j \geqq 0$ for all i. Denote

$$K_{ki} = K_{ki}(\mathbf{p}_1^j, \ldots, \mathbf{p}_{i-1}^j, \mathbf{p}_{i+1}^j, \ldots, \mathbf{p}_N^j),$$

the expected pay-off of player i if he plays his kth pure strategy. Let k_i be the smallest index for which

$$K_{k_i i} \geqq K_{ki}, \quad (k=1, \ldots, s_i)$$

holds. Let the indicator of $\mathbf{x}^j$ be 1 if either $p_{s_i i}^j = 0$ for all i, or $k_i = s_i$ for all indices i for which $p_{s_i i}^j > 0$. Otherwise the indicator of $\mathbf{x}^j$ is

$$1 + \sum_{u=0}^{i-1} t_u + k_i$$

where

$$i = \min \{l \mid p_{s_l} > 0 \text{ and } k_l \neq s_l\}.$$

By the algorithm described in the proof of Theorem 1 we determine a primitive set, each element of which has a different indicator. Let this be $\{\mathbf{x}^{j_1}, \ldots, \mathbf{x}^{j_n}\}$.

LEMMA 1. The first component $x_1^{j_l}$ of any vector $\mathbf{x}^{j_l}$, $(l=1, \ldots, n)$ is nonnegative.

Proof. Suppose on the contrary that $x_1^{j_l} < 0$ for some l, say for $l=1$. By definition the indicator of $\mathbf{x}^{j_1}$ is 1 and $x_1^{j_1} = -\frac{1}{D}$, $x_1^{j_l} = 0$, $(l \neq 1)$ since otherwise there would be more vectors with indicator 1 in the primitive set. We assert that there is at least one nonnegative vector in our primitive set. Had any vector of the primitive set a negative component (which is necessarily $-\frac{1}{D}$), each row would contain exactly one of them, since the columns have different indicators. But in this case Property (i) implies that all columns are nonpositive, which contradicts the definition of the primitive set.

Thus there exists a vector, say $\mathbf{x}^{j_2}$, which is nonnegative. Since D is a multiple of N and the first component of $\mathbf{x}^{j_2}$ is 0, therefore the vectors $\mathbf{p}_1^{j_2}, \ldots, \mathbf{p}_N^{j_2}$ associated with $\mathbf{x}^{j_2}$ are such that either $p_{s_i i}^{j_2}=0$ for all i or $p_{s_i i}^{j_2}<0$ for some i. In both cases the indicator of $\mathbf{x}^{j_2}$ is 1, which is a contradiction since $\mathbf{x}^{j_1}$ is the one labeled by indicator 1. ∎

LEMMA 2. If $\mathbf{x}^j$ is a nonnegative element of the primitive set, then the last components of the vectors $\mathbf{p}_1, \ldots, \mathbf{p}_N$ associated with it are nonnegative.

Proof. Assume on the contrary that

$$\mathbf{x}^{j_3} \geqq \mathbf{0}$$

and

$$p_{s_1 1}^{j_3} < 0 .$$

It follows readily that $x_k^{j_3}>0$ for some k, say for $k=2$. By Property (i) $x_2^{j_3}>0$ implies $x_2^{j_l}\geqq 0$ for all l. Thus the vector supplied with indicator 2 got labelled by Case 2 of the labelling rule.

Sum up now the rows of the vectors forming the primitive set from row indices 2 to s_1. Using the fact that D is divisible by N we get the following implications,

$$p_{s_1 1}^{j_3} < 0 \Rightarrow N \sum_{k=2}^{s_1} x_k^{j_3} \geqq 1 + \frac{N}{D} \Rightarrow N \sum_{k=2}^{s_1} x_k^{j_l} \geqq 1 ,$$

$$(l=1, \ldots, n) .$$

Hence $p_{s_1 1} \leqq 0$ for all nonnegative vectors of the primitive set, i.e., no nonnegative vector in the primitive set can be labelled with indicator 2 by Case 2 of the labelling rule. Consequently no vector has indicator 2, which is a contradiction. ∎

THEOREM 2. [65] If $D = rN$ and the positive integer r tends to infinity, then any cluster point of the nonnegative vectors of the corresponding primitive sets provides an equilibrium point of the game Γ.

Proof. By Property (i) the difference between any pair of vectors in the primitive set tends to $\mathbf{0}$ as $D \to \infty$. Since the set X is closed and bounded there exists a cluster point $\mathbf{x}^*$ of the sequence of vectors determined by our algorithm. We have to prove that the N-tuple $(\mathbf{p}_1^*, \ldots, \mathbf{p}_N^*)$ associated with it by the rule (6) is an equilibrium point.

It is obvious that $\mathbf{p}_1^*, \ldots, \mathbf{p}_N^*$ are probability vectors since they are cluster points of probability vectors.

Lemmas 1 and 2 and the fact that $\mathbf{x}^*$ is also a cluster point of vectors having indicator 1 imply that either

$$(7) \qquad p_{s_i i}^* = 0 \quad \text{and} \quad p_{ki}^* > 0 \quad \text{for some } k$$

or

$$p_{s_i i}^* > 0 \quad \text{and} \quad k_i = s_i$$

holds for all i.

Assume that the $r_{k^* i}$th component of $\mathbf{x}^*$ is positive. Then $\mathbf{x}^*$ is also a cluster point of nonnegative vectors with indicator $r_{k^* i}$ by Case 2 of the labelling rule. Hence taking into account (7) and the continuity of the pay-off functions $K_1, \ldots, K_N$ we have that $p_{ki}^* > 0$ implies

$$K_{ki}(\mathbf{p}_1^*, \ldots, \mathbf{p}_{i-1}^*, \mathbf{p}_{i+1}^*, \ldots, \mathbf{p}_N^*) \geqq$$

$$\geqq K_{vi}(\mathbf{p}_1^*, \ldots, \mathbf{p}_{i-1}^*, \mathbf{p}_{i+1}^*, \ldots, \mathbf{p}_N^*)$$

for any $v = 1, \ldots, s_i$, i.e. $(\mathbf{p}_1^*, \ldots, \mathbf{p}_N^*)$ is an equilibrium point. ▮

To illustrate the algorithm let us consider a 3-person game where each player has three pure strategies denoted by $A_1, A_2, A_3, B_1, B_2, B_3, C_1, C_2, C_3$ resp. The pay-off functions are given by the following tables.

Player I

	B_1C_1	B_1C_2	B_1C_3	B_2C_1	B_2C_2	B_2C_3	B_3C_1	B_3C_2	B_3C_3
A_1	1	−1	2	3	3	2	3	4	1
A_2	2	3	−2	2	4	2	3	2	1
A_3	1	2	3	4	3	−1	2	1	4

Player II

	A_1C_1	A_1C_2	A_1C_3	A_2C_1	A_2C_2	A_2C_3	A_3C_1	A_3C_2	A_3C_3
B_1	2	2	0	−1	2	1	2	2	2
B_2	0	4	−1	−1	0	−3	2	1	2
B_3	−1	1	1	2	2	−2	−1	2	−1

Player III

	A_1B_1	A_1B_2	A_1B_3	A_2B_1	A_2B_2	A_2B_3	A_3B_1	A_3B_2	A_3B_3
C_1	1	3	1	2	4	−2	1	3	0
C_2	1	1	4	3	1	1	−1	5	−3
C_3	2	0	2	0	1	4	3	3	2

Now $n=7$ and we choose $D=99$. The Scarf–Hansen algorithm produced the primitive set determined by the following matrix $\mathbf{M}'$

$$\mathbf{M}' = \begin{bmatrix} 66 & 67 & 67 & 67 & 67 & 67 & 67 \\ 0 & -1 & 0 & 0 & 0 & 0 & 0 \\ 0 & 0 & -1 & 0 & 0 & 0 & 0 \\ 33 & 33 & 33 & 32 & 33 & 33 & 33 \\ 0 & 0 & 0 & 0 & -1 & 0 & 0 \\ 0 & 0 & 0 & 0 & 0 & -1 & 0 \\ 0 & 0 & 0 & 0 & 0 & 0 & -1 \end{bmatrix}$$

$I(l)=(1, 2, 3, 4, 5, 6, 7)$.

The strategies computed from the first column

$$\mathbf{p}_1 = (0, 0, 1)$$

$$\mathbf{p}_2 = (1, 0, 0)$$

$$\mathbf{p}_3 = (0, 0, 1)$$

constitute an equilibrium point.

We can start the algorithm from other matrices than the one given by (5). The first row by $\mathbf{M}'$ can be changed by any other row. The matrix obtained in this way determines a primitive set starting out from which another equilibrium point might be detected. Unfortunately this cannot be guaranteed even if we know that there exist more equilibrium points. As of now no efficient method exists to determine all equilibrium points of an n-person finite game.

It is interesting to note that any n-person finite game can be reduced to a finite 3-person game [28]. Thus any method which performs efficiently on 3-person games is a viable tool for solving n-person games. However, no method is known so far which makes use of the fact that there are only 3 players.

6. The oligopoly game[1]

In this chapter we are going to study a special n-person game of economic origin. The oligopoly game can rightly be looked upon as a representative of games very closely attached to real economic problems since many essential features of a competitive situation present themselves here in a typical way. In addition, the methods used for analyzing this game also deserve attention as they illustrate how a special, problem-oriented approach can alleviate the solution of a rather complex problem. The general ideas and techniques applied might prove to be useful for the treatment of other n-person games of similar nature, too.

First we describe briefly the economic model of an oligopoly. There are N groups of manufacturers producing M different goods to be sold on a market. Denote $x_{ki}^{(m)}$ the amount of the mth good produced by the ith manufacturer of the kth group ($k=1, \ldots, N$; $i=1, \ldots, i_k$; $m=1, \ldots, M$; i_k is the number of members in the kth group.) Let $L_{ki}^{(m)}$ denote the upper bound for $x_{ki}^{(m)}$, P_m the price, and C_k the cost functions ($m=1, \ldots, M$; $k=1, \ldots, N$). Then the profit of group k is given by

$$(1) \qquad \sum_{m=1}^{M} \sum_{i=1}^{i_k} x_{ki}^{(m)} P_m\left(\sum_{k=1}^{N} \sum_{i=1}^{i_k} x_{ki}^{(1)}, \ldots \sum_{k=1}^{N} \sum_{i=1}^{i_k} x_{ki}^{(M)}\right) - $$

$$- C_k(x_{k1}^{(1)}, \ldots, x_{ki_k}^{(1)}, \ldots, x_{k1}^{(M)}, \ldots, x_{ki_k}^{(M)})$$

[1] This chapter is based mostly on F. Szidarovszky's works on the subject [179], [180], [181].

Introduce the following notation:

$$\mathbf{x}_k=(x_{k1}^{(1)}, \ldots, x_{ki_k}^{(1)}, \ldots, x_{k1}^{(M)}, \ldots, x_{ki_k}^{(M)}), \quad (k=1, \ldots, N)$$

$$\mathbf{x}=(\mathbf{x}_1, \ldots, \mathbf{x}_N)$$

$$S_k=\{\mathbf{x}_k \mid 0 \leqq x_{ki}^{(m)} \leqq L_{ki}^{(m)}, \quad (i=1, \ldots, i_k; m=1, \ldots, M)\}, \quad (k=1, \ldots, N).$$

Let $\varphi_k(\mathbf{x})$ denote the expression (1).

The N-person game $\Gamma=\{S_1, \ldots, S_N; \varphi_1, \ldots, \varphi_N\}$ is called the *generalized oligopoly game.* We are primarily interested in finding an equilibrium point of Γ, i.e., an N-tuple of strategies $\mathbf{x}^*=(\mathbf{x}_1^*, \ldots, \mathbf{x}_N^*)$ to satisfy the inequalities

$$\text{(2)} \qquad \varphi_k(\mathbf{x}_1^*, \ldots, \mathbf{x}_{k-1}^*, \mathbf{x}_k^*, \mathbf{x}_{k+1}^*, \ldots, \mathbf{x}_N^*) \geqq \geqq \varphi_k(\mathbf{x}_1^*, \ldots, \mathbf{x}_{k-1}^*, \mathbf{x}_k, \mathbf{x}_{k+1}^*, \ldots, \mathbf{x}_N^*)$$

$$\mathbf{x}_k \in S_k, \quad (k=1, \ldots, N).$$

6.1. THE REDUCTION PRINCIPLE

Consider for each $k=1, \ldots, N$ the programming problem

$$\text{(3)} \qquad C_k(\mathbf{x}_k) \to \min$$

$$0 \leqq x_{ki}^{(m)} \leqq L_{ki}^{(m)}, \quad (i=1, \ldots, i_k; m=1, \ldots, M).$$

$$\sum_{i=1}^{i_k} x_{ki}^{(m)} = s_k^{(m)}, \quad (m=1, \ldots, M)$$

where $0 \leqq s_k^{(m)} \leqq \sum_{i=1}^{i_k} L_{ki}^{(m)}$ are parameters. Assuming that the function C_k is a continuous programming problem, (3) has an optimal solution for any $s_k^{(m)}$ satisfying $0 \leqq s_k^{(m)} \leqq \sum_{i=1}^{i_k} L_{ki}^{(m)}$. Let $Q_k(s_k^{(1)}, \ldots, s_k^{(M)})$ denote the optimal objective function's value and consider the game $\bar{\Gamma}$ with strategy sets

(4) $$\bar{S}_k = \mathop{\times}_{m=1}^{M} \left[0, \sum_{i=1}^{i_k} L_{ki}^{(m)}\right], \quad (k=1, \ldots, N)$$

and pay-off functions

(5) $$\Psi_k(\mathbf{s}_1, \ldots, \mathbf{s}_N) = \sum_{m=1}^{M} s_k^{(m)} P_m\left(\sum_{\nu=1}^{N} s_\nu^{(1)}, \ldots, \sum_{\nu=1}^{N} s_\nu^{(M)}\right) - Q_k(s_k^{(1)}, \ldots, s_k^{(M)}), \quad (k=1, \ldots, N).$$

The following lemma is an immediate consequence of definition (2).

LEMMA 1.

(a) Let $\mathbf{x}^*$ be an equilibrium point of the oligopoly game Γ, and let us define $s_k^{(m)*} = \sum_{i=1}^{i_k} x_{ki}^{(m)*}$ for $k=1, \ldots, N$; $m=1, \ldots, M$. Then $\mathbf{s}^* = (s_1^{(1)*}, \ldots, s_1^{(M)*}; \ldots, s_N^{(1)*}, \ldots, s_N^{(M)*})$ is an equilibrium point of the reduced game $\bar{\Gamma}$.

(b) Let $\mathbf{s}^* = (s_1^{(1)*}, \ldots, s_1^{(M)*}, \ldots, s_N^{(1)*}, \ldots, s_N^{(M)*})$ be an equilibrium point of $\bar{\Gamma}$ and $\mathbf{x}_k^*$ an optimal solution of problem (3) with right-hand sides $s_k^{(m)*}$, $(m=1, \ldots, M)$ for $k=1, \ldots, N$. Then $\mathbf{x}^* = (\mathbf{x}_1^*, \ldots, \mathbf{x}_N^*)$ is an equilibrium point of the oligopoly game Γ.

On the basis of Lemma 1 we may assume that each group consists of only one player.

For the development of methods to solve various special cases of the oligopoly game we shall need a few results concerning the connection between the cost functions C_k and Q_k, $(k=1, \ldots, N)$.

LEMMA 2. If C_k is continuous, then so is Q_k.

LEMMA 3.

(a) If C_k is monotone (strictly monotone) increasing in the variables $x_{k1}^{(m)}, \ldots, x_{ki_k}^{(m)}$ for fixed m, $(m=1, \ldots, M)$, then Q_k is monotone (strictly monotone) increasing in $s_k^{(m)}$.

(b) If C_k is monotone (strictly monotone) increasing in all variables, then so is Q_k.

LEMMA 4. If C_k is convex (strictly convex), then so is Q_k.

These lemmas follow directly from well-known theorems of mathematical programming [107] and the definition of games Γ and $\bar{\Gamma}$. ▮

6.2. THE GENERAL MULTIPRODUCT CASE

Our primary goal in this paragraph is to establish the existence of an equilibrium point for the general multiproduct oligopoly game. To this end we first prove a lemma.

LEMMA 5. Let $\mathbf{g}: \mathbb{R}^M \to \mathbb{R}^M$ be a continuously differentiable, concave function defined on a convex subset G of the nonnegative orthant $\mathbb{R}^M_+$. Denote $\mathbf{J}$ the Jacobian of $\mathbf{g}$. If for any $\mathbf{x} \in G$ the matrix $\mathbf{J}(\mathbf{x}) + \mathbf{J}^T(\mathbf{x})$ is negative definite, then the function

$$h(\mathbf{x}) = \mathbf{x}\mathbf{g}(\mathbf{x})$$

is concave on G.

Proof. The gradient of $\mathbf{h}$ can easily be computed as

$$\nabla h(\mathbf{x}) = \mathbf{g}(\mathbf{x}) + \mathbf{x}\mathbf{J}(\mathbf{x}). \tag{5}$$

Since any component of $\mathbf{g}$ is concave, we have

$$\mathbf{g}(\mathbf{y}) - \mathbf{g}(\mathbf{x}) \leqq \mathbf{J}(\mathbf{x})(\mathbf{y} - \mathbf{x}) \tag{6}$$

for any $\mathbf{x}, \mathbf{y} \in G$. By the negative definiteness of the matrix $\mathbf{J}(\mathbf{x}) + \mathbf{J}^T(\mathbf{x})$ we get

$$0 \geqq \frac{1}{2}(\mathbf{y} - \mathbf{x})[\mathbf{J}(\mathbf{x}) + \mathbf{J}^T(\mathbf{x})](\mathbf{y} - \mathbf{x}) =$$

$$= (\mathbf{y} - \mathbf{x})\mathbf{J}^T(\mathbf{x})(\mathbf{y} - \mathbf{x}).$$

Using (6) we deduce

$$\mathbf{y}[\mathbf{g}(\mathbf{y}) - \mathbf{g}(\mathbf{x})] \leqq \mathbf{y}\mathbf{J}(\mathbf{x})(\mathbf{y} - \mathbf{x}) =$$

$$= (\mathbf{y} - \mathbf{x})\mathbf{J}^T(\mathbf{x})\mathbf{y} \leqq (\mathbf{y} - \mathbf{x})\mathbf{J}^T(\mathbf{x})\mathbf{x}.$$

Hence

$$(\mathbf{y}-\mathbf{x})\,[\mathbf{g}(\mathbf{x})+\mathbf{J}(\mathbf{x})\mathbf{x}] \geqq \mathbf{y}\mathbf{g}(\mathbf{y})-\mathbf{x}\mathbf{g}(\mathbf{x})$$

which means that h is concave. ∎

Let us now make the following assumptions.

(a) There exists a nonempty closed, convex set $D \subseteqq \mathbb{R}_+^M$ such that $P_m(s^{(1)}, \ldots, s^{(M)})=0$ for any $(s^{(1)}, \ldots, s^{(M)}) \notin D$ and $m=1, \ldots, M$.

(b) P_m is a continuously differentiable, concave function for any $m=1, \ldots, M$ and the matrix $\mathbf{J}(\mathbf{s})+\mathbf{J}^T(\mathbf{s})$ is negative semidefinite for any $\mathbf{s} \in D$ where $\mathbf{J}$ denotes the Jacobian of $\mathbf{p}=(P_1, \ldots, P_M)$.

(c) For any $(s^{(1)}, \ldots, s^{(M)}) \in D$ and $0 \leqq \tilde{s}^{(m)} \leqq s^{(m)}$, $(m=1, \ldots, M)$ the relation $(s^{(1)}, \ldots, \tilde{s}^{(m)}, \ldots, s^{(M)}) \in D$ holds.

(d) The cost functions C_k, $(k=1, \ldots, N)$ are continuous, convex and monotone increasing with respect to each variable on the set S_k.

THEOREM 1. [180] Under assumptions (a)–(d) the multiproduct oligopoly game has at least one equilibrium point.

Proof. The proof consists of several steps.

1. First we show that if

$$\mathbf{x}^* = (\mathbf{x}_1^*, \ldots, \mathbf{x}_N^*)$$

$$(\mathbf{x}_k^* = (x_k^{(1)*}, \ldots, x_k^{(M)*}), \quad k=1, \ldots, N)$$

is an equilibrium point of the oligopoly game Γ, then $\mathbf{s}^* \in D$, where

$$\mathbf{s}^* = \left(\sum_{k=1}^N x_k^{(1)*}, \ldots, \sum_{k=1}^N x_k^{(M)*}\right).$$

Suppose on the contrary that $\mathbf{s}^* \notin D$. Then at least one component of $\mathbf{s}^*$ is positive, i.e., $x_p^{(q)*} > 0$ for some p and q.

Since D is closed, there exists an $x_p^{(q)}$ to satisfy

$$0 \leqq x_p^{(q)} < x_p^{(q)*}$$

$$\left(\sum_{k=1}^N x_k^{(1)*}, \ldots, \sum_{\substack{k=1\\k\neq p}}^N x_k^{(q)*} + x_p^{(q)}, \ldots, \sum_{k=1}^N x_k^{(M)*}\right) \notin D.$$

Thus using property (d) we obtain

$$\varphi_p(\mathbf{x}) = -C_p(x_p^{(1)*}, \ldots, x_p^{(q)}, \ldots, x_p^{(M)*}) >$$
$$-C_p(\mathbf{x}_p^*) = \varphi_p(\mathbf{x}^*)$$

contradicting (2).

2. Let us consider the reduced game $\tilde{\Gamma} = \{\tilde{S}_1, \ldots, \tilde{S}_N, \varphi_1, \ldots, \varphi_N\}$ by redefining the strategy sets

$$\tilde{S}_k = S_k \cap D_k, \quad (k = 1, \ldots, N)$$

where D_k is the set of those $\mathbf{x}_k = (x_k^{(1)}, \ldots, x_k^{(M)})$ for which there exist vectors $\mathbf{x}'_l \in S_l$, $(l \neq k)$ such that $(\mathbf{x}'_1, \ldots, \mathbf{x}'_{k-1}, \mathbf{x}_k, \mathbf{x}'_{k+1}, \ldots, \mathbf{x}_N) \in D$. We will now prove that Γ and $\tilde{\Gamma}$ are strategically equivalent.

In step 1 we have already proved that any equilibrium point of Γ is an equilibrium point of $\tilde{\Gamma}$, too. Let now $\mathbf{x}^* = (\mathbf{x}_1^*, \ldots, \mathbf{x}_N^*)$ be an equilibrium point of $\tilde{\Gamma}$. Choose an arbitrary index k and a strategy vector $\mathbf{x}_k \in \tilde{S}_k$. If $\mathbf{x}_k \in S_k$, then inequality (2) is obviously satisfied for $\mathbf{x} = (\mathbf{x}_1^*, \ldots, \mathbf{x}_k, \ldots, \mathbf{x}_N^*)$. If $\mathbf{x}_k \notin S_k$, then $\mathbf{x} \notin D$ and

$$\varphi_k(\mathbf{x}_1^*, \ldots, \mathbf{x}_k, \ldots, \mathbf{x}_N^*) = -C_k(\mathbf{x}_k) < -C_k(\mathbf{0}) =$$
$$= \varphi_k(\mathbf{x}_1^*, \ldots, \mathbf{0}, \ldots, \mathbf{x}_N^*) \leqq \varphi_k(\mathbf{x}^*)$$

since $\mathbf{0} \in \tilde{S}_k$ by assumption (c). Thus $\mathbf{x}^*$ is an equilibrium point of Γ, too.

3. By Lemma 1 $\tilde{\Gamma}$ satisfies all the conditions of the Nikaido–Isoda theorem (Theorem 2 of Chapter 3) therefore it has at least one equilibrium point. ∎

It should be remarked that under assumptions (a)–(d) the uniqueness of the equilibrium point is not assured. Relatively simple counterexamples can be constructed to support this statement.

Introduce now the notation $L^{(m)} = \sum_{k=1}^{N} L_{k1}^{(m)}$ and let $s_k^{(m)} \in [0, L^{(m)}]$, $(k = 1, \ldots, N; m = 1, \ldots, M)$. For $\mathbf{x}_k, \mathbf{t}_k \in S_k$, $(k = 1, \ldots, N)$ $\mathbf{s} \in S^*$ define the function

$$\Psi_k(\mathbf{s}, \mathbf{x}_k, \mathbf{t}_k) = \sum_{m=1}^{M} t_k^{(m)} P_m(s^{(1)} - x_k^{(1)} + t_k^{(1)}, \ldots, s^{(M)} - x_k^{(M)} + t_k^{(M)}) - C_k(\mathbf{t}_k) \tag{7}$$

where

$$S^* = \mathop{\times}_{m=1}^{M} [0, L^{(m)}] .$$

Consider now the M-dimensional point-to-set mapping $\mathbf{s} \to X(\mathbf{s})$

$$X(\mathbf{s}) = \{\mathbf{u} \mid \mathbf{u} = \sum_{k=1}^{N} \mathbf{x}_k , \mathbf{x}_k \in X_k(\mathbf{s}), \quad (k=1, \ldots, N)\}$$

where

$$X_k(\mathbf{s}) = \{\mathbf{x}_k \mid \mathbf{x}_k \in S_k , \Psi_k(\mathbf{s}, \mathbf{x}_k , \mathbf{x}_k) \geqq \geqq \Psi_k(\mathbf{s}, \mathbf{x}_k , \mathbf{t}_k), \quad \text{for all } \mathbf{t}_k \in S_k\} .$$

Obviously $\mathbf{s} \in S^*$ implies $X(\mathbf{s}) \subset S^*$, thus X maps S^* into itself.

THEOREM 2. [180] The N-tuple of strategies $\mathbf{x}^* = (\mathbf{x}_1^*, \ldots, \mathbf{x}_N^*)$ is an equilibrium point of the multiproduct oligopoly game Γ if and only if $\mathbf{s}^* = \sum_{k=1}^{N} \mathbf{x}_k^*$ is a fixed point of the mapping X on S^* and $\mathbf{x}_k^* \in X_k(\mathbf{x}^*)$ for $k=1, \ldots, N$.

Proof. The assertion of the theorem is a direct consequence of the definition of an equilibrium point and the mapping X. ∎

The above theorem reduces the original $M \cdot N$ dimensional problem to an M-dimensional fixed-point problem which is a considerable advantage if M is much less than N. As we shall see it gives rise to an efficient iterative solution method in the special case $M=1$.

6.3. THE GENERAL LINEAR CASE

In this section we shall assume that all the price and cost functions involved in the multiproduct oligopoly game are linear. Although – according to our reduction principle set forth in Section 6.1 –

each group of players (manufacturers) can be assumed to consist of only one member, we shall treat the linear case directly, i.e., without invoking programming problem (3). This way we can avoid problems arising from nondifferentiability of the cost functions of the reduced game. (Unfortunately, differentiability of the original cost functions does not carry over to the reduced game.)

Consider the generalized oligopoly game Γ as defined in the opening paragraph of this chapter. Denote

$$L_k^{(m)} = \sum_{i=1}^{i_k} L_{ki}^{(m)}, \quad L^{(m)} = \sum_{k=1}^{N} L_k^{(m)}, \qquad (k=1, \ldots, N;\ m=1, \ldots, M)$$

and make the following additional assumptions:

(i) For $\mu=1, \ldots, M$ and $(s^{(1)}, \ldots, s^{(M)}) \notin D$ the price function $P_\mu(s^{(1)}, \ldots, s^{(M)})$ is identically zero, where $D \subseteqq \mathbb{R}^M$ is a closed, convex set.

(ii) The price functions are linear, i.e.,

$$P_\mu(s^{(1)}, \ldots, s^{(M)}) = \sum_{m=1}^{M} a_\mu^{(m)} s^{(m)} + b_\mu \tag{8}$$

for $\mu=1, \ldots, M$ and $(s^{(1)}, \ldots, s^{(M)}) \in D$.

(iii) $(s^{(1)}, \ldots, s^{(M)}) \in D$ and $0 \leqq \tilde{s}^{(m)} \leqq s^{(m)}$ implies $(s^{(1)}, \ldots, s^{(m-1)}, \tilde{s}^{(m)}, s^{(m+1)}, \ldots, s^{(M)}) \in D$ for any $m=1, \ldots, M$.

(iv) The cost functions are also linear, i.e.,

$$C_k(\mathbf{x}_k) = \sum_{i=1}^{i_k} \sum_{m=1}^{M} r_{ki}^{(m)} x_{ki}^{(m)} + b_k$$

where all the coefficients $r_{ki}^{(m)}$ are assumed to be positive.

(v) $(L^{(1)}, \ldots, L^{(M)}) \in D$.

(vi) The matrix $\mathbf{A} + \mathbf{A}^T$ is negative semidefinite, where $\mathbf{A} = (a_\mu^{(m)})$.

THEOREM 3. [179] Under assumption (i)–(vi) the multiproduct linear oligopoly game has at least one equilibrium point.

Proof. The proof goes exactly along the lines of Theorem 1, and is left to the reader as an exercise. ∎

Now we shall show that an equilibrium point of the linear oligopoly game under assumptions (i)–(vi) can be found by solving a quadratic programming problem.

By assumption (vi) the pay-off function φ_k is concave in $\mathbf{x}_k$. An N-tuple $\mathbf{x}^*=(\mathbf{x}_1^*, \ldots, \mathbf{x}_N^*)$ is an equilibrium point of the game if and only if

$$\Psi_{kj}^{(\mu)}(x_{kj}^{(\mu)})=\varphi_k(\mathbf{x}_1^*, \ldots, \mathbf{x}_k, \ldots, \mathbf{x}_N^*)$$

attains its maximum at $x_{kj}^{(\mu)*}$

$$\mathbf{x}_k=(x_{k1}^{(1)*}, \ldots, x_{kj}^{(\mu)}, \ldots, x_{kN}^{(M)*}).$$

Since $\Psi_{kj}^{(\mu)}$ is concave we have

$$\text{(9)} \qquad \Psi_{kj}^{(\mu)\prime}(x_{kj}^{(\mu)*}) \begin{cases} \leqq 0, & \text{if } x_{kj}^{(\mu)*}=0 \\ \geqq 0, & \text{if } x_{kj}^{(\mu)*}=L_{kj}^{(\mu)} \\ =0, & \text{if } 0<x_{kj}^{(\mu)*}<L_{kj}^{(\mu)}. \end{cases}$$

By the linearity of the price and cost functions, (9) can be written as

$$\text{(10)} \qquad \sum_{m=1}^{M} s_k^{(m)} a_m^{(\mu)} + \sum_{m=1}^{M} a_\mu^{(m)} s^{(m)*} + b_\mu - r_{kj}^{(\mu)}$$

$$\begin{cases} \leqq 0, & \text{if } x_{kj}^{(\mu)*}=0 \\ \geqq 0, & \text{if } x_{kj}^{(\mu)*}=L_{kj}^{(\mu)} \\ =0, & \text{if } 0<x_{kj}^{(\mu)*}<L_{kj}^{(\mu)} \end{cases}$$

where

$$s_k^{(m)}=\sum_{i=1}^{i_k} x_{ki}^{(m)}, \quad (k=1, \ldots, N;\ m=1, \ldots, M)$$

$$s^{(m)}=\sum_{k=1}^{N} s_k^{(m)}, \quad (m=1, \ldots, M)$$

Define the following nonnegative variables:

$$w_{kj}^{(\mu)}=L_{kj}^{(\mu)}-x_{kj}^{(\mu)} \geqq 0,$$

$$z_{kj}^{(\mu)}\begin{cases}=0, & \text{if} \quad x_{kj}^{(\mu)}>0\,,\\ \geqq 0, & \text{if} \quad x_{kj}^{(\mu)}=0\,,\end{cases}$$

$$v_{kj}^{(\mu)}\begin{cases}=0, & \text{if} \quad x_{kj}^{(\mu)}<L_{kj}^{(\mu)}\,,\\ \geqq 0, & \text{if} \quad x_{kj}^{(\mu)}=L_{kj}^{(\mu)}\,.\end{cases}$$

Then (10) can be written as

$$\text{(11)} \qquad \sum_{m=1}^{M} s_k^{(m)*} a_m^{(\mu)} + \sum_{m=1}^{M} a_\mu^{(m)} s^{(m)*} + $$
$$+ b_\mu - r_{kj}^{(\mu)} - v_{kj}^{(\mu)} + z_{kj}^{(\mu)} = 0\,.$$

For notational simplicity we define hypermatrices **C** and **B**

$$\mathbf{C}=\begin{bmatrix} \mathbf{C}_{11} & \mathbf{C}_{12} & \dots & \mathbf{C}_{1N} \\ \mathbf{C}_{21} & \mathbf{C}_{22} & \dots & \mathbf{C}_{2N} \\ \dots & \dots & \dots & \dots \\ \mathbf{C}_{N1} & \mathbf{C}_{N2} & \dots & \mathbf{C}_{NN} \end{bmatrix},$$

$$\mathbf{B}=\begin{bmatrix} \mathbf{B}_{11} & \mathbf{B}_{12} & \dots & \mathbf{B}_{1N} \\ \mathbf{B}_{21} & \mathbf{B}_{22} & \dots & \mathbf{B}_{2N} \\ \dots & \dots & \dots & \dots \\ \mathbf{B}_{N1} & \mathbf{B}_{N2} & \dots & \mathbf{B}_{NN} \end{bmatrix}$$

where all entries of the i_p by i_q matrix $\mathbf{C}_{pq}$ are equal to 1, $\mathbf{B}_{pq}=\mathbf{0}$ for $p\neq q$ and $\mathbf{B}_{pp}=\mathbf{C}_{pp}$, $(p=1, \dots, N;\ q=1, \dots, N)$. We also define the tensor-product of matrices $\mathbf{G}=(g_{ij})_{i,j=1}^{p,q}$ and $\mathbf{H}=(h_{ij})_{i,j=1}^{u,v}$ by

$$\mathbf{G}\otimes\mathbf{H}=\begin{bmatrix} g_{11}\mathbf{H} & g_{12}\mathbf{H} & \dots & g_{1q}\mathbf{H} \\ g_{21}\mathbf{H} & g_{22}\mathbf{H} & \dots & g_{2q}\mathbf{H} \\ \dots & \dots & \dots & \dots \\ g_{p1}\mathbf{H} & g_{p2}\mathbf{H} & \dots & g_{pq}\mathbf{H} \end{bmatrix}.$$

Let $\mathbf{P}=\mathbf{C}\otimes\mathbf{A}+\mathbf{B}\otimes\mathbf{A}^T$. Thus (11) can be written as

$$\text{(12)} \quad \begin{aligned} \mathbf{Px}+\mathbf{b}-\mathbf{r}-\mathbf{v}+\mathbf{z}&=\mathbf{0},\\ \mathbf{x}+\mathbf{w}&=\mathbf{l},\\ \mathbf{xz}=\mathbf{vw}=\mathbf{vz}&=\mathbf{0},\\ \mathbf{x}, \mathbf{v}, \mathbf{z}, \mathbf{w}&\geqq\mathbf{0} \end{aligned}$$

where $\mathbf{x}$, $\mathbf{r}$, $\mathbf{v}$, $\mathbf{z}$, $\mathbf{w}$, $\mathbf{l}$ are vectors consisting of components $x_{ki}^{(m)}$, $r_{ki}^{(m)}$, $v_{ki}^{(m)}$, $z_{ki}^{(m)}$, $w_{ki}^{(m)}$, $L_{ki}^{(m)}$ and

$$\mathbf{b}=(b_1, \ldots, b_M, \ldots, b_1, \ldots, b_M).$$

Now we can state the above results in a theorem.

THEOREM 4. [179] Under assumptions (i)–(vi) the n-tuple of strategies $\mathbf{x}_1^*, \ldots, \mathbf{x}_N^*$ is an equilibrium point of the multiproduct linear oligopoly game if and only if there can be found vectors $\mathbf{v}^*$, $\mathbf{z}^*$, $\mathbf{w}^*$ of proper dimension such that $(\mathbf{x}^*, \mathbf{v}^*, \mathbf{z}^*, \mathbf{w}^*)$ satisfies inequality system (12). ∎

The solution of inequality system (12) can generally be very difficult. However, if we assume the matrix $\mathbf{A}$ to be symmetric and negative (semi)definite, then (12) can be reduced to a quadratic programming problem. To establish this result we need a lemma.

LEMMA 6. If $\mathbf{A}$ is a symmetric negative semidefinite matrix, then $\mathbf{P}$ is negative semidefinite. In the special case $i_1=i_2=\ldots i_N=1$ the negative definiteness of $\mathbf{A}$ implies the same property of $\mathbf{P}$.

Proof. First we prove that $\mathbf{B}+\mathbf{C}$ is positive semidefinite and if $i_1=i_2=, \ldots, =i_N=1$, then it is positive definite. Let $\mathbf{u}=(u_{11}, \ldots, u_{1i_1}, \ldots, u_{N1}, \ldots, u_{Ni_N})$ be arbitrary. Then

$$\mathbf{u}(\mathbf{B}+\mathbf{C})\mathbf{u}=\left(\sum_{k=1}^{N}\sum_{i=1}^{i_k}u_{ki}\right)^2+\sum_{k=1}^{N}\left(\sum_{i=1}^{i_k}u_{ki}\right)^2$$

immediately implying the positive (semi)definiteness of $\mathbf{B}+\mathbf{C}$.

By the symmetricity of **A** we have

$$\mathbf{P} = \mathbf{C} \otimes \mathbf{A} + \mathbf{B} \otimes \mathbf{A}^T = (\mathbf{B} + \mathbf{C}) \otimes \mathbf{A}.$$

By a theorem of Egerváry ([47]) the eigenvalues of **P** are the products of those of **B** + **C** and **A**. The matrix **P** is also symmetric by definition. ∎

THEOREM 5. [179] In addition to assumptions (i)–(vi) let us suppose that **A** is a symmetric, negative-definite matrix. Then any equilibrium point of the multiproduct linear oligopoly game is an optimal solution of the following quadratic programming problem

$$(13) \qquad \frac{1}{2}\mathbf{x}\mathbf{P}\mathbf{x} + (\mathbf{b} - \mathbf{a})\mathbf{x} \to \max$$

$$\mathbf{0} \leqq \mathbf{x} \leqq \mathbf{1}.$$

Proof. It is easy to see that the Kuhn–Tucker optimality conditions of (13) are identical to inequality system (12) leaving the single orthogonality condition $\mathbf{v}\mathbf{z} = 0$ out of consideration. By Lemma 6, (13) is a concave programming problem, thus the Kuhn–Tucker conditions are sufficient for optimality. Therefore any solution of (12) is an optimal solution of (13). But, by Theorem 4, the equilibrium points of the multiproduct linear oligopoly game and the solutions of (12) coincide. ∎

Uniqueness of the equilibrium point can be stated under additional assumptions.

THEOREM 6. Assume that (i)–(vi) and $i_1 = \ldots i_N = 1$ hold. If **A** is a positive definite matrix, then the multiproduct linear oligopoly game has a unique equilibrium point.

Proof. Theorem 3 states that the multiproduct linear oligopoly game has at least one equilibrium point.

By Lemma 6 $\mathbf{P} = (\mathbf{B} + \mathbf{C}) \otimes \mathbf{A}$ is negative definite. Thus (13) is a strictly concave programming problem having a unique optimal solution which is an equilibrium point of the game under consideration by Theorem 4. ∎

It is well known that concave quadratic programming problems can be solved efficiently. (See [107].)

6.4. THE SINGLE-PRODUCT CASE

Throughout this section we suppose that $M=1$, i.e., there is only one product manufactured and sold on the market. First we shall assume that each group of manufacturers consists of only one member, i.e., $i_1=i_2=\ldots i_N=1$. Since our oligopoly game is a special case of the multiproduct game discussed in Section 6.2, Theorems 1 and 2 hold for the single-product game *a fortiori*. However the conditions under which these theorems are valid become simpler.

Let

$$\Psi_k(s, x_k, t_k)=t_k P(s-x_k+t_k)-C_k(t_k) \tag{14}$$

where P is the (single) price function, $x_k, t_k \in S_k=[0, L_k]$, $s \in \left[0, \sum_{k=1}^{N} L_k\right]$. (Here L_k denotes the upper bound of production for the kth player.) Let furthermore

$$X_k(s)=\{x_k \mid x_k \in S_k, \Psi_k(s, x_k, x_k) \geqq \Psi_k(s, x_k, t_k), \quad \text{for all} \quad t_k \in S_k\} \tag{15}$$

and

$$X(s)=\left\{u \mid u=\sum_{k=1}^{N} x_k, x_k \in X_k(s), \quad (k=1,2,\ldots,N)\right\}. \tag{16}$$

Notice that $s \in \left[0, \sum_{k=1}^{N} L_k\right]$ implies

$$X(s) \subset \left[0, \sum_{k=1}^{N} L_k\right],$$

thus the mapping $s \rightarrow X(s)$ is a one-dimensional point-to-set mapping. Theorem 2 takes now a special form.

THEOREM 7. [179] An N-tuple of strategies $(x_1^*, \ldots, x_N^*)$ is an equilibrium point of the single-product oligopoly game if and only if $s^* = \sum_{k=1}^{N} x_k^*$ is a fixed point of the mapping X and $x_k^* \in X_k(s^*)$, $(k=1, \ldots, N)$. Theorem 1 also becomes simpler in the single-product case.

THEOREM 8. [179] Assume that

(a) there exists a number $\xi > 0$ for which

(i) $P(s)=0$ if $s \geqq \xi$;

(ii) P is continuous, concave and strictly decreasing in the interval $[0, \xi]$;

(b) the cost functions are continuous, convex and strictly increasing in the interval $[0, L_k]$.

Then the single-product oligopoly game has at least one equilibrium point.

Proof. Let $\{P_n\}$ be a sequence of continuously differentiable, concave, monotone decreasing functions converging uniformly to P on the interval $[0, \xi]$. Let furthermore $P_n(s)=0$ for $s \geqq \xi$. Consider the game $\Gamma^{(n)}$ with the original strategy sets S_k, $(k=1, \ldots, N)$ and pay-off functions

$$\varphi_k^{(n)}(\mathbf{x}) = x_k P_n\left(\sum_{k=1}^{N} x_i\right) - C_k(x_k), \quad (k=1, \ldots, N).$$

Obviously $\Gamma^{(n)}$ satisfies conditions (a)–(d) of Theorem 1 by choosing $D=[0, \xi]$ and $S = \times_{k=1}^{N} [0, L_k]$. Let $\mathbf{x}^{(n)*}$ be an equilibrium point of $\Gamma^{(n)}$ and $\mathbf{x}^*$ be a cluster point of the bounded sequence $\{\mathbf{x}^{(n)*}\}$. Choose a subset of indices $\{n_l\}$ such that $\mathbf{x}^{(n_l)*} \to \mathbf{x}^*$. Then

$$\varphi_k^{(n_l)}(\mathbf{x}^{(n_l)*}) \geqq \varphi_k^{(n_l)}(x_1^{(n_l)*}, \ldots, x_k, \ldots, x_N^{(n_l)*})$$

for any $x_k \in S_k$ and $k=1, \ldots, N$. Letting $l \to \infty$ we get

$$\varphi_k(\mathbf{x}^*) \geqq \varphi_k(x_1^*, \ldots, x_k, \ldots, x_N^*)$$

$$\text{(for all } x_k \in S_k; \quad k=1, \ldots, N)$$

by the uniform convergence of the functions $\varphi_k^{(n)}$. Thus $\mathbf{x}^*$ is an equilibrium point of our original game Γ. ▮

Two remarks are in order here.

1. The uniqueness of the equilibrium point is not assured by the above theorem. A counterexample was given in Section 4.4.

2. The proof of Theorem 1 is based on the Nikaido–Isoda theorem which is an existence proof by nature. Therefore it does not provide any method for finding an equilibrium point.

In the following we are going to set up a numerical method to determine an equilibrium point of the single-product oligopoly game under the conditions specified in Theorem 8.

Let us take a fixed k, $(k=1, \ldots, N)$ and introduce the following notation

$$P_\delta^+(s)=\frac{P(s+\delta)-P(s)}{\delta}, \qquad (0\leqq s<s+\delta\leqq L),$$

$$P_\delta^-(s)=\frac{P(s)-P(s-\delta)}{\delta}, \qquad (0\leqq s-\delta<s\leqq L),$$

$$C_{k\delta}^+(t)=\frac{C_k(t+\delta)-C_k(t)}{\delta}, \qquad (0\leqq t<t+\delta\leqq L_k),$$

$$C_{k\delta}^-(t)=\frac{C_k(t)-C_k(t-\delta)}{\delta}, \qquad (0\leqq t-\delta<t\leqq L_k),$$

where $L=\min\left\{\xi, \sum_{i=1}^{N} L_i\right\}$. Define furthermore the functions

$$\varphi_\delta^+(s, t)=P(s+\delta)+tP_\delta^+(s)-C_{k\delta}^+(t), \tag{17}$$

$$\varphi_\delta^-(s, t)=P(s-\delta)+tP_\delta^-(s)-C_{k\delta}^-(t). \tag{18}$$

By the assumptions imposed on functions P and C_k we immediately obtain that for fixed δ, s and δ, t, φ_δ^+ and φ_δ^- are strictly decreasing functions of t and s resp. It is also easy to see that

1. $\varphi_{\delta_1}^-(s, t)\geqq\varphi_{\delta_2}^+(s, t)$ for any t and $\delta_1, \delta_2>0$,

2. for fixed s and t, function φ_δ^+ is strictly decreasing, while φ_δ^- is strictly increasing with respect to δ.

Now we prove a few lemmas which will be very useful for the numerical solution of the single-product oligopoly game.

LEMMA 7. The following three assertions hold true

(a) $0 \in X_k(s)$ if and only if

$$\varphi_\delta^+(s,0) \leqq 0, \quad (s+\delta \leqq L, \delta \leqq L_k),$$

(b) $L_k \in X_k(s)$ if and only if

$$\varphi_\delta^-(s,L_k) \geqq 0, \quad (0 \leqq s-\delta, \delta \leqq L_k),$$

(c) for any $0<t<L_k$ the relation $t \in X_k(s)$ holds if and only if

$$\varphi_{\delta_1}^-(s,t) \geqq 0 \geqq \varphi_{\delta_2}^+(s,t),$$

$$(0 \leqq s-\delta_1, \quad \delta_1 \leqq t, \quad s+\delta_2 \leqq L, \quad t+\delta_2 \leqq L_k).$$

Proof. By definition (15) the point $t \in [0, L_k]$ belongs to $X_k(s)$ if

$$(t-\delta_1)P(s-\delta_1) - C_k(t-\delta_1) \leqq tP(s) - C_k(t), \qquad (0 \leqq s-\delta_1, \ \delta_1 \leqq t),$$

$$(t+\delta_2)P(s+\delta_2) - C_k(t+\delta_2) \leqq tP(s) - C_k(t), \qquad (s+\delta_2 \leqq L, \ t+\delta_2 \leqq L_k).$$

Using definition (17) and (18) we get assertions (a)–(c) by elementary algebra. ∎

LEMMA 8. $X_k(s)$ is a closed interval. If the price-function P is differentiable at s, then $X_k(s)$ consists of a single point.

Proof. Assume that $0 \notin X_k(s)$ and $L_k \notin X_k(s)$. Then, by Lemma 7, there exist numbers Δ_1, $\Delta_2 > 0$, such that $\varphi_{\Delta_1}^+(s,0) > 0$ and $\varphi_{\Delta_2}^-(s,L_k) < 0$. By the continuity of $\varphi_{\Delta_1}^+$ there exists a $u > 0$, for which $\varphi_{\Delta_1}^+(s,u) > 0$. Since φ_δ^+ is decreasing in δ we have $\varphi_\delta^+(s,u) > 0$ for arbitrary $0 < \delta \leqq \Delta_1$. Thus $\varphi_\delta^-(s,u) > 0$, and $\varphi_{\delta_1}^-(s,u) > 0$, for any $\delta_1 > 0$, since φ_δ^- is increasing in δ. Consider now the set

$$(19) \qquad T = \{t \mid \varphi_\delta^-(s,t) \geqq 0 \quad \text{for any } \delta > 0\}.$$

By the continuity of φ_δ^- with respect to t set T is closed. Thus

$$t_0 = \sup\{t \mid t \in T\} \in T$$

and $u \in T$, $L_k \notin T$, which means that $0 < t_0 < L_k$.

We now prove that t_0 satisfies condition (c) in Lemma 7. Suppose it does not. Then $\varphi_\Delta^+(s, t_0) > 0$ for some $\Delta > 0$. Since φ_Δ^+ is a continuous function of t there exists a $u > 0$ for which $\varphi_\Delta^+(s, t_0 + u) > 0$. The function φ_δ^+ is decreasing as to δ, therefore $\varphi_\delta^+(s, t_0 + u) > 0$ for any $0 < \delta \leqq \Delta$ implying $\varphi_\delta^-(s, t_0 + u) > 0$.

By the monotonicity of φ_δ^- we have $\varphi_\delta^-(s, t_0 + u) > 0$ for any $\delta > 0$, which means $t_0 + u \in T$, contradicting the definition of t_0. Hence t_0 satisfies condition (c) of Lemma 7, which means $t_0 \in X_k(s)$.

Suppose now that $t_1 < t_2$ and $t_1 \in X_k(s)$, $t_2 \in X_k(s)$. Let $t_1 < t < t_2$. We shall prove that $t \in X_k(s)$, i.e., $X_k(s)$ is either a single point or an interval. Assume first that $t_1 = 0$, $t_2 < L_k$. Then for sufficiently small $\delta_1, \delta_2, \delta_3$ we have $\varphi_{\delta_1}^+(s, 0) \leqq 0$, $\varphi_{\delta_2}^-(s, t_2) \geqq 0 \geqq \varphi_{\delta_3}^+(s, t_2)$ by Lemma 7. By the monotonicity of φ_δ^- and φ_δ^+ we obtain $\varphi_{\delta_2}^-(s, t) > 0 > \varphi_{\delta_3}^+(s, t)$, i.e., $t \in X_k(s)$ by virtue of assertion (c) in Lemma 7. The proof for $t_1 > 0$, $t_2 = L_2$ goes along the same lines. Now we consider the case $0 < t_1 < t_2 < L_k$. By Lemma 7

$$\varphi_{\delta_1}^-(s, t_1) \geqq 0 \geqq \varphi_{\delta_2}^+(s, t_1),$$

$$\varphi_{\delta_3}^-(s, t_2) \geqq 0 \geqq \varphi_{\delta_4}^+(s, t_2)$$

for arbitrary $\delta_1, \delta_2, \delta_3, \delta_4$ satisfying $0 \leqq s - \delta_1$, $\delta_1 < t_1$, $s + \delta_2 \leqq L$, $t_1 + \delta_2 \leqq L_k$; $0 \leqq s - \delta_3$, $\delta_3 < t_2$, $s + \delta_4 \leqq L$, $t_2 + \delta_4 \leqq L_k$. Then by the monotonicity of φ_δ^+ and φ_δ^- we have $\varphi_{\Delta_1}^-(s, t) \geqq 0 \geqq f_{\Delta_2}^+(s, t)$ for $\Delta_1 \leqq \min\{\delta_1, \delta_3\}$, $\Delta_2 \leqq \min\{\delta_2, \delta_4\}$. Since φ_δ^- (φ_δ^+) are monotonically increasing (decreasing) functions of δ the previous inequality holds for any Δ_1, $\Delta_2 (0 \leqq s - \Delta_1$, $\Delta_1 \leqq t$, $s + \Delta_2 \leqq L$, $t + \delta_2 \leqq L_2)$ implying $t \in X_k(s)$ by assertion (c) of Lemma 7.

By the continuity of the price and cost functions $X_k(s)$ is closed, therefore it is a closed interval.

Finally we prove that $X_k(s)$ is a single point if P is differentiable at s. First we show that for $t_1<t_2$, and for sufficiently small δ_1, δ_2, the inequality $\varphi_{\delta_1}^+(s, t_1)>\varphi_{\delta_2}^-(s, t_2)$ holds.

By substitution we obtain

$$\varphi_{\delta_1}^+(s, t_1)-\varphi_{\delta_2}^-(s, t_2)=P(s+\delta_1)-P(s-\delta_2)+$$

$$+t_1\frac{P(s+\delta_1)-P(s)}{\delta_1}-t_2\frac{P(s)-P(s-\delta_2)}{\delta_2}-$$

$$-\frac{C_k(t_1+\delta_1)-C_k(t_1)}{\delta_1}+\frac{C_k(t_2)-C_k(t_2-\delta_2)}{\delta_2}.$$

Letting $\delta_1, \delta_2\to 0$ we get

$$(t_1-t_2)P'(s)+C_k'(t_2-0)-C_k'(t_1+0)\geqq$$

$$\geqq(t_1-t_2)P'(s)>0$$

implying $\varphi_{\delta_1}^+(s, t_1)>\varphi_{\delta_2}^-(s, t_2)$.

If $0\in X_k(s)$ for some s, then, by assertion (a) of Lemma 7, the monotonicity of φ_δ^+ and inequality $\varphi_{\delta_1}^-(s, t)>\varphi_{\delta_2}^+(s, t)$ imply $\varphi_{\delta_2}^+(s, t)<0$ for any t. Hence for small enough δ_2 we get $\varphi_{\delta_2}^-(s, t)<0$; i.e., neither of conditions (b) and (c) in Lemma 7 are satisfied.

If $L_k\in X_k(s)$, then similar argumentation leads to $\varphi_{\delta_2}^-(s, t)>0$ for any $t<L_k$ implying $\varphi_{\delta_1}^+(s, t)>0$ for sufficiently small δ_1. Therefore neither case (a) nor case (c) of Lemma 7 occurs.

It only remains to be shown that no distinct values of t satisfying condition (c) in Lemma 7 can exist. Assume on the contrary that for $t_1<t_2$ the inequalities

$$\text{(20)}\qquad \varphi_{\delta_1}^-(s, t_1)\geqq 0\geqq\varphi_{\delta_2}^+(s, t_1)$$

$$\varphi_{\delta_1}^-(s, t_2)\geqq 0\geqq\varphi_{\delta_2}^+(s, t_2)$$

hold for sufficiently small positive δ_1, δ_2. But we have already proved that $\varphi_{\delta_2}^+(s, t_1)>\varphi_{\delta_1}^-(s, t_2)$ if δ_1, δ_2 are small enough. Thus only one point of the interval $[0, L_k]$ can belong to $X_k(s)$, which was to be proved. ∎

LEMMA 9. If $s_1 < s_2$, $t_1 \in X_k(s_1)$, $t_2 \in X_k(s_2)$, then $t_1 \geqq t_2$.

Proof. We remind the reader that if P is differentiable at s and $t_1 < t_2$, then $\varphi^+_{\delta_1}(s, t_1) > \varphi^-_{\delta_2}(s, t_2)$ for small enough δ_1, δ_2 as it was shown in the proof of Lemma 8.

Assume now that the assertion of Lemma 9 does not hold, i.e., $s_1 < s_2$ and $t_1 < t_2$. Since P is concave there exists a point s between s_1 and s_2 where P is differentiable. If $0 < t_1 < t_2 < L_k$, then, by Lemma 7, we have

$$0 \leqq \varphi^-_\delta(s_2, t_2) < \varphi^-_\delta(s, t_2) <$$
$$< \varphi^+_\delta(s, t_1) < \varphi^+_\delta(s_1, t_1) \leqq 0$$

which is an obvious contradiction. Similar argumentation applies for cases (a) and (b) in Lemma 7. ∎

LEMMA 10. Under conditions of Theorem 8,

$$\sum_{k=1}^{N} x_k^* = \sum_{k=1}^{N} x_k^{**}$$

if $(x_1^*, \ldots, x_N^*)$ and $(x_1^{**}, \ldots, x_N^{**})$ are arbitrary equilibrium points of the single-product oligopoly game.

Proof. Denote

$$s^* = \sum_{k=1}^{N} x_k^*,$$

$$s^{**} = \sum_{k=1}^{N} x_k^{**}.$$

Assume $s^* \neq s^{**}$ or without loss of generality $s^* < s^{**}$. Since $x_k^* \in X_k(s^*)$, $x_k^{**} \in X_k(s^{**})$ we have by Lemma 9 $s^* \geqq s^{**}$, a contradiction. ∎

Introduce now the notation for $s \in [0, L]$

$$(21)\quad L_k(s)=\min\{t\,|\,t\in X_k(s)\},$$

$$\beta_k(s)=\max\{t\,|\,t\in X_k(s)\},\quad (k=1,\ldots,N)$$

$$\alpha(s)=\sum_{k=1}^{N}\alpha_k(s),$$

$$\beta(s)=\sum_{k=1}^{N}\beta_k(s).$$

THEOREM 9. [181] Under conditions of Theorem 8, $\mathbf{x}^*=(x_1^*,\ \ldots,\ x_N^*)$ is an equilibrium point of the single-product oligopoly game if and only if

(i) $s^*=\sum_{k=1}^{N}x_k^*$ is the unique solution of the inequality

$$(22)\quad \alpha(s^*)\leqq s^*\leqq\beta(s^*),$$

(ii) $x_k^*\in X_k(s^*)$, $(k=1,\ldots,N)$.

Proof. The theorem is a direct consequence of Theorem 7 and Lemma 10. ∎

COROLLARY 1. Under conditions of Theorem 8, the set of equilibrium points is the simplex determined by the inequality system

$$\alpha_k(s^*)\leqq x_k\leqq\beta_k(s^*),\quad (k=1,\ldots,N)$$

$$\sum_{k=1}^{N}x_k=s^*$$

where s^* is the unique solution of (22).

COROLLARY 2. If in addition to the assumptions specified in Theorem 8, P is differentiable at s^*, then the single-product oligopoly game has a unique equilibrium point. This assertion follows directly from Lemma 8 since $\alpha_k(s^*)=\beta_k(s^*)$ holds in this case for any $k=1,\ \ldots,\ N$.

Theorem 9 and the preceding lemmas provide a good basis for a numerical method to determine the equilibrium point of the single-product oligopoly game if P is differentiable and C_k is a differentiable strictly convex function, $(k=1, \ldots, N)$.

In this case the maximumpoint of $\Psi_k(s, x_k, t)$ as to t is unique and $X_k(s)$ is the unique solution of the equation $\left.\frac{\partial \Psi_k}{\partial t}\right|_{t=x_k}=0$. Thus $X_k(s)$ is a single-valued monotonically decreasing function of s for any $k=1, \ldots, N$. Hence the solution of the equation

$$\sum_{k=1}^{N} X_k(s)=s$$

is an easy task by numerical methods (e.g. interval bisection). If s^* is the unique solution of this equation, then $X_k(s^*)$, $(k=1, \ldots, N)$ is the equilibrium point we have searched for.

The general single-product case $i_k \geqq 1$, $(k=1, \ldots, N)$ can be reduced to the special case discussed above via the reduction principle of Section 6.1, since the continuity, monotonicity and convexity of the cost functions carry over to the reduced game, and the price function remains unaltered. Thus let us consider the game with strategy sets

$$(23) \qquad S_k=\mathop{\times}_{i=1}^{i_k} [0, L_{ki}], \quad (i \leqq k \leqq N)$$

and pay-off functions

$$(24) \qquad \varphi_k(\mathbf{x})=\left(\sum_{i=1}^{i_k} x_{ki}\right) P\left(\sum_{l=1}^{N} \sum_{j=1}^{i_l} x_{lj}\right)-C_k(x_{k1}, \ldots, x_{i_k}).$$

Assume that

(a) there exists a $\xi>0$, for which

(i) $P(s)=0$, if $s \geqq \xi$,

(ii) the price-function P is continuous, concave and monotonically decreasing in the interval $[0, \xi]$,

(b) the cost function C_k is continuous, convex and strictly monotonically increasing in each variable on the set S_k, $(k=1, \ldots, N)$.

THEOREM 10. [181] Under the above assumptions the general single product oligopoly game has an equilibrium point and the sum of equilibrium strategies s^* is uniquely determined. If in addition C_k is strictly convex for any $k=1, \ldots, N$ and P is differentiable at s^*, then the equilibrium point is unique.

Proof. If the reduction principle of Section 6.1 is applied for the game (23), (24), then the game thus obtained satisfies the conditions of Theorem 8. So there exists at least one equilibrium point and the sum of the equilibrium points is uniquely determined by Lemma 10. If P is differentiable at s^*, then by Corollary 2 of Theorem 9 the equilibrium point of the reduced game is unique. By the strict concavity of the function C_k programming problem (3) has also a unique solution, $(k=1, \ldots, N)$ which means that the original single-product oligopoly game has a unique equilibrium point. ∎

As an example let us consider the single-product oligopoly game given by the following data:

$$N=3$$

$$P(s)=\begin{cases} 2-2s-s^2, & \text{if } \quad 0 \leqq s \leqq \sqrt{3}-1 \\ 0, & \text{if } \quad s \geqq \sqrt{3}-1 \end{cases}$$

$$C_k(x)=kx^3+x, \quad (k=1, 2, 3)$$

$$L_k=1, \qquad (k=1, 2, 3).$$

This game obviously satisfies all the assumptions of Theorem 10 for any group formation. We have solved this game for all possible group formations:

(a) (1), (2), (3)
(b) (1, 2), 3;
(c) (1, 3), 2;

(d) 1, (2, 3);
(e) (1, 2, 3).

(We simply denoted the players by numbers 1, 2, 3.)

The equilibrium points and the pay-offs can be seen in the table below:

	a	*b*	*c*	*d*	*e*
x_1^*	0.1077	0.0880	0.0924	0.1317	0.0921
x_2^*	0.0986	0.0622	0.1211	0.0762	0.0651
x_3^*	0.0919	0.1135	0.0534	0.0622	0.0532
$\varphi_1(\mathbf{x}^*)$	0.0326	0.0348	0.0357	0.0486	0.0485
$\varphi_2(\mathbf{x}^*)$	0.0291	0.0246	0.0443	0.0286	0.0343
$\varphi_3(\mathbf{x}^*)$	0.0266	0.0414	0.0206	0.0233	0.0280
$\sum_{k=1}^{3} \varphi_k(\mathbf{x}^*)$	0.0883	0.1007	0.1006	0.1006	0.1108

It deserves mentioning that, in the linear case, the equilibrium points can be computed in closed form without recourse to quadratic programming.

Based on theorems and lemmas of this chapter the reader himself can derive the formulas for the equilibrium strategies.

7. Two-person games

GENERAL THEOREMS

According to the general definition of a game a two-person game Γ is given by strategy sets $\{\Sigma_1, \Sigma_2\}$ and pay-off functions $K_1 : \Sigma_1 \times \Sigma_2 \to \mathbb{R}$, $K_2 : \Sigma_1 \times \Sigma_2 \to \mathbb{R}$. We write briefly: $\Gamma = \{\Sigma_1, \Sigma_2; K_1, K_2\}$.

Although two-person games can always be viewed as special cases of n-person games yet it is worth studying their special properties: on the one hand because a great number of conflict situations emerging from practice can be modelled as two-person games, on the other hand because of their relative simplicity compared to n-person games ($n \geqq 3$). Therefore much more is known about the mathematical structure of two-person games; and computationally efficient methods are available for solving a number of important special cases.

Just as for n-person games, in the case of two-person games we mostly focus our attention on equilibrium points. Existence theorems and methods for finding equilibrium points will be our main concern.

As we already know a pair $\sigma_1^* \in \Sigma_1$, $\sigma_2^* \in \Sigma_2$ is called a (Nash) equilibrium point if

$$K_1(\sigma_1^*, \sigma_2^*) \geqq K_1(\sigma_1, \sigma_2^*) \quad \text{for all} \quad \sigma_1 \in \Sigma_1;$$
$$K_2(\sigma_1^*, \sigma_2^*) \geqq K_2(\sigma_1^*, \sigma_2) \quad \text{for all} \quad \sigma_2 \in \Sigma_2.$$

To begin with we prove a theorem needed very often later on.

THEOREM 1. [79] Let $f: X \times Y \to \mathbb{R}$ be a bounded function where X and Y are arbitrary sets. Then

$$\inf_{y \in Y} \sup_{x \in X} f(x, y) \geqq \sup_{x \in X} \inf_{y \in Y} f(x, y).$$

Proof. By the definition of the operator sup for any $y \in Y$ and $x \in X$ we get

$$\sup_{x \in X} f(x, y) \geqq f(x, y) .$$

Taking the infimum of both sides we have

$$\inf_{y \in Y} \sup_{x \in X} f(x, y) \geqq \inf_{y \in Y} f(x, y) .$$

Since on the right-hand side x is an arbitrary element of X we may write

$$\inf_{y \in Y} \sup_{x \in X} f(x, y) \geqq \sup_{x \in X} \inf_{y \in Y} f(x, y)$$

which is precisely what the theorem asserts. ∎

For two-person zero-sum games ($K_1 = -K_2 = f$) we shall very often use an equivalent definition of an equilibrium point. The equivalence is established in the following theorem.

THEOREM 2. [31] Let $f: \Sigma_1 \times \Sigma_2 \to \mathbb{R}$ be player I's pay-off function in the two-person zero-sum game $\Gamma = \{\Sigma_1, \Sigma_2, f\}$. If f is bounded, then Γ has an equilibrium point if and only if

$$\max_{\sigma_1 \in \Sigma_1} \inf_{\sigma_2 \in \Sigma_2} f(\sigma_1, \sigma_2) = \min_{\sigma_2 \in \Sigma_2} \sup_{\sigma_1 \in \Sigma_1} f(\sigma_1, \sigma_2) . \tag{1}$$

If (1) holds and v denotes the real number defined by either side o equation (1), then precisely those strategies $\sigma_1^0 \in \Sigma_1$ are optimal fo player I which satisfy

$$\inf_{\sigma_2 \in \Sigma_2} f(\sigma_1^0, \sigma_2) = v .$$

Similarly, a strategy $\sigma_2^0 \in \Sigma_2$ is optimal for player II if and only i

$$\sup_{\sigma_1 \in \Sigma_1} f(\sigma_1, \sigma_2^0) = v .$$

Furthermore, for any pair of optimal strategies (σ_1^0, σ_2^0) the equalit $f(\sigma_1^0, \sigma_2^0) = v$ holds.

Proof. Let us assume that (σ_1^0, σ_2^0) is a pair of optimal strategies for Γ and $f(\sigma_1^0, \sigma_2^0)=v$.

By the definition of an equilibrium point we have

$$(2) \qquad f(\sigma_1, \sigma_2^0) \leqq v \quad \text{for all} \quad \sigma_1 \in \Sigma_1 .$$

Thus

$$(3) \qquad \sup_{\sigma_1 \in \Sigma_1} f(\sigma_1, \sigma_2^0) \leqq v .$$

We assert that the function $\varphi(\sigma_2)= \sup_{\sigma_1 \in \Sigma_1} f(\sigma_1, \sigma_2)$ attains its minimum at σ_2^0. Suppose on the contrary that there is a $\bar{\sigma}_2 \in \Sigma_2$ for which

$$\sup_{\sigma_1 \in \Sigma_1} f(\sigma_1, \bar{\sigma}_2) < \sup_{\sigma_1 \in \Sigma_1} f(\sigma_1, \sigma_2^0) .$$

By the definition of an equilibrium point we obtain

$$v=f(\sigma_1^0, \sigma_2^0) \leqq f(\sigma_1^0, \bar{\sigma}_2) \leqq \sup_{\sigma_1 \in \Sigma_1} f(\sigma_1, \bar{\sigma}_2) < \\ < \sup_{\sigma_1 \in \Sigma_1} f(\sigma_1, \sigma_2^0)$$

which contradicts (2). Thus from (3) we can conclude

$$\min_{\sigma_2 \in \Sigma_2} \sup_{\sigma_1 \in \Sigma_1} f(\sigma_1, \sigma_2) \leqq v .$$

In the same way we can prove that

$$\max_{\sigma_1 \in \Sigma_1} \inf_{\sigma_2 \in \Sigma_2} f(\sigma_1, \sigma_2) \geqq v$$

and thus we get

$$\min_{\sigma_2 \in \Sigma_2} \sup_{\sigma_1 \in \Sigma_1} f(\sigma_1, \sigma_2) \leqq \max_{\sigma_1 \in \Sigma_1} \inf_{\sigma_2 \in \Sigma_2} f(\sigma_1, \sigma_2)$$

which together with Theorem 1 implies

$$(4) \qquad \min_{\sigma_2 \in \Sigma_2} \sup_{\sigma_1 \in \Sigma_1} f(\sigma_1, \sigma_2) = \max_{\sigma_1 \in \Sigma_1} \inf_{\sigma_2 \in \Sigma_2} f(\sigma_1, \sigma_2) = v .$$

In order to prove the sufficiency part of our theorem we assume that (4) holds. Let us choose σ_1^0 and σ_2^0 to satisfy

$$(5) \qquad \inf_{\sigma_2 \in \Sigma_2} f(\sigma_1^0, \sigma_2) = v$$

$$\sup_{\sigma_1 \in \Sigma_1} f(\sigma_1, \sigma_2^0) = v\,.$$

Hence

$$f(\sigma_1^0, \sigma_2^0) \geqq v$$

$$f(\sigma_1^0, \sigma_2^0) \leqq v$$

implying $f(\sigma_1^0, \sigma_2^0) = v$ and

$$f(\sigma_1^0, \sigma_2) \geqq v \geqq f(\sigma_1, \sigma_2^0)\,.$$

for all $\sigma_1 \in \Sigma_1$, $\sigma_2 \in \Sigma_2$; which means that σ_1^0, σ_2^0 is a pair of equilibrium strategies. ▮

REMARK 1. When proving the first part (the necessity) of the theorem we chose an arbitrary pair of equilibrium strategies. Thereby we proved the validity of the so called *interchangeability property*. To put it in precise terms this means that if (σ_1^0, σ_2^0) and (σ_1^*, σ_2^*) are equilibrium points, then (σ_1^0, σ_2^*) and (σ_1^*, σ_2^0) are also equilibrium points and

$$f(\sigma_1^0, \sigma_2^0) = f(\sigma_1^*, \sigma_2^*) = f(\sigma_1^0, \sigma_2^*) = f(\sigma_1^*, \sigma_2^0) = v\,.$$

REMARK 2. The real number $v = f(\sigma_1^0, \sigma_2^0)$, where (σ_1^0, σ_2^0) is an arbitrary equilibrium point, is called the *value of the game* Γ.

In order to be able to replace sup and inf by max and min in formula (1), the sets Σ_1, Σ_2 and the function f should meet additional requirements. (See the application of Theorem 2 to matrix games in Chapter 9.)

Theorem 2 can very well be interpreted, too. Player I can never get more than $\inf_{\sigma_2 \in \Sigma_2} f(\sigma_1, \sigma_2)$ if player II plays "against" him by choosing the "worst" σ_2. (Assuming the players to be "black-hearted" is natural since they play a zero-sum game.) Player I seeks to maximize this lower bound, i.e. to apply a strategy $\sigma_1^0 \in \Sigma_1$ for which

$$\max_{\sigma_1 \in \Sigma_1} \inf_{\sigma_2 \in \Sigma_2} f(\sigma_1, \sigma_2) = \inf_{\sigma_2 \in \Sigma_2} f(\sigma_1^0, \sigma_2)$$

holds.

Player II behaves in the same manner. He tries to keep player I's pay-off as low as possible by playing a strategy $\sigma_2^0 \in \Sigma_2$ which makes the equality

$$\min_{\sigma_2 \in \Sigma_2} \sup_{\sigma_1 \in \Sigma_1} f(\sigma_1, \sigma_2) = \sup_{\sigma_1 \in \Sigma_1} f(\sigma_1, \sigma_2^0)$$

to hold. If there exists an equilibrium point, then the above efforts of the players result in an "equilibrium state".

In case of an arbitrary game $\Gamma = \{\Sigma_1, \Sigma_2, f\}$ the numbers (if they exist)

$$\underline{v} = \max_{\sigma_1 \in \Sigma_1} \inf_{\sigma_2 \in \Sigma_2} f(\sigma_1, \sigma_2)$$

$$\bar{v} = \min_{\sigma_2 \in \Sigma_2} \sup_{\sigma_1 \in \Sigma_1} f(\sigma_1, \sigma_2)$$

are called the *lower and upper values* resp. of the game. By Theorem 1 $\underline{v} \leq \bar{v}$. Theorem 2 asserts that the lower and upper values of the game are equal if and only if the game has at least one pair of equilibrium strategies.

Certain generalizations of the equilibrium strategies and the value of a game are of theoretical and practical importance in the case of two-person games.

Let $\Gamma = \{\Sigma_1, \Sigma_2, f\}$ be a two-person zero-sum game with bounded pay-off function f. If

$$\sup_{\sigma_1 \in \Sigma_1} \inf_{\sigma_2 \in \Sigma_2} f(\sigma_1, \sigma_2) = \inf_{\sigma_2 \in \Sigma_2} \sup_{\sigma_1 \in \Sigma_1} f(\sigma_1, \sigma_2) = v,$$

then the real number v is called the *sup-inf value* of the game Γ.

The following theorem concerns the generalized equilibrium points associated with the sup-inf value.

THEOREM 3. [31] A real number v is a sup-inf value of the game $\Gamma = \{\Sigma_1, \Sigma_2, f\}$ if and only if for any positive ε both players have "ε optimal" strategies, i.e., there exist $\sigma_1^0(\varepsilon) \in \Sigma_1$, $\sigma_2^0(\varepsilon) \in \Sigma_2$ satisfying

$$f(\sigma_1^0(\varepsilon), \sigma_2) \geqq v - \varepsilon \quad \text{for all} \quad \sigma_2 \in \Sigma_2$$

$$f(\sigma_1, \sigma_2^0(\varepsilon)) \leqq v + \varepsilon \quad \text{for all} \quad \sigma_1 \in \Sigma_1 .$$

Proof. If $\sup_{\sigma_1 \in \Sigma_1} \inf_{\sigma_2 \in \Sigma_2} f(\sigma_1, \sigma_2) = v$, then for any $\varepsilon > 0$ there is a $\sigma_1^0(\varepsilon) \in \Sigma_1$ such that $\inf_{\sigma_2 \in \Sigma_2} f(\sigma_1^0(\varepsilon), \sigma_2) \geqq v - \varepsilon$. This implies $f(\sigma_1^0(\varepsilon), \sigma_2) \geqq v - \varepsilon$, for all $\sigma_2 \in \Sigma_2$. The existence of an "ε optimal" strategy for player II can be proved in the same way.

To prove the sufficiency let us assume that for any $\varepsilon > 0$ there is a $\sigma_1^0(\varepsilon) \in \Sigma_1$ to satisfy $f(\sigma_1^0(\varepsilon), \sigma_2) \geqq v - \varepsilon$, for all $\sigma_2 \in \Sigma_2$. Obviously

$$\inf_{\sigma_2 \in \Sigma_2} f(\sigma_1^0(\varepsilon), \sigma_2) \geqq v - \varepsilon$$

and a fortiori

$$\sup_{\sigma_1 \in \Sigma_1} \inf_{\sigma_2 \in \Sigma_2} f(\sigma_1, \sigma_2) \geqq v - \varepsilon .$$

Since ε is arbitrarily small we have

$$\text{(6)} \qquad \sup_{\sigma_1 \in \Sigma_1} \inf_{\sigma_2 \in \Sigma_2} f(\sigma_1, \sigma_2) \geqq v .$$

By similar reasoning we get

$$\text{(7)} \qquad \inf_{\sigma_2 \in \Sigma_2} \sup_{\sigma_1 \in \Sigma_1} f(\sigma_1, \sigma_2) \leqq v .$$

Inequalities (6), (7) and Theorem 1 imply

$$\text{(8)} \qquad \sup_{\sigma_1 \in \Sigma_1} \inf_{\sigma_2 \in \Sigma_2} f(\sigma_1, \sigma_2) = \inf_{\sigma_2 \in \Sigma_2} \sup_{\sigma_1 \in \Sigma_1} f(\sigma_1, \sigma_2) = v$$

which was to be proved. ∎

COROLLARY. If a strategy pair (σ_1^0, σ_2^0) is "ε optimal" for any $\varepsilon > 0$, then sup and inf in (8) can be replaced by max and min resp. Then by virtue of Theorem 2 (σ_1^0, σ_2^0) is an equilibrium point and v is the value of the game.

Analogously to the definition of the lower and upper values of a game we can define the *lower and upper sup-inf value* of a two-person zero-sum game. $\underline{v}$ and $\bar{v}$ are called the lower and upper sup-inf values resp. of a two-person zero-sum game $\Gamma=\{\Sigma_1, \Sigma_2, f\}$ if

$$\underline{v}=\sup_{\sigma_1\in\Sigma_1}\inf_{\sigma_2\in\Sigma_2} f(\sigma_1, \sigma_2)$$

$$\bar{v}=\inf_{\sigma_2\in\Sigma_2}\sup_{\sigma_1\in\Sigma_1} f(\sigma_1, \sigma_2).$$

If f is bounded, then $\underline{v}$ and $\bar{v}$ are finite. In this case we also have $\underline{v}\leqq\bar{v}$ and $\underline{v}=\bar{v}$ if and only if the game has a sup-inf value.

In Chapter 2 we already defined the mixed extension of an n-person game. This generalization is very useful for two-person games, as well. The game $\bar{\Gamma}=\{\bar{\Sigma}_1, \bar{\Sigma}_2, \bar{K}_1, \bar{K}_2\}$ is called the *mixed extension of the finite game*

$$\Gamma=\{\Sigma_1, \Sigma_2, K_1, K_2\}$$

$$(\Sigma_1=\{\sigma_1^{(1)}, \ldots, \sigma_1^{(m)}\}, \Sigma_2=\{\sigma_2^{(1)}, \ldots, \sigma_2^{(n)}\})$$

if $\bar{\Sigma}_1, \bar{\Sigma}_2$ are simplices of probability m and n-vectors resp. and

$$\text{(9)}\qquad \bar{K}_1(\mathbf{x}, \mathbf{y})=\sum_{i=1}^{m}\sum_{j=1}^{n} K_1(\sigma_1^{(i)}, \sigma_2^{(j)})x_i y_j,$$

$$\mathbf{x}\in\bar{\Sigma}_1, \mathbf{y}\in\bar{\Sigma}_2,$$

$$\bar{K}_2(\mathbf{x}, \mathbf{y})=\sum_{i=1}^{m}\sum_{j=1}^{n} K_2(\sigma_1^{(i)}, \sigma_2^{(j)})x_i y_j.$$

If Σ_1 and Σ_2 are rectangular regions of $\mathbb{R}^m$ and $\mathbb{R}^n$ resp., then $\bar{\Gamma}=\{\bar{\Sigma}_1, \bar{\Sigma}_2, \bar{K}_1, \bar{K}_2\}$ is the *mixed extension of the infinite game* $\Gamma=\{\Sigma_1, \Sigma_2, K_1, K_2\}$ if $\bar{\Sigma}_1$ and $\bar{\Sigma}_2$ are the families of distribution functions defined on Σ_1 and Σ_2 resp. and

$$\text{(10)}\qquad \bar{K}_1(f_1, f_2)=\int_{\Sigma_1}\int_{\Sigma_2} K_1(\sigma_1, \sigma_2)\mathrm{d}f_1(\sigma_1)\mathrm{d}f_2(\sigma_2),$$

$$\bar{K}_2(f_1, f_2)=\int_{\Sigma_1}\int_{\Sigma_2} K_2(\sigma_1, \sigma_2)\mathrm{d}f_1(\sigma_1)\mathrm{d}f_2(\sigma_2),$$

$$f_1\in\bar{\Sigma}_1, f_2\in\bar{\Sigma}_2.$$

The mixed extension can be defined for more general games, too, but this is beyond the scope of this book.

A comprehensive theory of two-person games has not been worked out yet. A few results can be established for general two-person games but these are useful for developing efficient methods only for special games. Therefore we deem it proper to turn our attention to special two-person games right away. However we are going to state some simple theorems for general two-person games, too. When proving these theorems we shall use notations making the generality of the reasoning apparent.

8. Bimatrix games

A finite two-person game is usually referred to as a *bimatrix game*[1] since it is completely determined by the pay-off matrices of the players. Let player I and player II have m and n pure strategies resp. Then the pay-off to player I and player II are a_{ij} and b_{ij} resp. if player I plays his ith while player II his jth strategy, $(i=1, \ldots, m;$ $j=1, \ldots, n)$. Then the game is given by the pair of matrices $(\mathbf{A}, \mathbf{B})$ where $\mathbf{A}=[a_{ij}]$, $\mathbf{B}=[b_{ij}]$. Primarily we are interested in the mixed extension of a bimatrix game $(\mathbf{A}, \mathbf{B})$, i.e., in the game where the strategy sets of the players are

$$\Sigma_1 = X_m = \{\mathbf{x} \mid \mathbf{x} \geqq \mathbf{0},\ \mathbf{1x} = 1\},$$

$$\Sigma_2 = Y_n = \{\mathbf{y} \mid \mathbf{y} \geqq \mathbf{0},\ \mathbf{1y} = 1\},$$

and the pay-off functions are given as

$$\begin{aligned} K_1(\mathbf{x}, \mathbf{y}) &= \mathbf{xAy}, \\ K_2(\mathbf{x}, \mathbf{y}) &= \mathbf{xBy}. \end{aligned} \qquad \mathbf{x} \in X_m,\ \mathbf{y} \in Y_n$$

Just as in case of n-person games finding equilibrium points and studying their structural properties will be our main concern. We remind the reader here of the definition of an equilibrium point in terms of bimatrix game $(\mathbf{A}, \mathbf{B})$. A pair $(\mathbf{x}^0, \mathbf{y}^0)$ where $\mathbf{x}^0 \in X_m$, $\mathbf{y}^0 \in Y_n$ is called an equilibrium point if

$$\mathbf{xAy}^0 \leqq \mathbf{x}^0\mathbf{Ay}^0, \quad \text{for all} \quad \mathbf{x} \in X_m$$

$$\mathbf{x}^0\mathbf{By} \leqq \mathbf{x}^0\mathbf{By}^0, \quad \text{for all} \quad \mathbf{y} \in Y_n.$$

[1] The term "bimatrix" was first used by Vorobyev [196].

By Theorem 3 of Chapter 3 any bimatrix game has at least one equilibrium point.

We remind the reader that as of now we are only dealing with non-cooperative games, therefore any form of communication, bargaining or agreement between the players is not provided by the rules.

A number of definitions and theorems in the following hold for more general two-person games. Although we give these definitions and theorems for a bimatrix game, the reader can easily see where the finiteness of the strategy sets is not important. To help recognize these cases we use matrix notation only if it is necessary.

8.1. BASIC DEFINITIONS AND SOME SIMPLE PROPERTIES OF BIMATRIX GAMES

Let us denote by X^0 and Y^0 the nonempty sets of player I's and player II's equilibrium strategies resp. Unfortunately X^0 and Y^0 are not necessarily convex and do not possess the *interchangeability property* either, i.e., $\mathbf{x} \in X^0$ and $\mathbf{y} \in Y^0$ do not imply that $\mathbf{x}, \mathbf{y}$ is an equilibrium point. The following example illustrates this:

$$(1) \qquad \mathbf{A} = \begin{bmatrix} 2 & -1 \\ -1 & 1 \end{bmatrix}, \qquad \mathbf{B} = \begin{bmatrix} 1 & -1 \\ -1 & 2 \end{bmatrix}.$$

Now $(\mathbf{e}_1, \mathbf{e}_1)$ and $(\mathbf{e}_2, \mathbf{e}_2)$ are equilibrium points but neither $(\mathbf{e}_1, \mathbf{e}_2)$ nor $(\mathbf{e}_2, \mathbf{e}_1)$ are such.

Two equilibrium points of a bimatrix game $(\mathbf{x}, \mathbf{y})$ and $(\mathbf{u}, \mathbf{v})$ are *equivalent* if $K_k(\mathbf{x}, \mathbf{y}) = K_k(\mathbf{u}, \mathbf{v})$, $(k = 1, 2)$. A bimatrix game is said to possess the *equivalency property* if any two equilibrium points are equivalent. It is easily seen that the game (1) does not have this property.

A bimatrix game is called *solvable* if it has the interchangeability and equivalency property. Then $K_1(\mathbf{x}^0, \mathbf{y}^0) = v_1$ and $K_2(\mathbf{x}^0, \mathbf{y}^0) = v_2$ are called the *values* of the game. ($(\mathbf{x}^0, \mathbf{y}^0)$ is an equilibrium point.)

Even in the class of solvable games we can find such games where an equilibrium can hardly be accepted as a "solution" since there is a nonequilibrium point which is more advantageous for both players. In the following game

$$\mathbf{A}=\begin{bmatrix}2 & 4\\ 1 & 3\end{bmatrix} \qquad \mathbf{B}=\begin{bmatrix}2 & 1\\ 4 & 3\end{bmatrix}$$

there is only one equilibrium point: $(\mathbf{e}_1, \mathbf{e}_1)$ but the point $(\mathbf{e}_2, \mathbf{e}_2)$ gives larger pay-off for both players.

A pair of strategies $(\mathbf{x}, \mathbf{y})$ is said *to dominate* the strategy pair $(\mathbf{u}, \mathbf{v})$ and is denoted by $(\mathbf{x}, \mathbf{y}) \succsim (\mathbf{u}, \mathbf{v})$, if

$$K_k(\mathbf{x}, \mathbf{y}) \geqq K_k(\mathbf{u}, \mathbf{v}), \quad (k=1, 2)$$

and the inequality is strict for at least one index k. If

$$K_k(\mathbf{x}, \mathbf{y}) > K_k(\mathbf{u}, \mathbf{v}), \quad (k=1, 2),$$

then $(\mathbf{x}, \mathbf{y})$ strictly dominates the pair $(\mathbf{u}, \mathbf{v})$ and we write

$$(\mathbf{x}, \mathbf{y}) \succ (\mathbf{u}, \mathbf{v}).$$

A strategy pair which is not dominated by any other pair of strategies is said to be a *dominant point*. If it is not dominated strictly by any strategy pair, then it is called a *weak dominant point*.

The existence of nondominant points is characteristic to nonzero-sum games.

THEOREM 1. [83] Any strategy pair of a zero-sum two-person game is dominant.

Proof. Let us assume on the contrary that $(\mathbf{x}, \mathbf{y}) \succsim (\mathbf{u}, \mathbf{v})$, that is

$$K_1(\mathbf{x}, \mathbf{y}) \geqq K_1(\mathbf{u}, \mathbf{v}),$$

$$K_2(\mathbf{x}, \mathbf{y}) \geqq K_2(\mathbf{u}, \mathbf{v})$$

and one of these inequalities is strict. Summing them up we get

$$0 = K_1(\mathbf{x}, \mathbf{y}) + K_2(\mathbf{x}, \mathbf{y}) > K_1(\mathbf{u}, \mathbf{v}) + K_2(\mathbf{u}, \mathbf{v}) = 0,$$

which is a contradiction. ∎

We call a solvable game *strictly solvable* if it has at least one dominant equilibrium strategy pair.

Among two-person games zero-sum games are of great importance. (See Chapter 9.) Therefore it is important to decide whether a two-person game can be reduced to a zero-sum game. The following theorem gives an easy-to-check criterion.

THEOREM 2. [83] Let $\Gamma=\{\Sigma_1, \Sigma_2, K_1, K_2\}$ be a two-person game. If there exists a positive number p and a real number q such that

$$K_2(\sigma_1, \sigma_2)=-pK_1(\sigma_1, \sigma_2)+q, \qquad \text{for all} \quad \sigma_1\in\Sigma_1, \sigma_2\in\Sigma_2,$$

then Γ is strategically-equivalent to the game $\Gamma_0=\{\Sigma_1, \Sigma_2, K_1\}$ provided both Γ and Γ_0 have at least one equilibrium point.

Proof. Let (σ_1^0, σ_2^0) be an equilibrium point of Γ_0. Then

$$K_1(\sigma_1, \sigma_2^0)\leqq K_1(\sigma_1^0, \sigma_2^0)\leqq K_1(\sigma_1^0, \sigma_2) \qquad \text{for all} \quad \sigma_1\in\Sigma_1, \sigma_2\in\Sigma_2.$$

Substituting

$$K_1(\sigma_1, \sigma_2)=\frac{1}{p}[-K_2(\sigma_1, \sigma_2)+q]$$

we get

$$-\frac{1}{p}K_2(\sigma_1^0, \sigma_2^0)+\frac{q}{p}\leqq-\frac{1}{p}K_2(\sigma_1^0, \sigma_2)+\frac{q}{p} \qquad \text{for all} \quad \sigma_2\in\Sigma_2$$

implying

$$K_2(\sigma_1^0, \sigma_2^0)\geqq K_2(\sigma_1^0, \sigma_2) \quad \text{for all} \quad \sigma_2\in\Sigma_2$$

which means that (σ_1^0, σ_2^0) is an equilibrium point of Γ, too.

Proving that any equilibrium point of Γ is that of Γ_0 goes similarly. ∎

COROLLARIES

1. If Γ is finite (a bimatrix game), then the condition of Theorem 2 can be formulated as $\mathbf{B}=-p\,\mathbf{A}+q[1]$ where $\mathbf{A}$ and $\mathbf{B}$ are the pay-off matrices.

2. The solution of *constant-sum games*, where $K_1(\sigma_1, \sigma_2)+$ $+K_2(\sigma_1, \sigma_2)=$Constant for all $\sigma_1 \in \Sigma_1, \sigma_2 \in \Sigma_2$ can be reduced to zero-sum games.

3. The games $\Gamma=\{\Sigma_1, \Sigma_2, K_1, K_2\}$, $\Gamma'=\{\Sigma_1, \Sigma_2, pK_1+q, rK_2+s\}$ are strategically-equivalent (p, q, r, s are constants, $p>0, r>0$). Therefore without loss of generality we may assume (if necessary) that the values of a solvable game are (0, 0).

In a nonzero-sum game the aim of a player is not to increase the loss of the opponent (rather to increase his own pay-off). Nevertheless for each player it is good to know how much he can assure himself no matter what his opponent does and what strategies he has to choose to assure that amount. We call these strategies *defensive strategies.* Thus $\bar{\mathbf{x}}$ and $\bar{\mathbf{y}}$ are defensive strategies if

$$\max_{\mathbf{x}} \inf_{\mathbf{y}} K_1(\mathbf{x}, \mathbf{y})= \inf_{\mathbf{y}} K_1(\bar{\mathbf{x}}, \mathbf{y})$$

$$\max_{\mathbf{y}} \inf_{\mathbf{x}} K_2(\mathbf{x}, \mathbf{y})= \inf_{\mathbf{x}} K_2(\mathbf{x}, \bar{\mathbf{y}}).$$

Similarly we can define *attacking strategies.* Strategies $\hat{\mathbf{x}}$ and $\hat{\mathbf{y}}$ are attacking strategies if

$$\min_{\mathbf{x}} \sup_{\mathbf{y}} K_2(\mathbf{x}, \mathbf{y})= \sup_{\mathbf{y}} K_2(\hat{\mathbf{x}}, \mathbf{y})$$

$$\min_{\mathbf{y}} \sup_{\mathbf{x}} K_1(\mathbf{x}, \mathbf{y})= \sup_{\mathbf{x}} K_1(\mathbf{x}, \hat{\mathbf{y}}).$$

When applying these strategies the players, disregarding their own benefit, want to hurt their opponent as badly as they can.

It is easy to see that for zero-sum games any defensive strategy is attacking at the same time and vice versa.

So far we have been assuming that any player takes into consideration the opponent's all feasible strategies no matter how unfavourable they are. Restricting the strategy sets by deleting certain "disadvantageous" strategies we can extend the class of solvable games.

Let $X^* \subseteqq X_m$ have the following properties:

1. If $X^* \neq X_m$, then for any $\mathbf{x} \in X_m - X^*$ there exists an $\bar{\mathbf{x}} \in X^*$ to satisfy

$$K_1(\bar{\mathbf{x}}, \mathbf{y}) \geqq K_1(\mathbf{x}, \mathbf{y}), \quad \text{for all} \quad \mathbf{y} \in Y_n$$

and the inequality is strict for at least one $\mathbf{y}$.

2. No proper subset of X^* has Property 1.

For player II, $Y^* \subseteqq Y_n$ is defined analogously.

THEOREM 3. [98] For bimatrix games the sets X^* and Y^* exist and are uniquely determined.

Proof. It is enough to prove the theorem for X^*. Let S be the intersection of all sets satisfying Property 1. We are going to prove that S is not empty. Let $\mathbf{x}_0 \in X_m$ be arbitrary and consider the following linear programming problem:

$$z = \mathbf{xA1} \to \max$$

$$\mathbf{xA} \geqq \mathbf{x}_0 \mathbf{A}$$

$$\mathbf{x} \in X_m .$$

It is easy to see that any optimal basic solution $\bar{\mathbf{x}}$ of this problem is an *efficient point*, i.e., there is no $\mathbf{x} \in X_m$ for which $\mathbf{xA} \geqq \bar{\mathbf{x}}\mathbf{A}$, $\mathbf{xA} \neq \bar{\mathbf{x}}\mathbf{A}$ hold. But this implies that $\bar{\mathbf{x}}$ should belong to any set having Property 1. Therefore $\bar{\mathbf{x}} \in S$.

Since S is the intersection of all sets with Property 1, therefore $S = X^*$, and the way we defined S assures its uniqueness, too.

We call X^* and Y^* *minimal strategy sets*. Let us consider now the game $\Gamma^* = \{X^*, Y^*, K_1, K_2\}$. If Γ^* is strictly solvable, then we say that the original game Γ is *weakly solvable*.

For instance game (1) is not strictly solvable but it is weakly solvable since $X^* = \{\mathbf{e}_1\}$, $Y^* = \{\mathbf{e}_1\}$ and $(\mathbf{e}_1, \mathbf{e}_1)$ is the only feasible strategy pair.

If we liked to reduce Γ^* further by defining subsets $X^{**} \subsetneqq X^*$, $Y^{**} \subsetneqq Y^*$ we would have to face the difficulty that the strategies in X^{**} are not necessarily "efficient" against strategies in $Y_n - Y^*$. The same holds for Y^{**}.

8.2. METHODS FOR SOLVING BIMATRIX GAMES

By solving a bimatrix game we mean finding an equilibrium point or all the equilibrium points of the game. For the characterization of equilibrium points the following theorem is very useful. Since the theorem is a straightforward consequence of the definition of equilibrium points we give it without proof. (The proof is an easy exercise for the reader.)

THEOREM 4. [103] For a pair $(\mathbf{x}^0, \mathbf{y}^0)$ to be an equilibrium point of the bimatrix game $(\mathbf{A}, \mathbf{B})$ it is necessary and sufficient that there exist real numbers α^0, β^0 such that $\mathbf{x}^0, \mathbf{y}^0, \alpha^0, \beta^0$ satisfy the following system of inequalities:

$$
\begin{aligned}
(2) \qquad \mathbf{xAy} - \alpha &= 0, \\
\mathbf{xBy} - \beta &= 0, \\
\mathbf{Ay} - \alpha\mathbf{1} &\leqq \mathbf{0}, \\
\mathbf{B}^T\mathbf{x} - \beta\mathbf{1} &\leqq \mathbf{0}, \\
\mathbf{1x} &= 1, \\
\mathbf{1y} &= 1, \\
\mathbf{x} &\geqq \mathbf{0}, \\
\mathbf{y} &\geqq \mathbf{0}.
\end{aligned}
$$

Among the equilibrium points of a bimatrix game a very important role is played by *extreme equilibrium points*. A quadruple

$(\mathbf{x}^0, \mathbf{y}^0, \alpha^0, \beta^0)$ is said to be an extreme equilibrium point if $(\mathbf{x}^0, \beta^0)$ and $(\mathbf{y}^0, \alpha^0)$ are extreme points of S and T resp. where

$$S=\{(\mathbf{x}, \beta) \mid \mathbf{B}^T\mathbf{x}-\beta\mathbf{1} \leqq \mathbf{0}, \mathbf{x} \in X_m\},$$

$$T=\{(\mathbf{y}, \alpha) \mid \mathbf{A}\mathbf{y}-\alpha\mathbf{1} \leqq \mathbf{0}, \mathbf{y} \in Y_n\}.$$

THEOREM 5. [84] Any equilibrium point of a bimatrix game can be expressed as a convex linear combination of the extreme equilibrium points.

Proof. Let $(\bar{\mathbf{x}}, \bar{\mathbf{y}}, \bar{\alpha}, \bar{\beta})$ be an equilibrium point. Consider the following linear programming problem

$$(3) \qquad \begin{aligned} & z(\mathbf{x}, \beta)=\mathbf{x}(\mathbf{A}+\mathbf{B})\bar{\mathbf{y}}-\beta-\bar{\alpha} \to \max \\ & \mathbf{x} \in X_m \\ & \mathbf{B}^T\mathbf{x}-\beta\mathbf{1} \leqq \mathbf{0}. \end{aligned}$$

We shall prove that $(\bar{\mathbf{x}}, \bar{\beta})$ is an optimal solution of (3). By Theorem 4 $(\bar{\mathbf{x}}, \bar{\beta})$ is feasible and $z(\bar{\mathbf{x}}, \bar{\beta})=0$. But

$$\mathbf{x}(\mathbf{A}+\mathbf{B})\bar{\mathbf{y}}-\beta-\bar{\alpha}=\bar{\mathbf{x}}(\mathbf{A}\bar{\mathbf{y}}-\bar{\alpha}\mathbf{1})+\bar{\mathbf{y}}(\mathbf{B}^T\mathbf{x}-\beta\mathbf{1}) \leqq 0$$

for any feasible $(\mathbf{x}, \beta)$, therefore $(\bar{\mathbf{x}}, \bar{\beta})$ is optimal. It is well known from the theory of linear programming that $(\bar{\mathbf{x}}, \bar{\beta})$ can be expressed as a convex linear combination of optimal extreme points of the feasible set of (3) which is S. Let the set of these optimal extreme points be U. By the optimality of $(\bar{\mathbf{x}}, \bar{\beta})$ we have

$$(4) \qquad \mathbf{x}(\mathbf{A}+\mathbf{B})\bar{\mathbf{y}}-\beta-\bar{\alpha}=0 \quad \text{for all} \quad (\mathbf{x}, \beta) \in U.$$

In the same way it can be shown that $(\bar{\mathbf{y}}, \bar{\alpha})$ is a convex linear combination of certain extreme points of T. Let us denote this set by V. Now (4) can be written as

$$(5) \qquad \mathbf{x}(\mathbf{A}\bar{\mathbf{y}}-\bar{\alpha}\mathbf{1})+\bar{\mathbf{y}}(\mathbf{B}^T\mathbf{x}-\beta\mathbf{1})=0 \quad \text{for all} \quad (\mathbf{x}, \beta) \in U.$$

Since $\mathbf{x} \geqq \mathbf{0}$, $\mathbf{A}\bar{\mathbf{y}}-\bar{\alpha}\mathbf{1} \leqq \mathbf{0}$, $\bar{\mathbf{y}} \geqq \mathbf{0}$, $\mathbf{B}^T\mathbf{x}-\beta\mathbf{1} \leqq \mathbf{0}$ it follows from (5) that

$$(6) \qquad \bar{\mathbf{y}}(\mathbf{B}^T\mathbf{x}-\beta\mathbf{1})=0 \quad \text{for all} \quad (\mathbf{x}, \beta) \in U.$$

Since $\bar{\mathbf{y}} = \sum_{i=1}^{r} \lambda_i \mathbf{y}_i$, $\left(\lambda_i > 0, \sum_{i=1}^{r} \lambda_i = 1, (\mathbf{y}_i, \alpha_i) \in V, (i = 1, \ldots, r)\right)$ we derive from (6)

$$\mathbf{y}_i(\mathbf{B}^T\mathbf{x} - \beta\mathbf{1}) = 0 \quad \text{for all} \quad (\mathbf{x}, \beta) \in U, \quad (i = 1, \ldots, r).$$

Similarly we can prove

$$\mathbf{x}_j(\mathbf{A}\mathbf{y} - \alpha\mathbf{1}) \quad \text{for all} \quad (\mathbf{x}_j, \beta_j) \in U, \ (\mathbf{y}, \alpha) \in V, \qquad (j = 1, \ldots, s).$$

But this means exactly that $(\mathbf{x}_j, \mathbf{y}_i)$, $(i = 1, \ldots, r; j = 1, \ldots, s)$ are all (extreme) equilibrium points and $(\bar{\mathbf{x}}, \bar{\beta})$, $(\mathbf{y}, \bar{\alpha})$ can be expressed as their convex linear combination. ∎

Theorem 5 gives rise to a method for determining all equilibrium points of a bimatrix game. All we have to do is find all extreme points of S and T (e.g., by complete description) and choose those satisfying

$$\text{(7)} \qquad \mathbf{x}(\mathbf{A} + \mathbf{B})\mathbf{y} - \alpha - \beta = 0.$$

It should be mentioned that complete description methods are computationally efficient only if m and n are not large. As an example let us find all extreme equilibrium strategies of game (1). First we have to determine all extreme points of the polyhedron given by the inequality systems.

$$\text{(8)} \qquad \begin{bmatrix} 1 & -1 \\ -1 & 2 \end{bmatrix} \begin{bmatrix} x \\ 1-x \end{bmatrix} - \beta \begin{bmatrix} 1 \\ 1 \end{bmatrix} \leqq \begin{bmatrix} 0 \\ 0 \end{bmatrix}$$

$$\text{(9)} \qquad \begin{bmatrix} 2 & -1 \\ -1 & 1 \end{bmatrix} \begin{bmatrix} y \\ 1-y \end{bmatrix} - \alpha \begin{bmatrix} 1 \\ 1 \end{bmatrix} \leqq \begin{bmatrix} 0 \\ 0 \end{bmatrix}.$$

The extreme points of (8) are

$$\mathbf{x}_1 = (0, 1) \qquad \beta_1 = 2$$

$$\mathbf{x}_2 = (1, 0) \qquad \beta_2 = 1$$

$$\mathbf{x}_3 = \left(\frac{3}{5}, \frac{2}{5}\right) \qquad \beta_3 = \frac{1}{5}.$$

Those of (9) are

$$\mathbf{y}_1 = (0, 1) \qquad \alpha_1 = 1$$

$$\mathbf{y}_2 = (1, 0) \qquad \alpha_2 = 2$$

$$\mathbf{y}_3 = \left(\frac{2}{5}, \frac{3}{5}\right) \qquad \alpha_3 = \frac{1}{5}.$$

Substituting all possible pairs of $(\mathbf{x}_i, \beta_i)$ and $(\mathbf{y}_j, \alpha_j)$ into (7) we can select the three extreme equilibrium points:

$$P_1 = (\mathbf{x}_1, \mathbf{y}_1)$$

$$P_2 = (\mathbf{x}_2, \mathbf{y}_2)$$

$$P_3 = (\mathbf{x}_3, \mathbf{y}_3).$$

Much more efficient methods exist if our aim is only to find at least one equilibrium point. The algorithm we are going to set forth is due to Majthay [101] and is based on complementary pivoting.[1] Its complexity is similar to that of the simplex method for linear programming.

By Theorem 2 we may assume that the m by n matrices of the game we are going to solve are negative: $\mathbf{A} < \mathbf{0}$, $\mathbf{B} < \mathbf{0}$. Then the real numbers in inequality system (2) are also negative: $\alpha < 0$, $\beta < 0$. Let us rewrite (2) in a more convenient form by introducing the notation:

$$\mathbf{G} = \mathbf{A}, \qquad \hat{\mathbf{G}} = \mathbf{B}^T$$

$$\mathbf{u} = \frac{1}{\alpha}\mathbf{A}\mathbf{y} - \mathbf{1} = \frac{1}{\alpha}\mathbf{G}\mathbf{y} - \mathbf{1}$$

[1] The first method of this type was given by Lemke and Howson [90], [91].

$$\hat{\mathbf{u}} = \frac{1}{\beta}\mathbf{B}^T\mathbf{x} - \mathbf{1} = \frac{1}{\beta}\hat{\mathbf{G}}\mathbf{y} - \mathbf{1}$$

$$\mathbf{z} = -\frac{1}{\beta}\mathbf{x}, \qquad \hat{\mathbf{z}} = -\frac{1}{\alpha}\mathbf{y}.$$

Inequality system (2) is equivalent to

(10) $\mathbf{u} + \mathbf{G}\hat{\mathbf{z}} = -\mathbf{1}$

(11) $\mathbf{u} \geqq \mathbf{0}, \quad \hat{\mathbf{z}} \geqq \mathbf{0}$

(12) $\hat{\mathbf{u}} + \hat{\mathbf{G}}\mathbf{z} = -\mathbf{1}$

(13) $\hat{\mathbf{u}} \geqq \mathbf{0}, \quad \mathbf{z} \geqq \mathbf{0}$

(14) $\mathbf{u}\mathbf{z} = 0, \quad \hat{\mathbf{u}}\hat{\mathbf{z}} = 0.$

If $\mathbf{u}$, $\hat{\mathbf{u}}$, $\mathbf{z}$, $\hat{\mathbf{z}}$ is a feasible solution to the above system, then taking $\alpha = -\frac{1}{\mathbf{1}\hat{\mathbf{z}}}$ and $\beta = -\frac{1}{\mathbf{1}\mathbf{z}}$ we get an equilibrium point

$$\mathbf{x} = -\beta\mathbf{z}, \quad \mathbf{y} = -\alpha\hat{\mathbf{z}}.$$

By simple substitution we can easily verify the above assertion.

Now we want to find a solution to the system (10)–(14). For the sake of easy reference quadruples ($\mathbf{u}$, $\hat{\mathbf{z}}$, $\hat{\mathbf{u}}$, $\mathbf{z}$) satisfying (10) and (12) are called *solutions.* If in addition they also satisfy (11) and (13) they are called *feasible solutions.* If a feasible solution satisfies (14), then it is said to be a *complementary solution.* Components of $\mathbf{u}$, $\mathbf{z}$ and $\hat{\mathbf{u}}$, $\hat{\mathbf{z}}$ with the same index are called *complementary variables* while the vectors of the matrices $\mathbf{H}$, $\hat{\mathbf{H}}$ belonging to them are referred to as *complementary pairs,* where

$$\mathbf{H} = [\mathbf{E}, \mathbf{G}],$$

$$\hat{\mathbf{H}} = [\hat{\mathbf{E}}, \hat{\mathbf{G}}].$$

By introducing the notation

$$\mathbf{v} = (\mathbf{u}, \hat{\mathbf{z}}), \quad \mathbf{h}_0 = -\mathbf{1}$$

$$\hat{\mathbf{v}} = (\hat{\mathbf{u}}, \mathbf{z}), \quad \hat{\mathbf{h}}_0 = -\mathbf{1}$$

system (10)–(13) can be written briefly as

$$\mathbf{H}\mathbf{v} = \mathbf{h}_0$$

$$\mathbf{v} \geqq \mathbf{0}$$

$$\hat{\mathbf{H}}\hat{\mathbf{v}} = \hat{\mathbf{h}}_0$$

$$\hat{\mathbf{v}} \geqq \mathbf{0}.$$

The rank of $\mathbf{H}$ and $\hat{\mathbf{H}}$ is m and n resp. Let $\mathbf{K}$ and $\hat{\mathbf{K}}$ be bases of the column space of $\mathbf{H}$ and $\hat{\mathbf{H}}$ resp.

$$\mathbf{K} = (\mathbf{h}_{i_1}, \ldots, \mathbf{h}_{i_m}),$$

$$\hat{\mathbf{K}} = (\hat{\mathbf{h}}_{\hat{i}_1}, \ldots, \hat{\mathbf{h}}_{\hat{i}_n})$$

and the index sets of the bases are

$$I = \{i_1, \ldots, i_m\},$$

$$\hat{I} = \{\hat{i}_1, \ldots, \hat{i}_n\}.$$

Let us denote by $\mathbf{D}$ and $\hat{\mathbf{D}}$ the coordinates of $(\mathbf{h}_0, \mathbf{H})$ and $(\hat{\mathbf{h}}_0, \hat{\mathbf{H}})$ resp. with respect to the bases $\mathbf{K}$ and $\hat{\mathbf{K}}$ resp.

$$\mathbf{D} = (\mathbf{d}_0, \mathbf{d}_1, \ldots, \mathbf{d}_{m+n}) = \begin{bmatrix} \boldsymbol{\delta}_{i_1} \\ \vdots \\ \boldsymbol{\delta}_{i_m} \end{bmatrix} = [d_{i,p}],$$

$$\hat{\mathbf{D}} = (\hat{\mathbf{d}}_0, \hat{\mathbf{d}}_1, \ldots, \hat{\mathbf{d}}_{m+n}) = \begin{bmatrix} \boldsymbol{\delta}_{\hat{i}_1} \\ \vdots \\ \boldsymbol{\delta}_{\hat{i}_n} \end{bmatrix} = [d_{\hat{i},p}].$$

A basic solution belonging to $\mathbf{K}$ and $\hat{\mathbf{K}}$ is defined by

$$v_r = \begin{cases} d_{r,0} & \text{if} \quad r \in I \\ 0 & \text{if} \quad r \notin I \end{cases}$$

$$\hat{v}_r = \begin{cases} \hat{d}_{\hat{r},0} & \text{if} \quad \hat{r} \in \hat{I} \\ 0 & \text{if} \quad \hat{r} \notin \hat{I}. \end{cases}$$

Matrices $\mathbf{D}$ and $\hat{\mathbf{D}}$ are called *l-positive* (lexicographically positive) if all of their rows are *l*-positive. A basis $\mathbf{K}$ (or $\hat{\mathbf{K}}$) is *l-feasible* if the matrix of coordinates with respect to it $\mathbf{D}$ (or $\hat{\mathbf{D}}$) is *l*-positive. A pair of bases $\mathbf{K}$, $\hat{\mathbf{K}}$ is said to be *l-feasible* if both $\mathbf{K}$ and $\hat{\mathbf{K}}$ are *l*-feasible.

If a nonbasic vector $\mathbf{h}_k$ can be drawn into the *l*-feasible basis $\mathbf{K}$ in such a manner that the new basis $\mathbf{K}^{(1)}$ is also *l*-feasible, then $\mathbf{K}^{(1)}$ is said to be a neighbour of $\mathbf{K}$.

The following assertion is well known from the theory of linear programming. A basis $\mathbf{K}$ has a neighbour containing the nonbasic column $\mathbf{h}_k$ if and only if $\mathbf{d}_k$ has at least one positive component. If there exists such a neighbour, then it is uniquely determined.

An *l*-feasible basis pair $\mathbf{H}^{(1)}$, $\hat{\mathbf{H}}^{(1)}$ is said to be a *neighbour* of the *l*-feasible basis pair $\mathbf{H}$, $\hat{\mathbf{H}}$ if either $\mathbf{H}^{(1)}=\mathbf{H}$ and $\hat{\mathbf{H}}^{(1)}$ is a neighbour of $\hat{\mathbf{H}}$ or $\hat{\mathbf{H}}^{(1)}=\hat{\mathbf{H}}$ and $\mathbf{H}^{(1)}$ is a neighbour of $\mathbf{H}$. The following assertion is an immediate consequence of the above definition: An *l*-feasible basis pair $\mathbf{H}$, $\hat{\mathbf{H}}$ has a neighbour obtainable by drawing the nonbasic vector $\mathbf{h}_k$ into $\mathbf{H}$ (or $\hat{\mathbf{h}}_k$ into $\hat{\mathbf{H}}$) if and only if $\mathbf{d}_k$ (or $\hat{\mathbf{d}}_k$) has at least one positive component. This neighbour is also uniquely determined.

Now we are able to set up the algorithm for finding an equilibrium point of the game $(\mathbf{A}, \mathbf{B})$.

Let our initial bases be $\mathbf{K}^{(-2)}=(\mathbf{e}_1, \ldots, \mathbf{e}_m)$, $\hat{\mathbf{K}}^{(-2)}=(\hat{\mathbf{e}}_1, \ldots, \hat{\mathbf{e}}_n)$, $\mathbf{D}^{(-2)}=(\mathbf{h}_0, \mathbf{H})$, $\hat{\mathbf{D}}^{(-2)}=(\hat{\mathbf{h}}_0, \hat{\mathbf{H}})$. Clearly, the basis pair $\mathbf{K}^{(-2)}$, $\hat{\mathbf{K}}^{(-2)}$ is not feasible. Let us define the basis pair $\mathbf{K}^{(-1)}$, $\hat{\mathbf{K}}^{(-1)}$ in the following manner: let $k=m+1$ and draw $\mathbf{h}_k$ into the basis $\mathbf{K}^{(-2)}$. The index h of the outgoing vector is given by

$$(15) \qquad \frac{1}{d_{h,k}^{(-2)}}\,\delta_h^{(-2)} = \operatorname*{lexmax}_{i \in I^{(-2)}} \frac{1}{d_{i,k}^{(-2)}}\,\delta_i^{(-2)}$$

("lexmax" stands for lexicographically maximal and it is well defined since $d_{ik}^{(-2)}<0$ by the assumption $\mathbf{G}=\mathbf{A}<\mathbf{0}$). (15) represents the regular pivot selection rule of the simplex method. Since the rows of $\mathbf{D}^{(-2)}$ are linearly independent the index h is uniquely determined. Let furthermore $\hat{\mathbf{K}}^{(-1)}=\hat{\mathbf{K}}^{(-2)}$.

In the next step let $\mathbf{K}^{(0)}=\mathbf{K}^{(-1)}$, $k=n+h$ and draw $\hat{\mathbf{h}}_k$ into $\hat{\mathbf{K}}^{(-1)}$. The index j of the outgoing vector is obtained similarly to (15):

$$\frac{1}{\hat{d}_{j,k}^{(-1)}}\hat{\boldsymbol{\delta}}_j^{(-1)}=\operatorname*{lexmax}_{i\in I^{(-1)}}\frac{1}{\hat{d}_{i,k}^{(-1)}}\hat{\boldsymbol{\delta}}_i^{(-1)}.$$

The index j is also uniquely determined. The basis pair $\mathbf{K}^{(0)}$, $\hat{\mathbf{K}}^{(0)}$ is l-feasible which is assured by the pivot selection rule we applied. By the rules of simplex transformation $\hat{\mathbf{d}}_j^{(0)}<\mathbf{0}$ and therefore the basis pair $\mathbf{K}^{(0)}$, $\hat{\mathbf{K}}^{(0)}$ does not have a neighbour obtainable by drawing $\hat{\mathbf{h}}_j$ into the basis.

Let us determine now the following sequence of feasible basis pairs:

$$(16)\qquad \mathbf{K}^{(0)},\hat{\mathbf{K}}^{(0)};\ \mathbf{K}^{(1)},\hat{\mathbf{K}}^{(1)};\ \mathbf{K}^{(2)},\hat{\mathbf{K}}^{(2)};\ \ldots$$

which is defined by the rules:

(a) If $\hat{\mathbf{K}}^{(q)}=\hat{\mathbf{K}}^{(q-1)}$ and $\mathbf{h}_j$ is the vector which has just left $\mathbf{K}^{(q-1)}$, then let $\mathbf{K}^{(q+1)}=\mathbf{K}^{(q)}$ and $\hat{\mathbf{K}}^{(q+1)}$ is obtained by drawing into $\hat{\mathbf{K}}^{(q)}$ the complementary pair $\hat{\mathbf{h}}_k$ of $\mathbf{h}_j$.

(b) If $\mathbf{K}^{(q)}=\mathbf{K}^{(q-1)}$ and $\hat{\mathbf{h}}_j$ has just left $\hat{\mathbf{K}}^{(q-1)}$, then $\hat{\mathbf{K}}^{(q+1)}=\hat{\mathbf{K}}^{(q)}$ and $\mathbf{K}^{(q+1)}$ is obtained by drawing the complementary pair $\mathbf{h}_k$ of $\hat{\mathbf{h}}_j$ into the basis $\mathbf{K}^{(q)}$.

By the application of the lexicographic pivot selection rule the sequence (16) is uniquely determined and each element of it is an l-feasible basis pair. If either $\mathbf{h}_{m+1}=\mathbf{g}_1$ or $\hat{\mathbf{h}}_1=\hat{\mathbf{e}}_1$ leaves the basis, then the algorithm terminates. The algorithm also terminates if no positive pivot can be found for some index q, $(q\geqq 0)$.

THEOREM 6. [101] Sequence (16) is finite.

Proof. Since the number of all possible basis pairs is finite we only have to prove that no basis pair occurs twice in sequence (16). If there were a basis pair occurring twice, then there would be a first one among them. By the construction of the sequence (16) there is exactly one complementary pair both in the basis and out of the basis. Each element of the sequence is a neighbour of its predecessor and it has at most two neighbours. The first element of the sequence has only one neighbour ($\hat{\mathbf{d}}_j^{(0)}$ cannot be drawn into the basis since it is negative). Thus the first of basis pairs occurring twice in the sequence has at least three neighbours which is a contradiction.▮

THEOREM 7. [101] The last element of sequence (16) provides a solution of the bimatrix game $(\mathbf{A}, \mathbf{B})$.

Proof. If the last element of sequence (16) is a complementary solution, then we have finished. This could only have happened if the last basis pair had been obtained by having pivoted out either $\mathbf{g}_1$ or $\hat{\mathbf{e}}_1$. All we have to do now is to prove that sequence (16) can terminate only in this way, i.e., that it cannot terminate because we could not find a positive pivot.

Let the last element be $\mathbf{K}$, $\hat{\mathbf{K}}$. This cannot be $\mathbf{K}^{(0)}$, $\hat{\mathbf{K}}^{(0)}$. By our stopping rule we know that $\mathbf{g}_1$ is in $\mathbf{K}$ and $\bar{\mathbf{e}}_1$ is in $\hat{\mathbf{K}}$. Let $\mathbf{e}_1, \ldots, \mathbf{e}_r$ $(0 \leqq r < m)$ be the unit vectors being in $\mathbf{K}$, while $\hat{\mathbf{e}}_1, \ldots, \hat{\mathbf{e}}_s$ $(1 \leqq s < n)$ those being in $\hat{\mathbf{K}}$. ($s < n$, since $\hat{\mathbf{K}}$ is feasible and $\hat{\mathbf{G}} < \mathbf{0}$.) Since there is only one complementary pair $(\mathbf{g}_1, \hat{\mathbf{e}}_1)$ in the basis the rest of the vectors in $\mathbf{K}$ form a subset of $\{\mathbf{g}_{s+1}, \ldots, \mathbf{g}_n\}$. The non-unit vectors in $\hat{\mathbf{K}}$ belong to the set $\{\hat{\mathbf{g}}_{r+1}, \ldots, \hat{\mathbf{g}}_m\}$. Thus we have listed $m+n+1$ vectors. Since a basis pair contains exactly $m+n$ vectors we know that one of them is not in the basis. We may assume that either $\mathbf{g}_{s+1}$ or $\hat{\mathbf{g}}_{r+1}$ is not a basis vector. In the first case

$$(17) \qquad \mathbf{K} = (\mathbf{e}_1, \ldots, \mathbf{e}_r, \mathbf{g}_1, \mathbf{g}_{s+2}, \mathbf{g}_n)$$

$$\hat{\mathbf{K}} = (\hat{\mathbf{e}}_1, \ldots, \hat{\mathbf{e}}_s, \hat{\mathbf{g}}_{r+1}, \ldots, \hat{\mathbf{g}}_m),$$

and in the second case

$$(18) \qquad \mathbf{K} = (\mathbf{e}_1, \ldots, \mathbf{e}_r, \mathbf{g}_1, \mathbf{g}_{s+1}, \ldots, \mathbf{g}_n)$$

$$\hat{\mathbf{K}} = (\hat{\mathbf{e}}_1, \ldots, \hat{\mathbf{e}}_s, \hat{\mathbf{g}}_{r+2}, \ldots, \hat{\mathbf{g}}_m).$$

In the first case the complementary pair being out of the basis is $\mathbf{g}_{s+1}, \hat{\mathbf{e}}_{s+1}$ while in the second case $\mathbf{e}_{r+1}, \hat{\mathbf{g}}_{r+1}$. By our assumption one of the vectors of the complementary pair has nonpositive coordinates with respect to the actual basis. So in the first case

$$\text{(19)}\quad \mathbf{g}_{s+1}=\sum_{i=1}^{r}\mathbf{e}_i d_{i,m+s+1}+\mathbf{g}_1 d_{m+1,m+s+1}+$$

$$+\sum_{i=m+s+2}^{m+n}\mathbf{g}_{i-m}d_{i,m+s+1},$$

$$\text{(20)}\quad \hat{\mathbf{e}}_{s+1}=\sum_{i=1}^{s}\hat{\mathbf{e}}_i\hat{d}_{i,s+1}+\sum_{i=n+r+1}^{n+m}\hat{\mathbf{g}}_{i-n}\hat{d}_{i,s+1}.$$

If in (19) each d_{ik} is nonpositive, then the second and third term is nonnegative. Since $\mathbf{g}_{s+1}$ is negative $r=m$ should hold which means $\mathbf{K}=\mathbf{K}^{(-2)}$ and we already know that this basis is not feasible. On the other hand if in (20) all $\hat{d}_{ik}$ are nonpositive, then $s=n-1$ implying $\mathbf{K}=\mathbf{K}^{(0)}$. Since the basis pair contains only one complementary pair (this is $\mathbf{g}_1$, $\hat{\mathbf{e}}_1$) $\hat{\mathbf{K}}=\hat{\mathbf{K}}^{(0)}$ which is a contradiction.

In the second case

$$\text{(21)}\quad \mathbf{e}_{r+1}=\sum_{i=1}^{r}\mathbf{e}_i d_{i,r+1}+\mathbf{g}_1 d_{m+1,r+1}+\sum_{i=m+s+1}^{m+n}\mathbf{g}_i d_{i,r+1},$$

$$\text{(22)}\quad \mathbf{g}_{r+1}=\sum_{i=1}^{s}\hat{\mathbf{e}}_i\hat{d}_{i,n+r+1}-\sum_{i=n+r+2}^{n+m}\hat{\mathbf{g}}_i\hat{d}_{i,n+r+1}.$$

Equality (21) is consistent only in case of $r=m-1$ implying $\mathbf{K}=\mathbf{K}^{(0)}$. Thus out of vectors $\hat{\mathbf{g}}_i$ at most $\mathbf{g}_h$ can occur in $\hat{\mathbf{K}}$ implying either $\hat{\mathbf{K}}=\hat{\mathbf{K}}^{(0)}$ or $\hat{\mathbf{K}}=\hat{\mathbf{K}}^{(-1)}$, both being impossible. Equality (22) can be consistent only if $s=n$ which also leads to a contradiction.

As we have seen termination of sequence (16) cannot be caused by the lack of positive pivot elements. Otherwise having pivoted out $\mathbf{g}_1$ or $\hat{\mathbf{e}}_1$ means that we reached a complementary solution. ▮

COROLLARY. Any bimatrix game has an equilibrium point.

Our procedure for finding an equilibrium point has been a constructive proof of the existence theorem (Theorem 3 of Chapter 3 for $n=2$).

As an example let us find an equilibrium point of the bimatrix game given by the pay-off matrices:

$$\mathbf{A}=\begin{bmatrix}1 & 1\\ 2 & 0\end{bmatrix}, \qquad \mathbf{B}=\begin{bmatrix}0 & -1\\ -1 & 1\end{bmatrix}.$$

By subtracting 3 from each element of **A** and **B** we get nonpositive matrices as required by our algorithm. The list of the basic and nonbasic variables as well as the coordinates with respect to the actual basis for the system (10)–(13) can best be arranged in a simplex table format. Thus our initial table is as follows (columns of **H** and $\hat{\mathbf{H}}$ are represented by the variables belonging to them).

Nonbasic variables

			$\hat{z}_1$	$\hat{z}_2$	z_1	z_2
Basic variables	u_1	-1	-2	-2	0	0
	u_2	-1	$\boxed{-1}$	-3	0	0
	$\hat{u}_1$	-1	0	0	-3	-4
	$\hat{u}_2$	-1	0	0	-4	$\boxed{-2}$

From this infeasible solution $(\mathbf{K}^{(-2)}, \hat{\mathbf{K}}^{(-2)})$ we get the first feasible basis pair $(\mathbf{K}^{(0)}, \hat{\mathbf{K}}^{(0)})$ by introducing the columns represented by $\hat{z}_1$ and z_2. (The pivots are marked by a square around them.)

		u_2	$\hat{z}_2$	z_1	$\hat{u}_2$
u_1	1	-2	$\boxed{4}$	0	0
$\hat{z}_1$	1	-1	3	0	0
$\hat{u}_1$	1	0	0	5	-2
z_2	$\frac{1}{2}$	0	0	2	$-\frac{1}{2}$

Since the nonbasic variable which has just left the basis is $\hat{u}_2$, its complementary pair $\hat{z}_2$ should enter the basis. The pivot is determined by the simplex pivot selection rule. The new table is the following:

		u_2	u_1	z_1	$\hat{u}_2$
$\hat{z}_2$	$\frac{1}{4}$	$-\frac{1}{2}$	$\frac{1}{4}$	0	0
$\hat{z}_1$	$\frac{1}{4}$	$\frac{1}{2}$	$-\frac{3}{4}$	0	0
$\hat{u}_1$	1	0	0	$\boxed{5}$	-2
z_2	$\frac{1}{2}$	0	0	2	$-\frac{1}{2}$

Since u_1 left the basis z_1 should enter it. According to the simplex pivot selection rule, $\hat{u}_1$ leaves the basis which is the stopping criterion of the algorithm. Thereby the table below provides a solution of the bimatrix game.

		u_2	u_1	$\hat{u}_1$	$\hat{u}_2$
$\hat{z}_2$	$\frac{1}{4}$	$-\frac{1}{2}$	$\frac{1}{4}$	0	0
$\hat{z}_1$	$\frac{1}{4}$	$\frac{1}{2}$	$-\frac{3}{4}$	0	0
z_1	$\frac{1}{5}$	0	0	$\frac{1}{5}$	$-\frac{2}{5}$
z_2	$\frac{1}{10}$	0	0	$-\frac{2}{5}$	$\frac{3}{10}$

By norming the vectors $\mathbf{z}$ and $\hat{\mathbf{z}}$ we obtain the equilibrium point:

$$\mathbf{x}^0 = \left(\frac{2}{3}, \frac{1}{3}\right),$$

$$\mathbf{y}^0 = \left(\frac{1}{2}, \frac{1}{2}\right).$$

Substituting into the inequality system (2), putting $\alpha = \mathbf{x}^0 \mathbf{A} \mathbf{y}^0 = 1$ and $\beta = \mathbf{x}^0 \mathbf{B} \mathbf{y}^0 = -\dfrac{1}{3}$, we can easily verify that $(\mathbf{x}^0, \mathbf{y}^0)$ is really an equilibrium point.

8.3. EXAMPLES

Conflict situations representable by nonzero-sum games, and in particular by bimatrix games come up frequently in business and economics. As an illustration we will present two examples which have been considerably simplified. However the basic ideas of game-theoretic modelling can well be demonstrated through them.

EXAMPLE 1. *A market model.* Let a one-product market be given with two major firms A and B of equal power. The quality of the product marketed by A and B is about the same. The price of the products is bounded from below and above. Thus if p and q denote prices charged by A and B resp., then

$$m \leqq p \leqq M$$

$$m \leqq q \leqq M ,$$

m and M denoting the lower and upper bounds resp.

The demand is supposed to depend only on the average market price and it is a monotonically decreasing function of it. Thus, assuming that the demand function $k(x)$ is differentiable, we have

$$k'\left(\frac{p+q}{2}\right) < 0$$

$$k\left(\frac{p+q}{2}\right) > 0, \quad (m \leqq p \leqq M,\ m \leqq q \leqq M) .$$

The market share of each firm is assumed to depend only on the difference between prices. Let $f(p-q)$ be the share of firm A of the overall demand as a function of $p-q$, the difference between the prices charged by A and B. Obviously the share of B is now $1-f(p-q)$. The function f is supposed to satisfy the following conditions:

1. $0 \leqq f(x) \leqq 1$ for $m-M \leqq x \leqq M-m$,
2. $f(x)$ is monotonically decreasing in the interval $[m-M, M-m]$,
3. $f(0) = \dfrac{1}{2}$.

Given the data specified above we can set up the pay-off functions for both players assuming that their goal is to maximize their sales:

$$(21)\quad F_A(p, q) = pf(p-q)k\left(\frac{p+q}{2}\right), \qquad (m \leqq p, q \leqq M)$$

$$F_B(p, q) = q[1-f(p-q)]k\left(\frac{p+q}{2}\right), \qquad (m \leqq p, q \leqq M).$$

Allowing only finitely many values for p and q a bimatrix game is determined by (21).

Let

$$k\left(\frac{p+q}{2}\right) = -(p+q)+20$$

$$f(p-q) = -\frac{1}{6}(p-q)+\frac{1}{2} \quad m=1, \quad M=4$$

$$p, q = 1, 2, 3, 4.$$

Now the pay-off matrices are obtained by simple calculation

$$\mathbf{A} = \begin{bmatrix} \boxed{9} & \frac{34}{3} & \frac{40}{3} & 15 \\ \frac{17}{3} & \boxed{16} & 20 & \frac{70}{3} \\ 8 & 15 & \boxed{21} & 26 \\ 0 & \frac{28}{3} & \frac{52}{3} & 24 \end{bmatrix},$$

$$\mathbf{B} = \begin{bmatrix} \boxed{9} & \frac{17}{3} & 8 & 0 \\ \frac{34}{3} & \boxed{16} & 15 & \frac{28}{3} \\ \frac{40}{3} & 20 & \boxed{21} & \frac{52}{3} \\ 15 & \frac{70}{3} & 26 & 24 \end{bmatrix}.$$

Player I controls the rows whereas player II controls the columns. There are three pure equilibrium points: $(\mathbf{e}_1, \mathbf{e}_2)$, $(\mathbf{e}_2, \mathbf{e}_2)$, $(\mathbf{e}_3, \mathbf{e}_3)$. Since the first two are dominated by $(\mathbf{e}_3, \mathbf{e}_3)$, they can be left out of consideration. The point $(\mathbf{e}_4, \mathbf{e}_4)$ dominates the point $(\mathbf{e}_3, \mathbf{e}_3)$ but it is not "stable" since each player can be better off by changing his strategy (decreasing his price while the other maintains his own price level). Therefore, if they want to play the strategy pair $(\mathbf{e}_4, \mathbf{e}_4)$, which is advantageous for both, they should come to an agreement. But this question leads us to the theory of cooperative games.

EXAMPLE 2. *An inventory model.* [24] A warehouse places an order for a certain commodity with a wholesaler every Friday. The amount ordered is to meet the demand of the next week. In the case of ordering above demand some of the goods remain unsold whereas ordering less than demand means an opportunity loss. Demand is a random variable with exponential distribution, having the density function:

$$f(x) = \lambda e^{-\lambda x}, \quad \lambda > 0 .$$

The unit inventory holding cost and profit of the commodity is b_1 and a_1 resp. Holding an inventory of y, the expected profit of the warehouse is given as

$$P(y) = a_1 \int_0^y x f(x)\,\mathrm{d}x + a_1 y \int_y^\infty f(x)\,\mathrm{d}x - b_1 y .$$

If demand is above inventory, then an additional order can be placed but it cannot exceed the safety stock held by the wholesaler. The profit gained by selling a unit of the goods received through the additional order is a_2 $(a_2 < a_1)$. Denoting by z the safety stock of the wholesaler we get the following formula for the expected profit of the warehouse:

$$M_1(y, z) = a_1 \int_0^y x f(x)\,\mathrm{d}x + a_1 y \int_y^{y+z} f(x)\,\mathrm{d}x +$$

$$+ a_2 \int_y^{y+z} (x-y) f(x)\,\mathrm{d}x + (a_1 y + a_2 z) \int_{y+z}^\infty f(x)\,\mathrm{d}x -$$

$$-b_1 y = \frac{1}{\lambda}[a_1(1-e^{-\lambda y}) + a_2(e^{-\lambda y}) -$$

$$-e^{-\lambda(y+z)})] - b_1 y.$$

If the wholesaler gets a profit a_3 for a unit sold to the warehouse and his inventory cost per unit is b_3, then his expected profit is:

$$M_2(y, z) = a_3 \int_y^{y+z} (x-y) f(x)\,dx +$$

$$+ a_3 z \int_{y+z}^{\infty} f(x)\,dx - b_2 z =$$

$$= \frac{1}{\lambda} a_3 (e^{-\lambda y} - e^{-\lambda(y+z)}) - b_2 z.$$

The situation can be looked upon as a two-person game with pay-off functions M_1 and M_2. If y and z can assume only a finite number of values, then we obtain a bimatrix game. Let

$$a_1 = 8, \quad a_2 = 4, \quad a_3 = 3,$$

$$b_1 = 3, \quad b_2 = 1, \quad \lambda = \frac{\ln 2}{100}.$$

Let the pure strategies (the possible values of y and z) be {0, 50, 100, 150, 200, 250} and {0, 50, 100, 150, 200, 300} resp. Then we get the following pay-off matrices (elements of the matrices are rounded).

Pay-off to the warehouse

y \ z	0	50	100	150	200	300
0	0	170	290	375	430	505
50	195	305	390	450	485	545
100	275	360	420	460	490	530
150	295	350	395	425	445	470
200	260	305	335	355	370	390
250	200	230	250	265	275	280

Pay-off to the wholesaler

y \ z	0	50	100	150	200	300
0	0	75	115	130	125	25
50	0	40	55	50	30	−30
100	0	15	10	−10	−40	−110
150	[0]	−5	−25	−50	−85	−165
200	0	−20	−45	−80	−120	−205
250	0	−30	−60	−100	−145	−235

There is only one pure equilibrium point, $y=150$; $z=0$. This means that the warehouse keeps an inventory of 150 while the wholesaler's safety stock is 0. This equilibrium point is hard to accept as a solution of the problem since it is dominated by a number of other points. Thus both players could benefit from cooperation which is not allowed in the theory of noncooperative games.

EXAMPLE 3. *The world oil market.* [167] In this example the world oil-market is modelled as a two-person non-zero sum game. Player 1 called OPIC represents the oil-importing nations and player 2 called OPEC represents the oil exporting nations. We have assumed that all major oil importing countries have formed a cartel and act collectively as one unit and this cartel is the sole market for the oil exported by OPEC which is also assumed to represent all the major oil exporting countries.

Suppose that OPIC needs a total of D million barrels of oil daily (mmbd) representing consumption required for a desired (possibly maximum) growth of their economy. A part of this requirement can be met by domestic production of oil. The demand for import-oil can be reduced by a number of ways, e.g. by finding new domestic sources, by working the existing oil wells harder, by using alternative sources of energy and by administrative tools such as rationing, taxes etc. These measures may result, however, in a decrease in OPIC's economic growth. In short, in our game-theoretic model the strategy for OPIC is to decide the quantity of oil imports. Formally, the strategy space of OPIC is denoted by

$$(22) \quad \Sigma_1 = \{x \in \mathbb{R}^1 \mid 0 \leqq x \leqq D\}.$$

Associated with a strategy $x \in \Sigma_1$ is a monetary cost to OPIC, denoted by $f(x)$, for restricting its imports to x mmbd. $f(x)$ does not include the cost of imports. A possible way of computing $f(x)$ is as follows.

Denote $h(y)$ the total cost in million dollars daily (mm $d) to ensure that the domestic production of oil (or other alternative sources of energy) is at least y mmbd. Also denote $g(z)$ the loss in mm$d in OPIC's GNP if the total oil (energy) consumption is restricted to z mmbd. Then we obtain

$$(23) \quad f(x) = \min_{0 \leqq y \leqq D-x} [h(y) + g(y+x)].$$

We will assume that $f(x)$ is a positive, nonincreasing function defined on Σ_1.

Let the total proven oil reserves of OPEC be R million barrels. It costs OPEC an average of c dollars per barrel to extract the oil from the fields.

The strategy for OPEC is to decide the price p of a barrel of oil exported to OPIC, i.e., formally

$$\Sigma_2 = \{p \in \mathbb{R} \mid c \leqq p\}.$$

Let $\Sigma = \Sigma_1 \times \Sigma_2$.

Now we define the pay-off function u_1, u_2 of each player. Neglecting political, psychological and other considerations we formulate the pay-off functions solely in terms of monetary gains (or losses). Thus OPIC's pay-off is

$$u_1(x, p) = -f(x) - px$$

for any $(x, p) \in \Sigma$. For OPEC the monetary pay-off is

(24) $u_2(x, p)=(p-c)x$

for any $(x, p)\in \Sigma$.

If we assume that OPEC is interested in obtaining the maximum revenue from its finite oil reserves of R million barrels and the parameters of the problem D, f, c remain constant with time we would have

$$(25)\quad u_2(x, p)=\begin{cases} 0 & \text{if } x=0 \\ \sum_{i=0}^{[\frac{R}{x}]} (p-c)x\beta^i & \text{if } x>0, \end{cases}$$

where $\left[\frac{R}{x}\right]$ denotes the largest integer not greater than $\frac{R}{x}$ and β is a discount factor. $u_2(x, p)$ represents the discounted sum of revenues that OPEC would earn from its total oil reserves of R million barrels given that outcome (x, p) is constant during this period (of $\left[\frac{R}{x}\right]$ days).

To determine the accurate nature of function f is a crucial point in this model and this task is by far not trivial. However, several studies have been made to estimate $f(x)$ for the USA. These studies indicate that $f(x)$ is really nonincreasing and is exponential in nature.

To determine the noncooperative solution (the Nash-equilibrium point(s)) of this game it is sufficient to assume that f is nonincreasing.

Let us first consider the short-term model.

$u_2(x, p)$ is defined by $(p-c)x$. Let (x^*, p^*) be an equilibrium point. By definition

$$u_2(x^*, p^*)\geqq u_2(x^*, p^*+\Delta p)$$

for all $\Delta p>0$. Thus

$$(p^*-c)x^*\geqq(p^*+\Delta p-c)x^*$$

i.e.

$$\Delta p x^* \leqq 0 \quad \text{for any} \quad \Delta p > 0 .$$

Hence, we have $x^* = 0$ and $u_1(0, p^*) = -f(0)$. Also

$$u_1(0+\Delta x, p^*) = -f(\Delta x) - p^* \Delta x .$$

Since

$$u_1(0, p^*) \geqq u_1(0+\Delta x, p^*)$$

holds for any $\Delta x > 0$, we have

$$-f(0) \geqq -f(\Delta x) - p^* \Delta x \quad \text{for any} \quad \Delta x > 0$$

which means

$$p^* \geqq \frac{f(0)-f(x)}{\Delta x} \quad \text{for any} \quad \Delta x > 0 .$$

Define

$$p_m = \sup_{0 < x \leqq D} \frac{f(0)-f(x)}{x}$$

then $(0, p^*)$ is an equilibrium point for each $p^* \geqq p_m$.

A similar analysis shows that if we apply pay-off function (25) we also get $(0, p^*)$ where $p^* \geqq p_m$ as an equilibrium point with the pay-off pair $(-f(0), 0)$.

By substitution we can obtain that $(-f(0), 0)$ is the pay-off for any equilibrium point, i.e., the game is solvable and $(-f(0), 0)$ is the solution. The noncooperative solution is for OPIC to import no oil at all and OPEC to charge a very high price. The Arab oil embargo in the winter of 1973 was an outcome of the game at its equilibrium, i.e., $x = 0, p \to \infty$. It was forced by OPEC in order to demonstrate the disadvantages of a noncooperative situation.

EXAMPLE 4. *The diffusion of new technology.* [143] In this example an attempt is made to analyze the diffusion of new technology in a game-theoretic framework. We consider an industry composed of two firms each using the current best-practice technology. When a cost reducing innovation comes up, each firm must determine when (if ever) to adopt it, based in part upon the discounted cost of implementing the new technology and in part upon the behaviour of the rival firm. On the other hand the costs associated with adoption may decline with the lengthening of the adjustment period. Thus the firm must weight the costs and benefits of delaying adoption, as well as take account of its rival's strategic behaviour.

In the following we will formalize this problem as a two-person nonzero-sum game and compute the Nash-equilibrium points.

Consider two firms each making equilibrium profits Π_0 denoted by (Π_0, Π_0). At time $t=0$, a technological improvement designed to reduce production costs is announced and offered for sale.

If firm 1 purchases the new technology before firm 2, then the profits are (Π_1, Π_2) during the period before firm 2 adopts the innovation, symmetrically, if firm 2 adopts first, then the profits are (Π_2, Π_1). After both firms have adopted the innovation, profits are (Π_3, Π_3).

Denoting the adoption dates of firms 1 and 2 by T_1 and T_2 resp. the profit opportunities described above can be summarized in a tabular form:

t	Firm	
	1	2
$0 \leqq t \leqq \min\{T_1, T_2\}$	Π_0	Π_0
$T_1 \leqq t \leqq T_2$	Π_1	Π_2
$T_2 \leqq t \leqq T_1$	Π_2	Π_1
$\infty > t \geqq \max\{T_1, T_2\}$	Π_3	Π_3

The following assumptions describe the relative magnitudes of the profits.

A1. $\Pi_i > 0 \; i=0, 1, 2, 3$.

A2. $\Pi_1 > \Pi_3 > \Pi_2$; $\Pi_1 > \Pi_0 > \Pi_2$.

A3. $\alpha = \Pi_1 - \Pi_0 + \Pi_2 - \Pi_3 > 0$.

A1. states that both firms make positive profits. A2. implies that profits to firm i are greatest when i has adopted the innovation but j

has not; next greatest profits occur either when both have adopted or when no firm has yet adopted; finally profit opportunities for i are worst when j has adopted the innovation but i has not. A3 states that the net value of being first $\alpha=(\Pi_1-\Pi_0)-(\Pi_3-\Pi_2)$ is positive.

Define $p(t)$ to be the discounted price of the innovation at time t. This includes all the costs of adjustment. We also assume that p is twice continuously differentiable at all positive t. Denote r the market rate of interest. Then the firms' pay-offs are defined as follows:

The pay-off to firm 1 is

$$(26)\quad f^1(T_1, T_2)=\begin{cases} g^1(T_1, T_2) & \text{if } T_1 \leqq T_2 \\ g^2(T_1, T_2) & \text{if } T_1 \geqq T_2 \end{cases}$$

where

$$g^1(T_1, T_2)=\int_0^{T_1} \Pi_0 e^{-rt}\,dt+\int_{T_1}^{T_2} \Pi_1 e^{-rt}\,dt+\int_{T_2}^{\infty} \Pi_3 e^{-rt}\,dt-p(T_1)$$

and

$$g^2(T_1, T_2)=\int_0^{T_2} \Pi_0 e^{-rt}\,dt+\int_{T_2}^{T_1} \Pi_2 e^{-rt}\,dt+\int_{T_1}^{\infty} \Pi_3 e^{-rt}\,dt-p(T_1).$$

By symmetry, the pay-off to firm 2 is simply $f^2(T_1, T_2)\equiv f^1(T_2, T_1)$. Without loss of generality it is sufficient to deal with firm 1's decision problem, the corresponding results for firm 2 can be deduced by symmetry.

It should be noted that while $f^1(T_1, T_2)$ is continuous in T_1 for fixed T_2, it is not differentiable at $T_1=T_2$. The left-hand derivative at T_2 is

$$g_1^1(T_2, T_2)=(\Pi_0-\Pi_1)e^{-rT_2}-p'(T_2)$$

while the right-hand derivative at T_2 is

$$g_1^2(T_2, T_2)=(\Pi_2-\Pi_3)e^{-rT_2}-p'(T_2).$$

By A3 $g_1^1(T_2, T_2)<g_1^2(T_2, T_2)$ for all $T_2\geqq 0$.

We will also impose an assumption on the behaviour of the cost function p.

A4. $p(t) \geqq 0$ and $p''(t) > r[\Pi_1 - \Pi_0]e^{-rt}$ for all $t \geqq 0$. Further assume that $\lim_{t\to\infty} p'(t) > 0$.

A4 states that there is some optimal adjustment period beyond which any further prolongation of the adjustment process begins to increase costs.

As a consequence of this assumption we can show that $g^1(T_1, T_2)$ and $g^2(T_1, T_2)$ are strictly concave in T_1 (for fixed T_2). To see this, note that $g^1_{11} = r(\Pi_1 - \Pi_0)e^{-rt} - p''(t) < 0$ by A4 and $g^2_{11} = r(\Pi_3 - -\Pi_2)e^{-rt} - p''(t) < 0$ by A4 and A3. (g^1_{11} and g^2_{11} denote second derivatives).

We need a series of lemmas in the sequel. We can readily prove two of them.

LEMMA 1. There exists a unique $\hat{T} \geqq 0$ and $\hat{\hat{T}} \geqq 0$ which maximize $g^1(T_1, T_2)$ and $g^2(T_1, T_2)$ resp., independently of T_2. Furthermore

(a) if $p'(0) < \Pi_2 - \Pi_3$, then $\hat{T} \geqq 0$ and $\hat{\hat{T}} > 0$,
(b) if $p'(0) \geqq \Pi_2 - \Pi_3$, then $\hat{T} = \hat{\hat{T}} = 0$.

Proof. For all $T_2 \geqq 0$ we have

$$\lim_{t\to\infty} g^1_1(t, T_2) = \lim_{t\to\infty} (\Pi_0 - \Pi_1)e^{-rt} - p'(t) < 0$$

by A4. By strict concavity and continuity of g^1, there exists a unique $\hat{T} \geqq 0$ which maximizes $g^1(T_1, T_2)$. $\hat{T}$ is defined (independently of T_2) by

$$(\Pi_0 - \Pi_1)e^{-r\hat{T}} - p'(\hat{T}) \leqq 0, \quad \hat{T} \geqq 0,$$

and

$$[(\Pi_0 - \Pi_1)e^{-r\hat{T}} - p'(\hat{T})]\hat{T} = 0.$$

If $p'(0) \geqq \Pi_2 - \Pi_3$, then $g^1_1(0, T_2) < 0$ for all $T_2 \geqq 0$, thus $\hat{T} = 0$.

Also by A4 we have

$$\lim_{t\to\infty} g^2_1(t, T_2) = \lim_{t\to\infty} (\Pi_2 - \Pi_3)e^{-rt} - p'(t) < 0$$

for all $T_2 \geqq 0$.

By strict concavity and continuity of g^2, there exists a unique $\hat{\hat{T}} \geqq 0$ which maximizes $g^2(T_1, T_2)$. $\hat{\hat{T}}$ is defined (independently of T_2) by

$$(\Pi_2 - \Pi_3)e^{-r\hat{\hat{T}}} - p'(\hat{\hat{T}}) \leqq 0, \quad \hat{\hat{T}} \geqq 0$$

and

$$[(\Pi_2 - \Pi_3)e^{-r\hat{\hat{T}}} - p'(\hat{\hat{T}})]\hat{\hat{T}} = 0.$$

If $p'(0) < \Pi_2 - \Pi_3$, then $g_1^2(0, T_2) > 0$ for all $T_2 \geqq 0$, thus $\hat{\hat{T}} > 0$. If $p'(0) \geqq \Pi_2 - \Pi_3$, then $g_1^2(0, T_2) \leqq 0$ for all $T_2 \geqq 0$ and strict concavity $(g_{11}^2 < 0)$ implies that $\hat{\hat{T}} = 0$.

LEMMA 2. If $p'(0) < \Pi_2 - \Pi_3$, then $\hat{\hat{T}} > \hat{T}$.

Proof. Recall that $g_1^2(t, T_2) > g_1^1(t, T_2)$ holds for any $t, T_2 \geqq 0$. By Lemma 1, $0 < \hat{\hat{T}} < \infty$ and $\hat{T} < \infty$. If $\hat{T} = 0$, then $\hat{\hat{T}} > \hat{T}$ is obvious. Suppose $\hat{T} > 0$. Now $g_1^2(\hat{T}, T_2) > g_1^1(\hat{T}, T_2) = 0$ holds for all $T_2 \geqq 0$. Since $g_1^2(\hat{\hat{T}}, T_2) = 0$ and $g^2(t, T_2)$ is strictly concave in t for fixed T_2, we have $\hat{\hat{T}} > \hat{T}$.

Now we define a two-person non-zero-sum game G with the firms as players as follows

$$G = \{S_1, S_2; f_1, f_2\}$$

where the strategy sets $S_i = \{[0, \infty)\}$ $(i = 1, 2)$ and f_1, f_2 are defined as in (26). We are interested in finding a pair of Nash-equilibrium strategies (T_1^N, T_2^N). We have

THEOREM 1

(a) If $p'(0) < \Pi_2 - \Pi_3$, then there exist two Nash-equilibria $(T_1^N, T_2^N) = (\hat{T}, \hat{\hat{T}})$ and $(T_1^N, T_2^N) = (\hat{\hat{T}}, \hat{T})$ where $\hat{T}, \hat{\hat{T}}$ are as defined in Lemma 1.

(b) If $p'(0) \geqq \Pi_2 - \Pi_3$, then there is a unique Nash-equilibrium point $(T_1^N, T_2^N) = (0, 0)$.

Proof. Assertion (b) is an immediate consequence of Lemma 1. The proof of assertion (a) is accomplished via a series of lemmas.

LEMMA 3. $g^1(T_1, T_2) \gtreqless g^2(T_1, T_2)$ as $T_1 \lesseqgtr T_2$.

Proof. Easy calculation shows that

$$g^1(T_1, T_2) - g^2(T_1, T_2) = \frac{\alpha}{r}(e^{-rT_1} - e^{-rT_2}),$$

which immediately implies the assertion of the lemma.

LEMMA 4. $g^2(\hat{\hat{T}}, \hat{T}) > g^1(\hat{T}, \hat{T})$.

Proof. $g^2(\hat{\hat{T}}, \hat{T}) > g^2(\hat{T}, \hat{T}) = g^1(\hat{T}, \hat{T})$ where the inequality follows from Lemma 1. The equality follows from Lemma 3.

LEMMA 5. $g^1(\hat{T}, \hat{\hat{T}}) > g^2(\hat{\hat{T}}, \hat{\hat{T}})$.

Proof. The same as for Lemma 4.

LEMMA 6. There exists a $\tilde{T}$ such that $\hat{T} < \tilde{T} < \hat{\hat{T}}$ and $g(\hat{T}, T_2) \lesseqgtr g^2(\hat{\hat{T}}, T_2)$ as $T_2 \lesseqgtr \tilde{T}$.

Proof. Define $\gamma(\hat{T}, \hat{\hat{T}}, T_2) = g^1(\hat{T}, T_2) - g^2(\hat{\hat{T}}, T_2)$. By Lemma 4, $\gamma(\hat{T}, \hat{\hat{T}}, \hat{T}) < 0$. By Lemma 5, $\gamma(\hat{T}, \hat{\hat{T}}, \hat{\hat{T}}) > 0$. Since $\frac{\partial \gamma}{\partial T_2} = \alpha e^{-rT_2} > 0$, i.e., γ is monotone increasing and continuous in the interval $(\hat{T}, \hat{\hat{T}})$ which implies the assertion of the lemma.

Define player i's best responses to a strategy of player j by

$$\Phi_i(T_j) = \{T_i \in S_i \mid f^i(T_i, T_j) \geqq f^i(T_i', T_j) \quad \text{for all} \quad T_i' \in S_i\} \qquad (i = 1, 2, j \neq i).$$

Now we can state Lemma 7.

LEMMA 7.

$$\Phi_1(T_2) = \begin{cases} \hat{\hat{T}} & \text{for} \quad T_2 < \tilde{T} \\ \{\hat{T}, \hat{\hat{T}}\} & \text{for} \quad T_2 = \tilde{T} \\ \hat{T} & \text{for} \quad T_2 > \tilde{T}. \end{cases}$$

Proof. We distinguish three cases (the reasons for the various relations are found in parentheses immediately below the relation itself).

CASE 1. $T_2 < \tilde{T}$. We have for all $T_1 \leqq T_2$

$$f^1(\hat{\hat{T}}, T_2) = g^2(\hat{\hat{T}}, T_2) > g^1(\hat{T}, T_2) \geqq g^1(T_1, T_2).$$

(26) (Lemma 6)(Lemma 1)

Also by (26) we have $g^1(T_1, T_2) = f^1(T_1, T_2)$.

On the other hand for $T_1 \geqq T_2$, $T_1 \neq \hat{\hat{T}}$ we have

$$f^1(\hat{\hat{T}}, T_2) = g^2(\hat{\hat{T}}, T_2) > g^2(T_1, T_2) = f^1(T_1, T_2).$$

(26) (Lemma 1) (26)

Thus for all $T_1 \neq \hat{\hat{T}}$ we have $f^1(\hat{\hat{T}}, T_2) > f^1(T_1, T_2)$. Hence $\Phi_1(T_2) = \hat{\hat{T}}$ for all $T_2 < \tilde{T}$.

CASE 2. $T_2 = \tilde{T}$. For all $T_1 \leqq \tilde{T}$, $T_1 \neq \hat{T}$ we have

$$f^1(\hat{T}, \tilde{T}) = g^1(\hat{T}, \tilde{T}) > g^1(T_1, \tilde{T}) = f^1(T_1, \tilde{T})$$

(26) (Lemma 1) (26)

and for $T_1 \geqq \tilde{T}$, $T_1 \neq \hat{\hat{T}}$ we have

$$f^1(\hat{\hat{T}}, \tilde{T}) = g^2(\hat{\hat{T}}, \tilde{T}) > g^2(T_1, \tilde{T}) = f^1(T_1, \tilde{T}).$$

(26) (Lemma 1) (26)

But since $g^1(\hat{T}, \tilde{T}) = g^2(\hat{\hat{T}}, \tilde{T})$ holds by Lemma 6, we have $f^1(\hat{T}, \tilde{T}) = f^1(\hat{\hat{T}}, \tilde{T}) > f^1(T_1, \tilde{T})$ for all $T_1 \notin \{\hat{T}, \hat{\hat{T}}\}$. Thus $\Phi_1(\tilde{T}) = \{\hat{T}, \hat{\hat{T}}\}$.

CASE 3. $T_2 > \tilde{T}$. For all $T_1 \leqq T_2$, $T_1 \neq \hat{T}$ we have

$$f^1(\hat{T}, T_2) = g^1(\hat{T}, T_2) > g^1(T_1, T_2) = f^1(T_1, T_2).$$

(26) (Lemma 1) (26)

For $T_1 \geqq T_2$ we deduce

$$f^1(\hat{T}, T_2) = g^1(\hat{T}, T_2) > g^2(\hat{\hat{T}}, T_2) \geqq g^2(T_1, T_2) = f^1(T_1, T_2).$$

(Lemma 6) (Lemma 1)

Hence for any $T_1 \neq \hat{T}$ the inequality $f^1(\hat{T}, T_2) > f^1(T_1, T_2)$ holds. Hence $\Phi_1(T_2) = \hat{T}$ for all $T_2 > \tilde{T}$ which concludes the proof of the lemma.

Now to prove assertion (a) of our theorem it is sufficient to observe that by symmetry

$$\Phi_2(T_1) = \begin{cases} \hat{\hat{T}} & \text{for} \quad T_1 < \tilde{T} \\ \{\hat{T}, \hat{\hat{T}}\} & \text{for} \quad T_1 = \tilde{T} \\ \hat{T} & \text{for} \quad T_1 > \tilde{T}. \end{cases}$$

It is clear that functions Φ_1 and Φ_2 intersect only at two distinct points, $(\hat{T}, \hat{\hat{T}})$ and $(\hat{\hat{T}}, \hat{T})$ which are obviously the Nash-equilibrium points. ∎

We remind the reader that to compute $\hat{T}$ and $\hat{\hat{T}}$ we only have to solve the equations

$$-(\Pi_1 - \Pi_0)e^{-r\hat{T}} - p'(\hat{T}) = 0$$

$$-(\Pi_3 - \Pi_2)e^{-r\hat{\hat{T}}} - p'(\hat{\hat{T}}) = 0.$$

It is also interesting to note that $\hat{T}$ and $\hat{\hat{T}}$ depend on the profit rates only through the relevant opportunity costs of delaying adoption one period $(\Pi_1 - \Pi_0)$ for $\hat{T}$ and $(\Pi_3 - \Pi_2)$ for $\hat{\hat{T}}$.

Before concluding this example let us say a few words about the economic implications of Theorem 1. In the degenerate case (b) adjustment costs do not decline sufficiently rapidly as to warrant waiting. Thus, regardless of rival behaviour both firms have to adopt immediately.

In case (a) if there is a net value of being first (Assumption A3), then at equilibrium one firm adopts early ($\hat{T}$), while the other adopts late at $\hat{\hat{T}}$. It is never a Nash-equilibrium for identical firms to bring the new technology on line at the same time. Hence even in the case of identical firms and complete certainty there will be a diffusion of the innovation over time.

9. Matrix games

If a bimatrix game is zero-sum, then the pay-off of both players is completely determined by the pay-off matrix of one of the players, say by that of player I. Denote by **A** player I's pay-off matrix. Then player II's pay-off is $-\mathbf{A}$. For convenience we will always look at a matrix game from player I's point of view, i.e., the entries of matrix **A** will denote his "gains" and player II's "losses" for pure strategy pairs. Just as we did for bimatrix games we are going to concentrate on the mixed extension of the matrix game **A**. Unless otherwise indicated when speaking of a matrix game we think of its mixed extension.

Matrix games are models of antagonistic situations. Interests of the players are entirely opposed, their "gains" can only be increased at the expense of the other. A great number of parlour games (two-person card games, chess etc.), and certain competitive economic and military situations fall into this class. In addition, the theory of cooperative games, decision theory and other fields of mathematics make use of the theory and methods developed for matrix games.

9.1. EQUILIBRIUM AND THE MINIMAX PRINCIPLE

THEOREM 1. Any matrix game has at least one equilibrium strategy pair.

This theorem is an immediate consequence of Theorem 3 of Chapter 3. Theorem 1 is one of the earliest and most significant results of game theory first proved by John von Neumann. Therefore we call it Neumann's theorem.

If we put $v^0 = \mathbf{x}^0\mathbf{A}\mathbf{y}^0$ we have for any equilibrium strategy pair $(\mathbf{x}^0, \mathbf{y}^0)$

$$\text{(1)} \qquad \mathbf{x}\mathbf{A}\mathbf{y}^0 \leqq \mathbf{v}^0 \leqq \mathbf{x}^0\mathbf{A}\mathbf{y} \quad \text{for all} \quad \mathbf{x} \in X_m, \mathbf{y} \in Y_n .$$

v^0 is the value of game **A** and it can be interpreted as the expected pay-off to player I if the players play the strategies $\mathbf{x}^0$, $\mathbf{y}^0$ resp. Theorem 1 and (1) imply that any matrix game possesses the interchangeability property and therefore $\bar{\mathbf{x}}\bar{\mathbf{A}}\bar{\mathbf{y}} = v^0$ for any equilibrium point $(\bar{\mathbf{x}}, \bar{\mathbf{y}})$.

Special properties of the strategy sets and the pay-off function enable us to sharpen Theorem 2 of Chapter 7 for matrix games.

THEOREM 2. [31] For a strategy pair $(\mathbf{x}^0, \mathbf{y}^0)$ to be an equilibrium point and for the real number v to be the value of the matrix game **A**, it is necessary and sufficient to satisfy the equations

$$\text{(2)} \qquad \max_{\mathbf{x} \in X_m} \min_{\mathbf{y} \in Y_n} \mathbf{x}\mathbf{A}\mathbf{y} = \min_{\mathbf{x} \in X_m} \max_{\mathbf{y} \in Y_n} \mathbf{x}\mathbf{A}\mathbf{y} = v ,$$

$$\text{(3)} \qquad \min_{\mathbf{y} \in Y_n} \mathbf{x}^0\mathbf{A}\mathbf{y} = v ;$$

$$\text{(4)} \qquad \max_{\mathbf{x} \in X_m} \mathbf{x}\mathbf{A}\mathbf{y}^0 = v .$$

Proof. Our theorem follows directly from Theorem 2 of Chapter 7. Since X_m and Y_n are compact subsets of finite dimensional Euclidean spaces and the pay-off function $f(\mathbf{x}, \mathbf{y}) = \mathbf{x}\mathbf{A}\mathbf{y}$ is continuous therefore sup and inf in Theorem 2 of Chapter 7 can be changed to max and min. ∎

Theorem 2 is often referred to as the *minimax principle* and it can be interpreted in the same way as Theorem 2 of Chapter 7.

9.2. THE SET OF EQUILIBRIUM STRATEGIES

Let X^0 and Y^0 be the sets of equilibrium strategies of player I and II resp. Because of the interchangeability property, any pair $(\mathbf{x}^0, \mathbf{y}^0)$, $\mathbf{x}^0 \in X^0$, $\mathbf{y}^0 \in Y^0$ is an equilibrium point. The following theorem gives a geometric characterization of X^0 and Y^0.

THEOREM 3. [79] The set of equilibrium strategies is a convex polyhedron.

Proof. We first prove that for a strategy pair $(\mathbf{x}^0, \mathbf{y}^0)$ to be an equilibrium point and for a real number v^0 to be the value of the game it is necessary and sufficient to satisfy the following linear inequality system:

$$(5) \qquad \begin{aligned} &\mathbf{x} \geqq \mathbf{0}, \quad \mathbf{y} \geqq \mathbf{0} \\ &\mathbf{1x} = 1 \\ &\mathbf{1y} = 1 \\ &\mathbf{A}^T\mathbf{x} - \mathbf{1}v \geqq \mathbf{0} \\ &\mathbf{Ay} - \mathbf{1}v \leqq \mathbf{0}. \end{aligned}$$

Let $\mathbf{x}^0$, $\mathbf{y}^0$, v^0 satisfy system (5). Then $\mathbf{A}^T\mathbf{x}^0 - \mathbf{1}v^0 \geqq \mathbf{0}$, $\mathbf{Ay}^0 - \mathbf{1}v^0 \leqq \mathbf{0}$. Premultiplying the first inequality by an arbitrary probability vector $\mathbf{y}$, and postmultiplying the second one by an arbitrary $\mathbf{x} \in X_m$, we get

$$(6) \qquad \mathbf{xAy}^0 \leqq v^0 \leqq \mathbf{x}^0\mathbf{Ay} \quad \text{for all} \quad \mathbf{x} \in X_m\ \mathbf{y} \in Y_n,$$

which means that $(\mathbf{x}^0, \mathbf{y}^0)$ is an equilibrium point and v^0 is the value.

On the other hand if $(\mathbf{x}^0, \mathbf{y}^0)$ is an equilibrium point and v^0 is the value of the game, then (6) implies $\mathbf{x}^0\mathbf{Ay} \geqq v^0$ for all $\mathbf{y} \in Y_n$. In particular $\mathbf{x}^0\mathbf{Ae}_j \geqq v^0$, for unit vectors $\mathbf{e}_j$, $(j = 1, \ldots, n)$, or written in another way $\mathbf{A}^T\mathbf{x}^0 - \mathbf{1}v^0 \geqq \mathbf{0}$. The validity of $\mathbf{Ay}^0 - \mathbf{1}v^0 \leqq \mathbf{0}$ can be proved in the same way. Thus $(\mathbf{x}^0, \mathbf{y}^0, v^0)$ satisfies (5).

Holding v fixed system (5) determines a convex polyhedron which is precisely $X^0 \times Y^0$. ∎

COROLLARY. X^0 and Y° are convex polyhedrons in $\mathbb{R}^m$ and $\mathbb{R}^n$ resp.

Theorem 3 provides a method (not efficient at any rate) for finding all equilibrium points of a matrix game. All we have to do is to determine the vertices of the convex polyhedron defined by (5) since any equilibrium point can be expressed as a convex linear combination of them. For enumerating vertices of a convex polyhedron there are several methods available (see, e.g. [107]).

Since X^0 and Y^0 are convex polyhedrons, their extreme points (vertices) deserve special attention. Extreme points of X^0 and Y^0 are called *extreme equilibrium points* of the matrix game **A**. The following two theorems relate the extreme equilibrium points to certain nonsingular submatrices of **A**.

THEOREM 4. [79] Let $\mathbf{x}^0$ and $\mathbf{y}^0$ be extreme equilibrium strategies of the matrix game **A** and $v^0>0$ be the value of the game. Then there is a nonsingular submatrix **M** of **A** to satisfy

$$v^0 = \frac{1}{\mathbf{1}\mathbf{M}^{-1}\mathbf{1}} \tag{7}$$

$$\mathbf{x}^0 = v^0\mathbf{1}\mathbf{M}^{-1} \tag{8}$$

$$\mathbf{y}^0 = v^0\mathbf{M}^{-1}\mathbf{1}\,.[1] \tag{9}$$

Proof. We prove this theorem in Chapter 12 by using the theory of linear programming. Without recourse to linear programming the proof is much more complicated. See, e.g. [79]. ∎

REMARK 1. The matrix **M** in Theorem 4 is usually referred to as the *kernel* of **A**. The kernel is not necessarily unique.

2. In view of the transformation theorem (Theorem 2 of Chapter 8), the condition $v^0>0$ can always be met.

[1] To simplify the cumbersome notation we also denote by $\mathbf{x}^0$ and $\mathbf{y}^0$ the vectors obtained by deleting certain 0 components of the equilibrium strategies $\mathbf{x}^0$ and $\mathbf{y}^0$.

THEOREM 5. [79] If for a pair of equilibrium strategies $(\mathbf{x}^0, \mathbf{y}^0)$ and the value v^0 equations (7), (8), (9) hold for some nonsingular submatrix $\mathbf{M}$ of $\mathbf{A}$, then $\mathbf{x}^0$ and $\mathbf{y}^0$ are extreme equilibrium strategies.

Proof. Suppose on the contrary that

$$\mathbf{x}^0 = \frac{1}{2}(\mathbf{x}^1 + \mathbf{x}^2),$$

where $\mathbf{x}^1 \in X^0$, $\mathbf{x}^2 \in X^0$, $\mathbf{x}^1 \neq \mathbf{x}^2$. Let $\mathbf{M}$ be the matrix for which

$$(10) \qquad \mathbf{x}^0\mathbf{M} = v\mathbf{1}\,.$$

Since $\mathbf{x}^1 \in X^0$, $\mathbf{x}^2 \in X^0$ (deleting the corresponding zero components)

$$(11) \qquad \begin{aligned} \mathbf{x}^1\mathbf{M} &\geqq v\mathbf{1} \\ \mathbf{x}^2\mathbf{M} &\geqq v\mathbf{1}\,. \end{aligned}$$

Because of (10) equalities should hold in (11). But this implies $(\mathbf{x}^1 - \mathbf{x}^2)\mathbf{M} = \mathbf{0}$. By the nonsingularity of $\mathbf{M}$ we have $\mathbf{x}^1 = \mathbf{x}^2$ contradicting our assumption.

Proving that $\mathbf{y}^0$ is an extreme point of Y^0 goes along the same lines. ∎

Based on Theorems 4 and 5 by enumerating every quadratic submatrix of $\mathbf{A}$ we can determine all extreme equilibrium strategies. For any quadratic submatrix $\mathbf{M}$ we have to check whether

1. it is non-singular;
2. vectors defined by (8), (9) satisfy (5).

Of course we have to make sure that the value of the game is positive.

It is clear that the above enumeration method is efficient only for small matrices. For games with large matrices computationally efficient methods should be applied. Such methods are subjects of Chapter 12.

To conclude this section we prove a simple theorem which can very often be used for finding solutions to actual matrix games.

THEOREM 6. [79] Let $\mathbf{y}^0$ be an optimal strategy of player II for which $y_k^0 > 0$ (y_k^0 is the kth entry of $\mathbf{y}^0$). Then, for any optimal strategy $\mathbf{x}^0$ of player I,

$$\mathbf{x}^0\mathbf{A}\mathbf{e}_k = v^0,$$

where v^0 is the value of the game.

Proof. Let $\mathbf{x}^0$ be an optimal strategy of player I. Since the unit vector $\mathbf{e}_j$ ($j=1, \ldots, n$) is a strategy of player II, we have

$$\mathbf{x}^0\mathbf{A}\mathbf{e}_j \geqq v^0, \quad (j=1, \ldots, n). \tag{12}$$

Assume now that, for $j=k$, inequality (12) is strict. Multiplying both sides of (12) by y_k^0, we get

$$y_k^0\mathbf{x}^0\mathbf{A}\mathbf{e}_k > y_k^0 v^0.$$

For $j \neq k$ we have

$$y_j^0\mathbf{x}^0\mathbf{A}\mathbf{e}_j \geqq y_j^0 v^0.$$

Adding up these inequalities we obtain

$$\mathbf{x}^0\mathbf{A}\mathbf{y}^0 > v^0,$$

which is a contradiction. Therefore for $j=k$ in (12)

$$\mathbf{x}^0\mathbf{A}\mathbf{e}_k = v^0. \qquad \blacksquare$$

Of course Theorem 6 remains true if we change the role of the players.

10. Symmetric games

A general two-person zero-sum game $\Gamma=\{\Sigma_1, \Sigma_2, f\}$ is called *symmetric*, if $\Sigma_1=\Sigma_2$ and $f(\sigma_1, \sigma_2)= -f(\sigma_2, \sigma_1)$ for any $\sigma_1 \in \Sigma_1$ and $\sigma_2 \in \Sigma_2$. Since the players' strategy sets coincide we can denote the game simply by $\Gamma=\{\Sigma, f\}$ where Σ is the players' common strategy set. If, in addition, game Γ is finite – i.e. it is a matrix game – then it is symmetric if and only if its pay-off matrix $\mathbf{A}$ is skew-symmetric, i.e., $\mathbf{A}=-\mathbf{A}^T$.

In a sense a symmetric game is the "fairest", as demonstrated by the following theorem.

THEOREM 1. [79] Let $\Gamma=\{\Sigma, f\}$ be a symmetric game having at least one equilibrium point. Then

(a) the value of Γ is 0,

(b) the equilibrium strategy set of both players is the same,

(c) for a strategy $\sigma_0 \in \Sigma$ to be optimal it is necessary and sufficient to satisfy

$$f(\sigma, \sigma_0) \leqq 0 \quad \text{for all} \quad \sigma \in \Sigma .$$

Proof. (a) By the definition of symmetricity $f(\sigma, \sigma)=0$ for any $\sigma \in \Sigma$. Suppose now that the value of Γ is positive: $v>0$. Let (σ_0, τ_0) be an equilibrium point. Then, by the definition of an equilibrium point,

$$0<v=f(\sigma_0, \tau_0) \leqq f(\sigma_0, \sigma) \quad \text{for all} \quad \sigma \in \Sigma .$$

Thus $f(\sigma_0, \sigma_0)>0$ which is a contradiction. The assumption $v<0$ leads to a contradiction in the same way. Thus $v=0$.

(b) Let σ_0 be an optimal strategy of player I. Then

$$f(\sigma_0, \sigma) \geqq v = 0 \quad \text{for all} \quad \sigma \in \Sigma .$$

Multiplying both sides of this inequality by -1 and taking into consideration that Γ is symmetric we get

$$f(\sigma, \sigma_0) \leqq 0 = v \quad \text{for all} \quad \sigma \in \Sigma$$

which means that σ_0 is an optimal strategy of player II, as well. The proof that player II's optimal strategies are those of player I's goes along the same lines.

(c) The assertion directly follows from (a). ∎

COROLLARY. If Γ is a symmetric matrix game with pay-off matrix $\mathbf{A}$, then we can put forward assertion (c) of Theorem 1 in the following way: for $\mathbf{x}_0$ to be an optimal strategy it is necessary and sufficient to satisfy the linear inequality system

$$(1) \qquad \mathbf{A}\mathbf{x} \leqq \mathbf{0} .$$

Substituting $v=0$ and using $\mathbf{A} = -\mathbf{A}^T$ we immediately get (1) from the inequality system (5) of Chapter 9.

As we have seen a symmetric game is "fair" since both players have the same set to choose their strategies from and none of them "loses" if playing "rationally". Significance of symmetric games is further enhanced by the fact that any two-person zero-sum game can be transformed into a symmetric one. We first give a very simple transformation which is easily applicable to general two-person zero-sum games.

Let $\Gamma = \{\Sigma_1, \Sigma_2, f\}$ be a zero-sum two-person game having at least one equilibrium point. Define a symmetric game $H = \{\Sigma_1 \times \Sigma_2, g\}$ where $g[(\sigma, \tau), (\sigma', \tau')] = f(\sigma, \tau') - f(\sigma', \tau)$ for any $(\sigma, \tau) \in \Sigma_1 \times \Sigma_2$, $(\sigma', \tau') \in \Sigma_1 \times \Sigma_2$. The reader can very easily prove that H is indeed symmetric.

THEOREM 2. [79] If (σ_0, τ_0) is an equilibrium point of Γ, then (σ_0, τ_0) is an optimal strategy of both players in H. If $[(\bar{\sigma}, \bar{\tau}), (\bar{\bar{\sigma}}, \bar{\bar{\tau}})]$ is an equilibrium point of H, then $(\bar{\sigma}, \bar{\bar{\tau}})$ is an equilibrium point of Γ.

Proof. If (σ_0, τ_0) is an equilibrium point of Γ, then

$$f(\sigma, \tau_0) \leqq f(\sigma_0, \tau_0) \leqq f(\sigma_0, \tau)$$
$$\text{for all} \quad \sigma \in \Sigma_1, \tau \in \Sigma_2 .$$

Hence

$$f(\sigma, \tau_0) - f(\sigma_0, \tau) = g[(\sigma, \tau), (\sigma_0, \tau_0)] \leqq 0$$
$$\text{for all} \quad (\sigma, \tau) \in \Sigma_1 \times \Sigma_2 .$$

Applying assertion (c) of Theorem 1, we see that (σ_0, τ_0) is an equilibrium point of H.

To prove the second part of our theorem, assume that $[(\bar\sigma, \bar\tau), (\bar{\bar\sigma}, \bar{\bar\tau})]$ is an equilibrium point of H. Then

$$\left.\begin{array}{l} f(\bar\sigma, \tau) - f(\sigma, \bar\tau) \geqq 0 \\ f(\sigma, \bar{\bar\tau}) - f(\bar{\bar\sigma}, \tau) \leqq 0 . \end{array}\right. \quad \text{for all} \quad (\sigma, \tau) \in \Sigma_1 \times \Sigma_2$$

Hence

$$f(\sigma, \bar{\bar\tau}) \leqq f(\bar{\bar\sigma}, \tau) \leqq f(\bar\sigma, \bar{\bar\tau})$$
$$f(\bar\sigma, \tau) \geqq f(\bar{\bar\sigma}, \bar\tau) \geqq f(\bar\sigma, \bar{\bar\tau}) .$$

Then we have

$$f(\sigma, \bar{\bar\tau}) \leqq f(\bar\sigma, \bar{\bar\tau}) \leqq f(\bar\sigma, \tau) \quad \text{for all} \quad \sigma \in \Sigma_1, \tau \in \Sigma_2 ,$$

which means that $(\bar\sigma, \bar{\bar\tau})$ is an equilibrium point of Γ. ∎

We now present another procedure of symmetrization directly applied for matrix games. As a result of this transformation the size of the matrix is usually much smaller than it would be by applying the transformation we have dealt with previously. (Note that by Theorem 2, if the matrix of game Γ is of size $m \times n$, then that of game H is $mn \times mn$.)

Without loss of generality we may assume that $\mathbf{A} > \mathbf{0}$. Define a matrix $\mathbf{S}$ by

$$\mathbf{S}=\begin{bmatrix} \mathbf{0} & \mathbf{A} & -\mathbf{1} \\ -\mathbf{A}^T & \mathbf{0} & \mathbf{1} \\ \mathbf{1} & -\mathbf{1} & 0 \end{bmatrix}.$$

Since **S** is skew-symmetric the matrix game determined by **S** is symmetric. Note that **S** is of size $(m+n+1)\times(m+n+1)$. Let

$$\mathbf{z}=(u_1, \ldots, u_m, v_1, \ldots, v_n, \lambda)=(\mathbf{u}, \mathbf{v}, \lambda)$$

be an equilibrium strategy of **S**. Then, by the corollary to Theorem 1, we have

$$\text{(2)} \qquad \begin{aligned} \mathbf{A}\mathbf{v} &\leqq \lambda\mathbf{1} \\ \mathbf{u}\mathbf{A} &\geqq \lambda\mathbf{1} \\ \mathbf{1}\mathbf{u}-\mathbf{1}\mathbf{v} &\leqq 0 . \end{aligned}$$

If λ were 0, then $\mathbf{A}\mathbf{v}\leqq\mathbf{0}$ and $\mathbf{A}>\mathbf{0}$ would imply $\mathbf{v}=\mathbf{0}$, $\mathbf{u}=\mathbf{0}$, which is impossible since **z** is a probability vector. Therefore $\lambda>0$. On the other hand $\lambda<1$ since otherwise $\mathbf{v}=\mathbf{0}$, $\mathbf{u}=\mathbf{0}$ and $\mathbf{u}\mathbf{A}\geqq\lambda\mathbf{1}>\mathbf{0}$ would not hold. By Theorem 6 of Chapter 9 $\lambda>0$ implies $\mathbf{1}\mathbf{u}=\mathbf{1}\mathbf{v}$.

Thus

$$\mathbf{1}\mathbf{u}=\mathbf{1}\mathbf{v}=\frac{1-\lambda}{2}>0 .$$

Denoting $a=\frac{1-\lambda}{2}$, $\mathbf{x}^0=\frac{1}{a}\mathbf{u}$, $\mathbf{y}^0=\frac{1}{a}\mathbf{v}$, $v=\frac{\lambda}{a}$, (2) implies

$$\begin{aligned} \mathbf{A}\mathbf{y}^0 &\leqq v\mathbf{1} \\ \mathbf{x}^0\mathbf{A} &\geqq v\mathbf{1} . \end{aligned}$$

But this means that $(\mathbf{x}^0, \mathbf{y}^0)$ is an equilibrium point of **A** and v is the value of the game determined by **A**. Thereby we have shown that by finding an equilibrium strategy of the symmetric game **S** we can easily get an equilibrium point of **A**, too.

On the other hand, if $(\mathbf{x}^0, \mathbf{y}^0)$ is an equilibrium point of $\mathbf{A}$ and the value of the game is v ($v \geqq 0$), then

$$\mathbf{z}^0 = \frac{1}{2+v}(\mathbf{x}^0, \mathbf{y}^0, v)$$

is an equilibrium strategy of S. This assertion can easily be proved by substituting $\mathbf{z}^0$ into the linear inequality system $\mathbf{Sz} \leqq \mathbf{0}$.

As we have seen any matrix game can be transformed into a symmetric one without increasing the size of the matrix considerably. Therefore any method capable of finding an equilibrium point of a symmetric matrix game is suitable for the solution of an arbitrary matrix game, too.

11. Connection between matrix games and linear programming

Let us consider the following pair of linear programming problems (primal and dual)

$$\text{Primal:}\quad \mathbf{x} \geqq \mathbf{0} \qquad\qquad \text{Dual:}\quad \mathbf{u} \geqq \mathbf{0}$$

$$\mathbf{A}\mathbf{x} \leqq \mathbf{b} \qquad\qquad \mathbf{u}\mathbf{A} \geqq \mathbf{c}$$

$$\mathbf{c}\mathbf{x} \to \max \qquad\qquad \mathbf{u}\mathbf{b} \to \min .$$

From the parameters of these problems we can construct a skew-symmetric matrix:

$$\mathbf{P} = \begin{bmatrix} \mathbf{0} & \mathbf{A} & -\mathbf{b} \\ -\mathbf{A}^T & \mathbf{0} & \mathbf{c} \\ \mathbf{b} & -\mathbf{c} & \mathbf{0} \end{bmatrix}.$$

The symmetric game determined by $\mathbf{P}$ always has an equilibrium strategy. Let $\mathbf{z}$ be one of them:

$$\mathbf{z} = (\mathbf{r}, \mathbf{s}, \lambda).$$

THEOREM 1. [79] If $\lambda > 0$, then $\mathbf{x} = \frac{1}{\lambda}\mathbf{s}$ and $\mathbf{u} = \frac{1}{\lambda}\mathbf{r}$ are optimal solutions to the primal and dual linear programming problems resp.

Proof. By Theorem 1 of Chapter 10, $\mathbf{P}\mathbf{z} < \mathbf{0}$. This means

$$(1) \qquad \begin{aligned} \mathbf{A}\mathbf{x} &\leqq \mathbf{b} \\ -\mathbf{A}^T\mathbf{u} &\leqq -\mathbf{c} \\ \mathbf{b}\mathbf{u}-\mathbf{c}\mathbf{x} &\leqq \mathbf{0}\,. \end{aligned}$$

By the weak duality theorem of linear programming,

$$\mathbf{c}\mathbf{x} \leqq \mathbf{b}\mathbf{u}$$

which together with the last inequality of (1) gives

$$\mathbf{c}\mathbf{x} = \mathbf{b}\mathbf{u}\,,$$

which is a necessary and sufficient condition of optimality by the strong duality theorem of linear programming (see, e.g. [107]). ∎

THEOREM 2. [79] If $\mathbf{x}$ and $\mathbf{u}$ are optimal strategies of the primal and dual linear programming problems resp., then the symmetric game determined by $\mathbf{P}$ has an equilibrium strategy the last component of which is positive.

Proof. Let $\lambda = \dfrac{1}{\mathbf{1}\mathbf{x}+\mathbf{1}\mathbf{u}+1}$ and $\mathbf{s}=\lambda\mathbf{x}$, $\mathbf{r}=\lambda\mathbf{u}$. Then $\mathbf{z}=(\mathbf{r}, \mathbf{s}, \lambda)$ satisfies the inequality system $\mathbf{P}\mathbf{z}\leqq\mathbf{0}$, implying by Theorem 1 of Chapter 10 that $\mathbf{z}$ is an equilibrium strategy of $\mathbf{P}$. ∎

We can put Theorems 1 and 2 together. For a pair of primal-dual linear programming problems to be solvable it is necessary and sufficient that the associated symmetric game $\mathbf{P}$ has an equilibrium strategy with a positive last component.

We have seen that linear programming problems can be handled as special matrix games. In this sense the theory of games (matrix games) is more general than that of linear programming. On the other hand as we will see in the next chapter, linear programming provides the most efficient means of solving a matrix game. Thus the relationship between matrix games and linear programming is two-sided.

12. Methods for solving general matrix games

In Chapter 9 we have seen that by enumerating the vertices of the polyhedron defined by inequality system (5) of Chapter 9 we can find all extreme equilibrium strategies of a matrix game. Since for matrices of considerable size this approach is highly inefficient methods that are not supposed to go over all basic solutions of an inequality system are favoured from a computational point of view. Out of these we are going to deal with three basically different approaches.

12.1. SOLUTION OF MATRIX GAMES BY LINEAR PROGRAMMING

Let **A** be a matrix game. Without loss of generality we may assume that the value of the game is positive. Let us consider the following pair of primal-dual linear programming problems:

(1) Primal: $\mathbf{y} \geqq \mathbf{0}$, $\mathbf{A}\mathbf{y} \leqq \mathbf{1}$, $\mathbf{1}\mathbf{y} \rightarrow \max;$ Dual: $\mathbf{x} \geqq \mathbf{0}$, $\mathbf{A}^T\mathbf{x} \geqq \mathbf{1}$, $\mathbf{1}\mathbf{x} \rightarrow \min.$

THEOREM 1. [178] If $(\bar{\mathbf{x}}, \bar{\mathbf{y}})$ is an equilibrium point of the matrix game **A**, and v is the value of the game ($v > 0$), then $\mathbf{y}^0 = \frac{1}{v}\bar{\mathbf{y}}$ and $\mathbf{x}^0 = \frac{1}{v}\bar{\mathbf{x}}$ are optimal solutions to the primal and dual resp. The optimal objective function's value is $z = \frac{1}{v}$.

Proof. Since $(\bar{\mathbf{x}}, \bar{\mathbf{y}})$ is an equilibrium point of $\mathbf{A}$ we have

$$\text{(2)} \qquad \mathbf{A}\bar{\mathbf{y}} \leqq v\mathbf{1}, \qquad \mathbf{A}^T\bar{\mathbf{x}} \geqq v\mathbf{1}.$$

Thus $\mathbf{y}^0$ and $\mathbf{x}^0$ are feasible solutions to the primal and dual resp. By the strong duality theorem of linear programming $\mathbf{y}^0$ and $\mathbf{x}^0$ are optimal solutions to the primal and dual resp. since

$$\mathbf{1x}^0 = \mathbf{1y}^0 = \frac{1}{v}. \qquad \blacksquare$$

THEOREM 2. [178] If $\mathbf{y}^0$ and $\mathbf{x}^0$ are optimal solutions to the primal and dual resp. with positive objective function's value z, $(z > 0)$, then $\bar{\mathbf{x}} = \frac{1}{z}\mathbf{x}^0$ and $\mathbf{y}^0 = \frac{1}{z}\mathbf{y}^0$ is a pair of equilibrium strategies of the matrix game $\mathbf{A}$. The value of the game is $v = \frac{1}{z}$.

Proof. Since $\mathbf{y}^0$ and $\mathbf{x}^0$ are optimal solutions to the primal and dual resp. we have by the strong duality theorem of linear programming

$$\mathbf{1y}^0 = \mathbf{1z}^0 = z > 0.$$

Thus $\bar{\mathbf{x}}$ and $\bar{\mathbf{y}}$ are probability vectors. Since

$$\mathbf{A}\bar{\mathbf{y}} \leqq \frac{1}{z}\mathbf{1}$$

$$\mathbf{A}^T\bar{\mathbf{x}} \geqq \frac{1}{z}\mathbf{1},$$

therefore $\left(\bar{\mathbf{x}}, \bar{\mathbf{y}}, \frac{1}{z}\right)$ satisfies inequality system (5) of Chapter 9. $\blacksquare$

Similar theorems hold for the case when the value of the game is negative. Then the associated linear programs look like this:

$$(3) \qquad \text{Primal:} \quad \mathbf{y} \geqq \mathbf{0} \qquad\qquad \text{Dual:} \quad \mathbf{x} \geqq \mathbf{0}$$
$$\mathbf{A}\mathbf{y} \leqq -\mathbf{1} \qquad\qquad \mathbf{A}^T\mathbf{x} \geqq -\mathbf{1}$$
$$-\mathbf{1}\mathbf{y} \to \max \qquad\qquad -\mathbf{1}\mathbf{x} \to \min .$$

Adjusting the proofs of Theorems 1 and 2 is an easy exercise for the reader.[1]

In view of Theorems 1 and 2, the equilibrium strategies of an arbitrary matrix game can be found by linear programming. Before actually solving (1) we have to add a suitable constant (if necessary) to each entry of the matrix **A** so that the matrix and thereby the value of the game be positive. By Theorem 2 of Chapter 8 this transformation does not affect the equilibrium strategies.

Computational experience available up to now indicates that, for the solution of general matrix games, linear programming is the most efficient method. Of course, other methods for structured game matrices may prove to be better in certain special cases.

To conclude this section we make use of the theory of linear programming in another way, too. When we proved Theorem 1, we saw that the equilibrium points of **A** satisfy inequality system (2). Since $v > 0$, following the lines of the proof of Theorem 1, we see that the optimal extreme points of the primal and dual linear programs of (1) are precisely the scalar multiples of the extreme equilibrium strategies. From the theory of linear programming we know that, for any extreme point $\mathbf{y}'$ and $\mathbf{x}'$, there is a nonsingular submatrix of **A**, say $\mathbf{A}_{11}$, such that

$$\mathbf{y}' = \begin{bmatrix} \mathbf{A}_{11}^{-1}\,\mathbf{1} \\ \mathbf{0} \end{bmatrix}$$

$$\mathbf{x}' = \begin{bmatrix} \mathbf{1}\,\mathbf{A}_{11}^{-1} \\ \mathbf{0} \end{bmatrix}$$

[1] This same method can also be derived from the results of Section 4.2.

and the optimal objective function's value is $\mathbf{1}\ \mathbf{A}_{11}^{-1}\ \mathbf{1}$. Now the value of the game is $v = \dfrac{1}{\mathbf{1}\ \mathbf{A}_{11}^{-1}\ \mathbf{1}}$ by Theorem 2. Thereby we have proved in the simplest possible way Theorem 4 of Chapter 9.

12.2. METHOD OF FICTITIOUS PLAY

Let us imagine two players engaged in playing a matrix game repeatedly. They do not know their optimal strategies but they apply a policy for choosing a particular pure strategy in a play which is based on the following intuitive consideration. Each player chooses a strategy assuring maximal pay-off provided the opponent's future actions will resemble the past.

Let, e.g. $\mathbf{x}_1$ be an initial strategy of player I in the matrix game $\mathbf{A}$ in which player II is maximizing his pay-off, i.e., the function $\varphi(\mathbf{y}) = -\mathbf{x}_1\mathbf{A}\mathbf{y}$ subject to $\mathbf{y} \geqq \mathbf{0}$, $\mathbf{1}\mathbf{y} = 1$. $\varphi(\mathbf{y})$ attains its maximum at a pure strategy $\mathbf{y}_1 = \mathbf{e}_{j_1}$ for which

$$\min_j \mathbf{x}_1\mathbf{A}\mathbf{e}_j = \mathbf{x}_1\mathbf{A}\mathbf{e}_{j_1}.$$

In the second run player I picks a strategy $\mathbf{x}_2 = \mathbf{e}_{i_2}$ to satisfy

$$\max_i \mathbf{e}_i\mathbf{A}\mathbf{y}_1 = \mathbf{e}_{i_2}\mathbf{A}\mathbf{y}_1.$$

Now player II applies a strategy which is optimal against his opponent's "average" behaviour, i.e., against $\bar{\mathbf{x}}_2 = \frac{1}{2}(\mathbf{x}_1 + \mathbf{x}_2)$. Therefore $\mathbf{y}_2 = \mathbf{e}_{j_2}$, where

$$\min_j \bar{\mathbf{x}}_2\mathbf{A}\mathbf{e}_j = \bar{\mathbf{x}}_2\mathbf{A}\mathbf{e}_{j_2}.$$

Generally in the kth play $\mathbf{x}_k = \mathbf{e}_{i_k}$, $\mathbf{y}_k = \mathbf{e}_{j_k}$, where

$$\max_i \mathbf{e}_i\mathbf{A}\bar{\mathbf{y}}_{k-1} = \mathbf{e}_{i_k}\mathbf{A}\bar{\mathbf{y}}_{k-1}$$

$$\min_j \bar{\mathbf{x}}_k\mathbf{A}\mathbf{e}_j = \bar{\mathbf{x}}_k\mathbf{A}\mathbf{e}_{j_k}$$

$$\bar{\mathbf{x}}_k = \frac{1}{k}\sum_{p=1}^{k} \mathbf{x}_p, \quad \bar{\mathbf{y}}_{k-1} = \frac{1}{k-1}\sum_{p=1}^{k-1} \mathbf{y}_p .$$

The above scheme can very well be simulated on a computer and is often referred to as "fictitious play". In this context a question arises naturally: do the strategies $\bar{\mathbf{x}}_k$, $\bar{\mathbf{y}}_k$ obtained by the realization of the above procedure approximate the optimal strategies for large enough k, and if so, how "accurate" is this approximation?

The answer to this question involves rather lengthy and cumbersome argumentation. In order to make the presentation and the new notations clear let us redefine the procedure of fictitious play in a slightly different way. (The players choose their strategies simultaneously in an iteration.)

Let Γ be a finite, two-person, zero-sum game defined by matrix $\mathbf{A}$ of size $m_1 \times m_2$. Let $\mathbf{u}^{(0)} \in \mathbb{R}^{m_1}$, $\mathbf{v}^{(0)} \in \mathbb{R}^{m_2}$ be two initial vectors. The upper index is an iteration counter. For $k \geqq 1$ let i_k denote the index of the maximal component of $\mathbf{u}^{(k-1)}$ and j_k that of the minimal component of $\mathbf{v}^{(k-1)}$. (Ties can be broken arbitrarily.) Define

$$\text{(4)} \qquad \begin{aligned} \mathbf{u}^{(k)} &= \mathbf{u}^{(k-1)} + \mathbf{c}_{j_k}, \\ \mathbf{v}^{(k)} &= \mathbf{v}^{(k-1)} + \mathbf{r}_{i_k}, \end{aligned}$$

where $\mathbf{c}_{j_k}$, $\mathbf{r}_{i_k}$ denote the j_kth column and the i_kth row resp. of matrix $\mathbf{A}$. Denote $\sigma_p^{(k)}$ and $\tau_p^{(k)}$ the number of times we added the pth column and the pth row resp. of matrix $\mathbf{A}$ to $\mathbf{u}^{(0)}$ and $\mathbf{v}^{(0)}$ resp. The vectors

$$\text{(5)} \qquad \begin{aligned} \mathbf{x}^{(k)} &= \frac{1}{k}(\tau_1^{(k)}, \ldots, \tau_{m_1}^{(k)}), \\ \mathbf{y}^{(k)} &= \frac{1}{k}(\sigma_1^{(k)}, \ldots, \sigma_{m_2}^{(k)}), \end{aligned}$$

are obviously strategy (probability) vectors and

$$\text{(6)} \qquad \begin{aligned} \mathbf{u}^{(k)} &= \mathbf{u}^{(0)} + k\mathbf{A}\mathbf{y}^{(k)}, \\ \mathbf{v}^{(k)} &= \mathbf{v}^{(0)} + k\mathbf{x}^{(k)}\mathbf{A}. \end{aligned}$$

It is easy to see that if $\mathbf{u}^{(0)}=\mathbf{0}$, $\mathbf{v}^{(0)}=\mathbf{0}$, then $\mathbf{x}^k$ is the average of optimal pure strategies of player I applied against player II's strategies $\mathbf{y}^{(t)}$, $(t \leqq k-1)$. So this procedure is indeed a realization of fictitious play. Of course, the same holds when we change the role of the players.

Before dealing with the convergence of the procedure we introduce a few new notations and prove a series of lemmas.

Let

$$\mathbf{u}^{(k)}=(u_1^{(k)}, \ldots, u_{m_1}^{(k)}),$$

$$\mathbf{v}^{(k)}=(v_1^{(k)}, \ldots, v_{m_2}^{(k)}),$$

$$\Delta_{u,v}^{(k)}=\max_i u_i^{(k)} - \min_j v_j^{(k)},$$

$$\Delta_{v,u}^{(k)}=\max_j v_j^{(k)} - \min_i u_i^{(k)},$$

$$\Delta_{u,u}^{(k)}=\max_i u_i^{(k)} - \min_i u_i^{(k)},$$

$$\Delta_{v,v}^{(k)}=\max_j v_j^{(k)} - \min_j v_j^{(k)}.$$

LEMMA 1. If $\max_i u_i^{(0)} = \min_j v_j^{(0)}=0$, then $\Delta_{v,u}^{(k)} \geqq 0$.

Proof. Let v^* be the value of the matrix game $\mathbf{A}^T$. Then by (6) we have

$$\frac{\min_i u_i^{(k)}}{k} \leqq \frac{\max_i u_i^{(0)}}{k} + \min_i \mathbf{e}_i \mathbf{A} \mathbf{y}^{(k)}.$$

Denoting by $(\mathbf{y}^*, \mathbf{x}^*)$ an equilibrium point of the matrix game $\mathbf{A}^T$ we get

$$\text{(7)} \qquad \frac{\min_i u_i^{(k)} - \max_i u_i^{(0)}}{k} \leqq \min_i \mathbf{y}^{(k)} \mathbf{A}^T \mathbf{e}_i \leqq \mathbf{y}^{(k)} \mathbf{A}^T \mathbf{x}^* \leqq$$

$$\leqq \mathbf{y}^* \mathbf{A}^T \mathbf{x}^* = v^*.$$

In exactly the same way we can prove that

$$\frac{\max_j v_j^{(k)} - \min_j v_j^{(0)}}{k} \geqq v^*. \tag{8}$$

Then (7) and (8) together imply $\Delta_{v,u}^{(k)} \geqq 0$ provided $\max_i u_i^{(0)} = \min_j v_j^{(0)} = 0$. ∎

LEMMA 2. Let v be the value of the matrix game $\mathbf{A}$. Then

$$\frac{\max_i u_i^{(k)} - \min_i u_i^{(0)}}{k} \geqq v \geqq \frac{\min_j v_j^{(k)} - \max_j v_j^{(0)}}{k}. \tag{9}$$

Proof. The proof goes similarly to that of the previous lemma. Denoting by $(\bar{\mathbf{x}}, \bar{\mathbf{y}})$ an equilibrium point of the matrix game $\mathbf{A}$ we obtain by (6)

$$\frac{\max_i u_i^{(k)} - \min_i u_i^{(0)}}{k} \geqq \max_i \mathbf{e}_i \mathbf{A} \mathbf{y}^{(k)} \geqq \bar{\mathbf{x}} \mathbf{A} \mathbf{y}^{(k)} \geqq$$

$$\geqq \bar{\mathbf{x}} \mathbf{A} \bar{\mathbf{y}} = v.$$

The other side of the inequality in (9) can be proved similarly. ∎

LEMMA 3. Let $\hat{\mathbf{u}}^{(k)}$, $\hat{\mathbf{v}}^{(k)}$ denote the vectors obtained from the iteration given by (6) with the initial vectors $\hat{\mathbf{u}}^{(0)} = \mathbf{u}^{(0)} - \alpha \mathbf{1}$ and $\hat{\mathbf{v}}^{(0)} = \mathbf{v}^{(0)} - \beta \mathbf{1}$. Then for $k \geqq 1$

$$\hat{\mathbf{u}}^{(k)} = \mathbf{u}^{(k)} - \alpha \mathbf{1} \tag{10}$$

$$\hat{\mathbf{v}}^{(k)} = \mathbf{v}^{(k)} - \beta \mathbf{1}$$

$$\Delta_{u,v}^{(k)} - \Delta_{\hat{u},\hat{v}}^{(k)} = \alpha - \beta. \tag{11}$$

Proof. Subtracting a constant from each component of $\mathbf{u}^{(0)}$ and $\mathbf{v}^{(0)}$ does not affect the index of maximal and minimal elements.

Thus these indices are the same for initial vectors $\hat{\mathbf{u}}^{(0)}$ and $\hat{\mathbf{v}}^{(0)}$ as for $\mathbf{u}^{(0)}$ and $\mathbf{v}^{(0)}$. Thus the assertions of our lemma directly follow from (6). ∎

LEMMA 4. Let k, m_1, m_2 be positive integers satisfying $m_1+m_2 \geqq 4$, $k \geqq 2^{(m_1+m_2-1)(m_1+m_2-2)}$. Then

$$\left(2^{m_1+m_2-3}-2^{\frac{(m_1+m_2-3)^2}{m_1+m_2-2}}\right)k^{\frac{m_1+m_2-3}{m_1+m_2-2}}>1 . \tag{12}$$

Proof. Obviously

$$1>\frac{1}{\sqrt{2}}+\frac{1}{16} \geqq \frac{1}{2^{\frac{m_1+m_2-3}{m_1+m_2-2}}}+\frac{1}{2^{(m_1+m_2)(m_1+m_2-3)}} .$$

By simple rearrangement we get

$$1<\left(1-2^{-\frac{m_1+m_2-3}{m_1+m_2-2}}\right)2^{(m_1+m_2)(m_1+m_2-3)}=$$

$$=\left(2^{m_1+m_2-3}-2^{\frac{(m_1+m_2-3)^2}{m_1+m_2-2}}\right)\cdot 2^{(m_1+m_2-1)(m_1+m_2-3)} \leqq$$

$$\leqq\left(2^{m_1+m_2-3}-2^{\frac{(m_1+m_2-3)^2}{m_1+m_2-2}}\right)k^{\frac{m_1+m_2-3}{m_1+m_2-2}} .$$ ∎

Now we are able to state the convergence theorem of fictitious play which gives a (rough) estimation of the speed of convergence, as well.

THEOREM 3. [158] If $\max_i u_i^{(0)} = \min_j v_j^{(0)} = 0$, then for $k \geqq 1$

$$\frac{\Delta_{u,v}^{(k)}}{k} \leqq a\frac{2^{m_1+m_2}}{k^{\frac{1}{m_1+m_2-2}}} \tag{13}$$

where $a = \max_{i,j} |a_{ij}|$.

Proof. The proof goes by induction on m_1+m_2. If $m_1+m_2=2$, i.e., $m_1=m_2=1$, then

$$u_1^{(k)}=v_1^{(k)}=k\alpha$$

where $\mathbf{A}=[\alpha]$. Thus $\Delta_{u,v}^{(k)}=0$ and (13) holds. Assume now that (13) holds for any $m_1+m_2<r$, $(r\geqq 3)$ and consider the case $m_1+m_2=r$.

Let $k=2, 3, \ldots$ and $0<T<k$ (T is integer). First we prove that one of the following inequalities holds:

(i) $\Delta_{u,v}^{(k)}\leqq 4aT$,

(ii) $\Delta_{u,v}^{(k)}-\Delta_{u,v}^{(k-T)}\leqq a2^{m_1+m_2-1}T^{1-\frac{1}{m_1+m_2-3}}$, where for $m_1+m_2=3$ the right-hand side is 0 by definition.

Suppose that (i) does not hold, i.e., $\Delta_{u,v}^{(k)}>4aT$. Then by the definition of $\Delta_{u,v}^{(k)}$, $\Delta_{u,u}^{(k)}$, $\Delta_{v,v}^{(k)}$ and Lemma 1 we get $\Delta_{u,u}^{(k)}+\Delta_{v,v}^{(k)}>4aT$. Thus either $\Delta_{u,u}^{(k)}>2aT$ or $\Delta_{v,v}^{(k)}>2aT$. We may assume $\Delta_{u,u}^{(k)}>2aT$ since for $\Delta_{v,v}^{(k)}>2aT$ the reasoning is similar. Thus

$$\text{(14)} \qquad \max_i u_i^{(k)}-\min_i u_i^{(k)}>2aT.$$

Suppose that $u_{i_1}^{(k)}=\max_i u_i^{(k)}$ and $u_{i_2}^{(k)}=\min_i u_i^{(k)}$. Since no component of $\mathbf{u}$ changes by more than a in an iteration, (14) implies $u_{i_1}^{(k-s)}-u_{i_2}^{(k-s)}>0$ for $0\leqq s\leqq T$. Thus the i_2th component of $\mathbf{u}$ never is maximal during T iterations preceding the kth one. Let $\mathbf{A}^*$ be the matrix stemming from $\mathbf{A}$ by deleting its i_2th row. Omitting the i_2th component of $\mathbf{u}$ we get a vector denoted by $\mathbf{u}^*$. Let us now apply the method of fictitious play to matrix $\mathbf{A}^*$ with initial vectors $\hat{\mathbf{u}}^{(k-T)*}$, $\hat{\mathbf{v}}^{(k-T)}$, where

$$\hat{\mathbf{u}}^{(k-T)*}=\mathbf{u}^{(k-T)*}-\left[\max_i u_i^{(k-T)*}\right]\mathbf{1},$$

$$\hat{\mathbf{v}}^{(k-T)}=\mathbf{v}^{(k-T)}-\left[\min_j v_j^{(k-T)}\right]\mathbf{1}.$$

Then by applying Lemma 3 and keeping in mind that the i_2th component of $\mathbf{u}^{(k-s)}$, $(0\leqq s\leqq T)$ never is maximal we obtain

$$\Delta^{(T)}_{\hat{u},\hat{v}} = \max_i \hat{u}_i^{(k)*} - \min_j \hat{v}_j^{(k)} =$$

$$= \max_i u_i^{(k)*} - \max_i u_i^{(k-T)} - \min_j v_j^{(k)} + \min_j v_j^{(k-T)} =$$

$$= \Delta^{(k)}_{u,v} - \Delta^{(k-T)}_{u,v}.$$

By induction we now have

$$\frac{\Delta^{(T)}_{\hat{u},\hat{v}}}{T} = \frac{\Delta^{(k)}_{u,v} - \Delta^{(k-T)}_{u,v}}{T} \leqq \frac{a2^{m_1+m_2-1}}{T^{\frac{1}{m_1+m_2-3}}},$$

which is precisely inequality (ii).

Coming back to the main line of our proof let us assume first that $k \leqq 2^{(m_1+m_2-1)(m_1+m_2-2)}$. Since no component of $\mathbf{u}^{(k)}$ and $\mathbf{v}^{(k)}$ changes by more than a in an iteration we have

$$\Delta^{(k)}_{u,v} \leqq 2ak = 2ak^{\frac{1}{m_1+m_2-2}} k^{1-\frac{1}{m_1+m_2-2}} \leqq$$

$$\leqq 2a2^{m_1+m_2-1} k^{1-\frac{1}{m_1+m_2-2}} =$$

$$= a2^{m_1+m_2} k^{1-\frac{1}{m_1+m_2-2}},$$

which is the assertion of our theorem.

Therefore we only need to consider the case when $k > 2^{(m_1+m_2-1)(m_1+m_2-2)}$. Let us take a fixed positive integer T. For the sake of easy reference an iteration index k will be called regular if (i) holds for it and irregular if (ii) is valid.

The definition of a implies $\Delta^{(k)}_{u,v} \leqq 2ak$. Therefore k is regular if $k \leqq T$. Denote by q the integral part of $\frac{k}{T}$. Then there is at least one regular index among numbers $k, k-T, k-2T, \ldots, k-qT$ (k is trivially one of them). Let $k-\tau T$ be the largest of them. Thus

$$\Delta^{(k)}_{u,v} = \sum_{t=1}^{\tau} [\Delta^{(k-(t-1)T)}_{u,v} - \Delta^{(k-tT)}_{u,v}] + \Delta^{(k-\tau T)}_{u,v} \leqq$$

$$\leqq (a2^{m_1+m_2-1} T^{1-\frac{1}{m_1+m_2-3}})\tau + 4aT.$$

Since $\tau \leqq \frac{k}{T}$ and $k > T > 0$ we get

$$(15) \qquad \frac{\Delta_{u,v}^{(k)}}{k} \leqq a\left(2^{m_1+m_2-1} T^{-\frac{1}{m_1+m_2-3}} + \frac{4T}{k}\right).$$

Now we consider two cases.

1. $m_1 + m_2 = 3$. For $k \leqq 2^{(m_1+m_2-1)(m_1+m_2-2)} = 4$ we have already proved our theorem. Suppose $k > 4$ and let $T = 2$. Then (15) implies

$$\frac{\Delta_{u,v}^{(k)}}{k} \leqq \frac{8a}{k},$$

which is exactly what our theorem says.

2. $m_1 + m_2 \geqq 4$. By Lemma 4 there exists a positive integer $T = T(k)$ to satisfy

$$(16) \qquad 2^{\frac{(m_1+m_2-3)^2}{m_1+m_2-2}} k^{\frac{m_1+m_2-3}{m_1+m_2-2}} \leqq T(k) \leqq 2^{m_1+m_2-3} k^{\frac{m_1+m_2-3}{m_1+m_2-2}}.$$

Since we have assumed $k > 2^{(m_1+m_2-1)(m_1+m_2-2)}$ the following inequalities hold:

$$\text{(a)} \qquad T(k) < k.$$

This follows from

$$T(k) \leqq 2^{m_1+m_2-3} k^{\frac{m_1+m_2-3}{m_1+m_2-2}} =$$

$$= \frac{1}{4}\left(2^{(m_1+m_2-2)(m_1+m_2-1)} k^{m_1+m_2-3}\right)^{\frac{1}{m_1+m_2-2}} <$$

$$< \frac{1}{4}\left(k^{m_1+m_2-2}\right)^{\frac{1}{m_1+m_2-2}} = \frac{1}{4} k < k.$$

$$\text{(b)} \qquad 2^{m_1+m_2-1} T(k)^{-\frac{1}{m_1+m_2-3}} \leqq \frac{4T(k)}{k}.$$

From (16) we have

$$T(k) \geqq 2^{\frac{(m_1+m_2-3)^2}{m_1+m_2-2}} k^{\frac{m_1+m_2-3}{m_1+m_2-2}}.$$

Therefore

$$T(k)^{\frac{m_1+m_2-2}{m_1+m_2-3}} = T(k)^{\frac{1}{m_1+m_2-3}+1} \geqq 2^{m_1+m_2-3}k.$$

By rearranging the terms we obtain the desired inequality.

(c) $$\frac{8T(k)}{k} \leqq 2^{m_1+m_2}k^{-\frac{1}{m_1+m_2-2}}.$$

This is an immediate consequence of inequality (16). Now by choosing $T=T(k)$ we get from inequality (15)

$$\frac{\Delta_{u,v}^{(k)}}{k} \leqq a\left(2^{m_1+m_2-1}T(k)^{-\frac{1}{m_1+m_2-3}} + \frac{4T(k)}{k}\right) \leqq$$

$$\leqq a\left(\frac{4T(k)}{k} + \frac{4T(k)}{k}\right) =$$

$$= \frac{a\,8T(k)}{k} \leqq 2^{m_1+m_2}k^{-\frac{1}{m_1+m_2-2}}a,$$

which was to be proved. ∎

COROLLARY 1. Dropping the restriction $\max_i u_i^{(0)} = \min_j v_j^{(0)} = 0$ we have

(17) $$\frac{\Delta_{u,v}^{(k)}}{k} = a\frac{2^{m_1+m_2}}{k^{\frac{1}{m_1+m_2-2}}} + \frac{\Delta_{u,v}^{(0)}}{k},$$

which directly follows from Theorem 3 and Lemma 3 by setting $\alpha = \max_i u_i^{(0)}$, $\beta = \min_j v_j^{(0)}$.

COROLLARY 2. Let

$$\bar{v}^{(k)} = \frac{\max_i u_i^{(k)} - \min_i u_i^{(0)}}{k},$$

$$\underline{v}^{(k)} = \frac{\min_j v_j^{(k)} - \max_j v_j^{(0)}}{k}.$$

Then by Lemma 2, $\underline{v}^{(k)} \leqq v \leqq \bar{v}^{(k)}$ and (17) implies $\bar{v}^{(k)} - \underline{v}^{(k)} \to 0$ if $k \to \infty$. Therefore, for large enough k, either of the numbers $\bar{v}^{(k)}$ and $\underline{v}^{(k)}$ is an approximation of the game value. The error of the approximation can be made arbitrarily small by letting k go to infinity. The estimation of the error provided by Corollary 1 is rather rough. Computational experience indicates that much faster convergence can be achieved in practice.

It is conjectured that the rate of convergence is of order $\sqrt{k}$. By the method of fictitious play we not only approximate the game value but the optimal strategy vectors as well.

THEOREM 4. [145] Any cluster point of the strategy sequences $\{\mathbf{x}^{(k)}\}$, $\{\mathbf{y}^{(k)}\}$ obtained in the course of fictitious play is an optimal strategy of the matrix game **A**.

Proof. Let $\mathbf{x}^0, \mathbf{y}^0$ be a pair of cluster points of the sequences $\{\mathbf{x}^{(k)}\}$, $\{\mathbf{y}^{(k)}\}$ resp. and let $\{k_t\}$ be an index set such that

$$\lim_{t \to \infty} \mathbf{x}^{(k_t)} = \mathbf{x}^0,$$

$$\lim_{t \to \infty} \mathbf{y}^{(k_t)} = \mathbf{y}^0.$$

Then by (6) and (7) we have

$$v = \lim_{t \to \infty} \max_i u_i^{(k_t)} = \lim_{t \to \infty} \max_i \mathbf{e}_i \mathbf{A} \mathbf{y}^{(k_t)} = \max_i \mathbf{e}_i \mathbf{A} \mathbf{y}^0,$$

$$v = \lim_{t \to \infty} \min_j v_j^{(k_t)} = \lim_{t \to \infty} \min_j \mathbf{x}^{(k_t)} \mathbf{A} \mathbf{e}_j = \min_j \mathbf{x}^0 \mathbf{A} \mathbf{e}_j.$$

Furthermore $\mathbf{1}\mathbf{x}^0 = \mathbf{1}\mathbf{y}^0 = 1$. By Theorem 3 of Chapter 9 $(\mathbf{x}^0, \mathbf{y}^0)$ is an equilibrium point and v is the value of the game. ▮

Since the sequences $\{\mathbf{x}^{(k)}\}$, $\{\mathbf{y}^{(k)}\}$ are bounded, they have at least one cluster point. In this way Theorem 4 provides a constructive proof for the existence theorem of matrix games (von Neumann's theorem, Theorem 1 of Chapter 9).

As an example let us apply the method of fictitious play to the matrix game

$$\mathbf{A}=\begin{bmatrix} 2 & 1 & 0 \\ 2 & 0 & 3 \\ -1 & 3 & 3 \end{bmatrix},$$

with the initial vectors $\mathbf{u}^{(0)}=\mathbf{v}^{(0)}=\mathbf{0}$. After 100 iterations the maximum of $\underline{v}^{(k)}$ is 1.235 294, ($k=85$) and the minimum of $\bar{v}^{(k)}$ is 1.308 824, ($k=68$). Thus the value of **A** should fall into the interval [1.235 294, 1.308 824]. Comparing the length of this interval 0.07353 to the error term of Theorem 3 we see that the estimation

$$\frac{3.2^6}{100^{\frac{1}{4}}} \approx 60.715 \qquad (a=3,\ k=100)$$

is very rough.

Details of the computation are summarized in Table I. After 100 iterations the approximate equilibrium point of the game is

$$\mathbf{x}^{(0)} \approx (0.445\,545,\quad 0.287\,129,\quad 0.267\,327),$$

$$\mathbf{y}^{(0)} \approx (0.386\,139,\quad 0.435\,644,\quad 0.178\,218).$$

The exact optimal strategies are

$$\mathbf{x}^0=\left(\frac{4}{7},\frac{4}{21},\frac{5}{21}\right), \qquad \left(v=\frac{9}{7}\right)$$

$$\mathbf{y}^0=\left(\frac{3}{7},\frac{3}{7},\frac{1}{7}\right)$$

and the maximal error committed in approximating the components is 0.125 883.

Table I

k	$\bar{v}^{(k)}$	$\underline{v}^{(k)}$	$x_1^{(k)}$	$x_2^{(k)}$	$x_3^{(k)}$	$y_1^{(k)}$	$y_2^{(k)}$	$y_3^{(k)}$
1	3.000 000	−2.000 000	1.000 000	0.000 000	0.000 000	0.500 000	0.000 000	0.500 000
2	3.000 000	−1.000 000	0.666 667	0.333 333	0.000 000	0.333 333	0.000 000	0.666 667
3	3.000 000	0.000 000	0.500 000	0.500 000	0.000 000	0.250 000	0.250 000	0.500 000
4	2.250 000	0.000 000	0.400 000	0.600 000	0.000 000	0.200 000	0.400 000	0.400 000
5	2.400 000	0.000 000	0.333 333	0.500 000	0.166 667	0.166 667	0.500 000	0.333 333
10	1.500 000	0.300 000	0.181 818	0.363 636	0.454 545	0.454 545	0.363 636	0.181 818
20	1.500 000	0.800 000	0.190 476	0.571 429	0.238 095	0.523 810	0.380 952	0.095238
30	1.566 667	1.066 667	0.322 581	0.387 097	0.290 323	0.354 839	0.580 645	0.064 516
40	1.475 000	1.050 000	0.414 634	0.292 683	0.292 683	0.512 195	0.439 024	0.048 780
50	1.560 000	1.200 000	0.529 412	0.235 294	0.235 294	0.568 627	0.392 157	0.039 216
60	1.466 667	1.166 667	0.606 557	0.196 721	0.196 721	0.475 410	0.491 803	0.032 787
68	1.308 824	1.029 412	0.652 174	0.173 913	0.173 913	0.420 290	0.434 783	0.144 928
70	1 357 143	1.042 857	0.633 803	0.169 014	0.197 183	0.408 451	0.422 535	0.169 014
80	1.487 500	1.187 500	0.555 556	0.185 185	0.259 259	0.407 407	0.370 370	0.222 222
85	1.517 647	1.235 294	0.523 256	0.232 558	0.244 186	0.441 860	0.348 837	0.209 302
90	1.455 556	1.177 778	0.494 505	0.274 725	0.230 769	0.417 582	0.384 615	0.197 802
96	1.427 083	1.135 417	0.463 918	0.298 969	0.237 113	0.391 753	0.422 680	0.185 567
97	1.443 299	1.154 639	0.459 184	0.295 918	0.244 898	0.387 755	0.428 571	0.183 673
98	1.459 184	1.173 469	0.454 545	0.292 929	0.252 525	0.383 838	0.434 343	0.181 818
99	1.474 747	1.191 919	0.450 000	0.290 000	0.260 000	0.380 000	0.440 000	0.180 000
100	1.490 000	1.200 000	0.445 545	0.287 129	0.267 327	0.386 139	0.435 644	0.178 218

12.3. VON NEUMANN'S METHOD

The method of fictitious play, as we have seen, is of discrete character. John von Neumann proposed a method for solving matrix games which is based on the same idea (i.e. adjusting strategies according to the opponent's behaviour observed during previous runs of the game). But unlike fictitious play both players react to the changes in the opponent's strategy continuously over time.

Without loss of generality we may assume that the matrix game $\mathbf{A}$ under consideration is symmetric, i.e., $\mathbf{A}=-\mathbf{A}^T$. ($\mathbf{A}$ is an n by n matrix.) We know from Chapter 10 that the solution of any matrix game can be reduced to finding equilibrium points of a symmetric matrix game. Let player II's strategy $\mathbf{y}(t)$ be a function of parameter t (time) and suppose $t \geqq 0$.

Let us now define the following functions:

$$(18) \qquad u_i(\mathbf{y}(t))=\mathbf{e}_i\mathbf{A}\mathbf{y}(t),$$

$$\varphi(u_i)=\max\{0, u_i\}, \quad (i=1, \ldots, n)$$

$$\Phi(\mathbf{y}(t))=\sum_{i=1}^{n} \varphi(u_i(\mathbf{y}(t))),$$

$$\Psi(\mathbf{y}(t))=\sum_{i=1}^{n} \varphi^2(u_i(\mathbf{y}(t))).$$

We first prove a lemma.

LEMMA 5

$$\sqrt{\Psi(\mathbf{y}(t))} \leqq \Phi(\mathbf{y}(t)) \leqq \sqrt{n\Psi(\mathbf{y}(t))}.$$

Proof. The left-hand side of the inequality obviously holds, since

$$\Psi(\mathbf{y}(t))=\sum_{i=1}^{n} \varphi^2(u_i(\mathbf{y}(t)) \leqq [\varphi(u_i(\mathbf{y}(t))]^2=\Phi^2(\mathbf{y}(t)).$$

The right-hand side of the inequality follows from the Cauchy–Schwarz inequality; since

$$\Phi(\mathbf{y}(t)) = \sum_{j=1}^{n} 1 \cdot \varphi(u_i(\mathbf{y}(t))) \leqq$$

$$\leqq \left[\sum_{j=1}^{n} 1\right]^{\frac{1}{2}} \left[\sum_{j=1}^{n} \varphi^2(u_i(\mathbf{y}(t)))\right]^{\frac{1}{2}} = \sqrt{n\Psi(y(t))}\,. \quad \blacksquare$$

Let $\mathbf{y}^0$ be an initial strategy of player II and let us consider the differential equation system:

$$\text{(19)} \quad \left.\begin{aligned} y_j'(t) &= \varphi(u_j(\mathbf{y}(t))) - \Phi(\mathbf{y}(t))\,y_j(t) \\ y_j(0) &= y_j^0 \end{aligned}\right\}, \quad (j=1, \ldots, n).$$

Since the right-hand side of equation system (19) is continuous it has at least one solution. Let $\mathbf{y}(t)$ be a solution of (19).

THEOREM 5. [79] Any cluster point of the sequence $\{\mathbf{y}(t_k)\}$ – where $t_k \to \infty$ – is an optimal strategy of the symmetric matrix game $\mathbf{A}$. Furthermore there exists a constant C, such that

$$\mathbf{e}_i \mathbf{A} \mathbf{y}(t_k) \leqq \frac{\sqrt{n}}{C+t_k}, \qquad (i=1, \ldots, n).$$

Proof. We first prove that $\mathbf{y}(t)$ is a probability vector. Let us suppose on the contrary that there is an index j and a real number $t_1 > 0$ such that $y_j(t_1) < 0$. Let

$$t_0 = \sup\{t \mid 0 < t < t_1,\ y_j(t) \geqq 0\}\,.$$

Since $y_j(t)$ is continuous and $y_j(0) \geqq 0$ we have $y_j(t_0) = 0$ and for any τ satisfying $t_0 < \tau \leqq t_1$ the inequality $y_j(\tau) < 0$ holds. By definition $\varphi(u_j) \geqq 0$, $\Phi(\mathbf{y}(t)) \geqq 0$ and therefore

$$y_j'(\tau) = \varphi(u_j(\mathbf{y}(\tau))) - \Phi(\mathbf{y}(\tau))\,y_j(\tau) \geqq 0, \quad t_0 < \tau \leqq t_1\,.$$

By Lagrange's theorem there is a τ, $(t_0 < \tau \leqq t_1)$ such that

$$y_j(t_1) = y_j(t_0) + y_j'(\tau)\,(t_1 - t_0) \geqq y_j(t_0) = 0\,,$$

which is a contradiction. Next we will prove that $\sum_{j=1}^{n} y_j(t)=1$. Simple calculation shows that

$$\left(1-\sum_{j=1}^{n} y_j(t)\right)'=-\sum_{j=1}^{n} y_j'(t)=-\sum_{j=1}^{n} \varphi(u_j(\mathbf{y}(t))+$$

$$+\Phi(\mathbf{y}(t))\sum_{j=1}^{n} y_j(t)=-\Phi(\mathbf{y}(t))\left(1-\sum_{j=1}^{n} y_j(t)\right).$$

In other words, the function $f(t)=1-\sum_{j=1}^{n} y_j(t)$ satisfies the differential equation

$$f'(t)=-\Phi(\mathbf{y}(t))f(t)$$

with the initial condition $f(0)=0$. It can easily be verified that this equation has a unique solution: $f(t)\equiv 0$, which means that the components of the nonnegative vector $\mathbf{y}(t)$ sum to 1, i.e., $\mathbf{y}(t)$ is a probability vector.

Proceeding with the proof of our theorem let us assume that $\varphi(u_i(\mathbf{y}(t)))>0$ for some $t\geqq 0$. Then

$$\text{(20)}\quad \varphi'(u_i(\mathbf{y}(t)))=\sum_{j=1}^{n} a_{ij}y_j'(t)=\sum_{j=1}^{n} a_{ij}[\varphi(u_j(\mathbf{y}(t)))-$$

$$-\Phi(\mathbf{y}(t))y_j(t)]=\sum_{j=1}^{n} a_{ij}\varphi(u_j(\mathbf{y}(t)))-$$

$$-\Phi(\mathbf{y}(t))\varphi(u_i(\mathbf{y}(t)).$$

Multiplying both sides of the equation by $\varphi(u_i)$ and summing up for $i=1, \ldots, n$ we immediately get

$$\text{(21)}\quad \sum_{i=1}^{n} \varphi(u_i(\mathbf{y}(t)))\varphi'(u_i(\mathbf{y}(t)))=$$

$$=\sum_{i=1}^{n}\sum_{j=1}^{n} a_{ij}\varphi(u_i(\mathbf{y}(t)))\varphi(u_j(\mathbf{y}(t)))-$$

$$-\Phi(\mathbf{y}(t))\sum_{i=1}^{n} \varphi^2(u_i(\mathbf{y}(t))).$$

The first term of the right-hand side is 0 by the skew symmetricity of matrix **A**. Thus

$$(22)\quad \sum_{i=1}^{n} \varphi(u_i(\mathbf{y}(t)))\varphi'(u_i(\mathbf{y}(t)))=$$

$$=\frac{1}{2}\Psi'(\mathbf{y}(t))=-\Phi(\mathbf{y}(t))\Psi(\mathbf{y}(t)).$$

This equation does hold even in the case $\varphi(u_i(\mathbf{y}(t)))=0$ (except for the break-points where $\varphi'(u_i(\mathbf{y}(t)))$ may not exist) since relation (21) remains valid.

Assume now that there is a positive t for which $\Phi(\mathbf{y}(t))=0$. Then (22) implies that $\Phi(\mathbf{y}(\tau))=0$ for any $\tau \geqq t$. Thus $\mathbf{Ay}(\tau)\leqq \mathbf{0}$ and $\mathbf{y}(\tau)$ is an optimal strategy of both players by Theorem 1 of Chapter 10.

If $\Phi(\mathbf{y}(t))>0$ for any $t\geqq 0$, then $\Psi(\mathbf{y}(t)>0$ and Lemma 5 together with (22) imply

$$\frac{1}{2}\Psi'(\mathbf{y}(t))\leqq -\Psi^{\frac{3}{2}}(\mathbf{y}(t))$$

or

$$\frac{1}{2}\Psi'(\mathbf{y}(t))\Psi^{-\frac{3}{2}}(\mathbf{y}(t))\leqq -1.$$

Integrating both sides from 0 to t we get

$$-\Psi^{-\frac{1}{2}}(\mathbf{y}(t))+c\leqq -t$$

or

$$\Psi^{\frac{1}{2}}(\mathbf{y}(t))\leqq \frac{1}{c+t},$$

where $c=\Psi(\mathbf{y}(0))^{-\frac{1}{2}}$. Then by Lemma 5 for $i=1, \ldots, n$ we obtain the inequalities

$$(23)\quad u_i(\mathbf{y}(t))\leqq \varphi(u_i(\mathbf{y}(t)))\leqq \Phi(\mathbf{y}(t))\leqq$$

$$\leqq \sqrt{n}\sqrt{\Psi(\mathbf{y}(t))}\leqq \frac{\sqrt{n}}{c+t},$$

which remains valid even for the break-points of φ by the continuity of the function u_i. Thus (23) holds for any $t \geqq 0$. Taking a sequence $\{\mathbf{y}(t_k)\}$ where $t_k \to \infty$, for any cluster point $\mathbf{y}^*$, we have $\mathbf{A}\mathbf{y}^* \leqq \mathbf{0}$ which renders $\mathbf{y}^*$ to be an equilibrium point of the symmetric game $\mathbf{A}$ by Theorem 1 of Chapter 10. ∎

Differential equation system (19) can be interpreted in the following way. Let us assume that

$$\varphi(u_j(\mathbf{y}(t))) > 0 .$$

If player II plays strategy $\mathbf{y}(t)$ player I can assure himself positive pay-off by playing the pure strategy $\mathbf{e}_j$. Player II by increasing $y_j(t)$ to 1 (while keeping his strategy a probability vector), can push down player I's pay-off to 0 since

$$\mathbf{e}_j\mathbf{A}\mathbf{e}_j = a_{jj} = 0$$

by the skew symmetricity of $\mathbf{A}$. Therefore it is player II's interest to increase $y_j(t)$. This is exactly what the jth equation in (19) describes. (On the right-hand side of this equation the second term is bound to keep $\mathbf{y}(t)$ a probability vector.)

As an example let us solve the game given by

$$\mathbf{A} = \begin{bmatrix} 2 & 1 & 0 \\ 2 & 0 & 3 \\ -1 & 3 & 3 \end{bmatrix} .$$

Since $\mathbf{A}$ is not skew-symmetric we have first to transform it as we described it in Chapter 10. By adding 2 to each element of $\mathbf{A}$ we arrive at the symmetric game given by the skew-symmetric matrix

$$\tilde{\mathbf{A}} = \begin{bmatrix} 0 & 0 & 0 & 4 & 3 & 2 & -1 \\ 0 & 0 & 0 & 4 & 2 & 5 & -1 \\ 0 & 0 & 0 & 1 & 5 & 5 & -1 \\ -4 & -4 & -1 & 0 & 0 & 0 & 1 \\ -3 & -2 & -5 & 0 & 0 & 0 & 1 \\ -2 & -5 & -5 & 0 & 0 & 0 & 1 \\ 1 & 1 & 1 & -1 & -1 & -1 & 0 \end{bmatrix}.$$

We have solved the differential equation system (19) numerically starting from the initial vector $\mathbf{y}(0)=\mathbf{e}_1$ with step-size $h=0.01$ on the interval [0, 100]. The details of the computation are given in Table II. Taking $\mathbf{y}(100)$ as an approximation (using the notation of Chapter 10) we have $\lambda \approx 0.627\,163$,

$$a = \frac{1-\lambda}{2} \approx 0.186\,418\,5\,,$$

$$\mathbf{x}^0 \approx (0.563\,619, \quad 0.232\,359, \quad 0.241\,988)\,,$$

$$\mathbf{y}^0 \approx (0.485\,258, \quad 0.361\,633, \quad 0.115\,144)\,.$$

The value of the game (after having subtracted 2)

$$v = \frac{\lambda}{a} - 2 \approx 1.364\,274\,.$$

Comparing these numbers with the exact solution (see p. 162), we see that the error of v is 0.078 56 and the maximal error of the components of the optimal strategies is 0.066 938.

Table II

t	u_1	u_2	u_3	v_1	v_2	v_3	λ
0.0	1.000 000	0.000 000	0.000 000	0.000 000	0.000 000	0.000 000	0.000 000
1.0	0.498 258	0.000 000	0.000 000	0.000 000	0.000 000	0.000 000	0.501 742
10.0	0.068 436	0.059 580	0.082 426	0.070 307	0.030 623	0.068 195	0.620 432
20.0	0.032 549	0.114 997	0.067 851	0.062 994	0.078 965	0.032 435	0.610 209
30.0	0.045 135	0.097 306	0.053 180	0.081 088	0.079 824	0.020 397	0.623 070
40.0	0.088 452	0.068 093	0.049 293	0.087 253	0.079 949	0.014 274	0.612 688
50.0	0.088 789	0.051 342	0.058 903	0.072 439	0.094 930	0.010 762	0.622 835
60.0	0.112 434	0.039 897	0.045 773	0.093 828	0.077 980	0.008 363	0.621 724
70.0	0.113 662	0.041 508	0.038 620	0.076 930	0.077 060	0.025 900	0.626 321
80.0	0.104 693	0.040 207	0.044 144	0.081 142	0.079 014	0.022435	0.628 366
90.0	0.107 290	0.038 990	0.053 894	0.081 531	0.077 974	0.025 644	0.614 677
96.0	0.103 772	0.041 103	0.048 589	0.085 743	0.070 298	0.023 119	0.627 375
97.0	0.101 853	0.040 943	0.047 691	0.089 990	0.069 537	0.022 692	0.627 894
98.0	0.106 032	0.043 789	0.046 551	0.093 348	0.069 567	0.022 150	0.618 563
99.0	0.106 578	0.043 938	0.045 759	0.091 760	0.068 383	0.021 773	0.621 809
99.7	0.105 517	0.043 500	0.045 304	0.090 847	0.067 703	0.021 556	0.625 573
99.8	0.105 368	0.043 439	0.045 239	0.090 718	0.067 607	0.021 526	0.626 104
99.9	0.105 218	0.043 377	0.045 175	0.090 589	0.067 511	0.021 495	0.626 635
100.0	0.105 069	0.043 316	0.045 111	0.090 461	0.067 415	0.021 465	0.627 163

13. Some special games and methods

In the previous chapter we dealt with methods capable of solving arbitrary matrix games. In practice, however, we often come across games having some special features enabling us to use simpler methods to solve them and/or their size can be considerably reduced. Thereby they are more amenable to numerical treatment. In this chapter we are going to discuss some special games and methods of this sort.

13.1 MATRICES WITH SADDLE-POINTS

The simplest matrix games are those having an equilibrium point in pure strategies. In this case there are indices i_0, j_0 such that $\mathbf{x}^0 = \mathbf{e}_{i_0}$ and $\mathbf{y}^0 = \mathbf{e}_{j_0}$ are equilibrium strategies. By Theorem 3 of Chapter 9 we have

$$\mathbf{x}^0\mathbf{A} = \mathbf{e}_{i_0}\mathbf{A} \geqq v\mathbf{1} = a_{i_0 j_0}\mathbf{1},$$

$$\mathbf{A}\mathbf{y}^0 = \mathbf{A}\mathbf{e}_{j_0} \leqq v\mathbf{1} = a_{i_0 j_0}\mathbf{1}.$$

This means that $a_{i_0 j_0}$ is minimal in the i_0th row and maximal in the j_0th column. We call $a_{i_0 j_0} = v$ the *saddle point* of the matrix game **A**.

Conceptionally, pure strategies provide the most acceptable solutions particularly in situations where the game cannot be played several times. When analyzing practical game situations the first thing we have to do is check for saddle-points. This can very easily be done by inspection.

Unfortunately if we pick "randomly" from all the possible $m \times n$ game matrices, then it is very improbable that we will get a matrix with a saddle-point. The following theorem makes this statement precise.

THEOREM 1. [61] Let **A** be an m by n matrix each element of which is a random variable of the same continuous distribution. Then the probability that **A** has a saddle-point is

$$P(m, n) = \frac{m!\, n!}{(m+n-1)!}.$$

Proof. Consider the events

E: **A** has a saddle-point,
$E(r, s)$: a_{rs} is a saddle-point of **A**.
F: **A** has all its elements distinct.

Our continuity assumption ensures $P(F)=1$, so that

$$P(m, n) = P(E) = P(E \cap F).$$

Since the value of a game is unique if a_{rs} and a_{tu} are both saddle-points, then $a_{rs} = a_{tu}$. Thus $E \cap F$ is the disjoint union of the events $E(r, s) \cap F$, and so

$$P(m, n) = \sum_{r=1}^{m} \sum_{s=1}^{n} P[E(r, s) \cap F].$$

By symmetry $P[E(r, s) \cap F]$ has the same value for all pairs (r, s) and thus

$$P(m, n) = mnP[E(1, 1) \cap F].$$

The $(m+n-1)!$ possible orderings of the elements of the first row and column are clearly equiprobable and exactly $(m-1)!(n-1)!$ of them make a_{11} a saddle-point. Therefore

$$P[E(1, 1) \cap F] = \frac{(m-1)!(n-1)!}{(m+n-1)!},$$

implying the validity of our theorem. ∎

13.2. DOMINANCE RELATIONS

Let us consider the matrix game

$$(1)\qquad \mathbf{A}=\begin{bmatrix} 2 & 2 & 2 & 2 & 2 & 2 \\ 6 & 4 & 2 & 4 & 3 & 3 \\ 6 & 5 & 3 & 5 & 4 & 4 \\ 6 & 9 & 9 & -3 & 3 & 4 \\ 6 & 5 & 6 & 1 & 4 & 4 \\ 6 & 5 & 5 & 0 & 4 & 4 \end{bmatrix}.$$

If player I is rational, then he will not use his first pure strategy since playing the third one assures him larger pay-off independently of the strategy choice of player II. Therefore the first row of $\mathbf{A}$ can be omitted without restricting player I's possibilities for choosing an optimal strategy.

We will now set up simple reduction rules for game matrices based on ideas similar to the one applied above.

A vector $\mathbf{a}$ is said to dominate $\mathbf{b}$ if $\mathbf{a} \geqq \mathbf{b}$ and $\mathbf{a}$ strictly dominates $\mathbf{b}$ if $\mathbf{a} > \mathbf{b}$.

THEOREM 2. [79] Assume that one of the rows of $\mathbf{A}$, say the last one, is dominated by a convex linear combination of the remaining rows. Omit the last row and let $\mathbf{A}'$ be the matrix game obtained in this way, v' be the value, $X^{0\prime}$, $Y^{0\prime}$ be the equilibrium strategy sets of the game $\mathbf{A}'$. Then

(a) $v' = v$,
(b) $Y^{0\prime} = Y^0$, $\bar{X}^0 \subseteqq X^0$ where $\bar{X}^0 = \{\mathbf{x}' \mid \mathbf{x}' = (\mathbf{x}, 0),\ \mathbf{x} \in X^{0\prime}\}$,
(c) if the dominance is strict, then $\bar{X}^0 = X^0$.

(v denotes the value, X^0, Y^0 the sets of equilibrium strategies of the game $\mathbf{A}$.)

Proof. Suppose the last row $\mathbf{r}_m$ of $\mathbf{A}$ is dominated by a convex linear combination of the rows $\mathbf{r}_1, \mathbf{r}_2, \ldots, \mathbf{r}_{m-1}$, i.e.,

$$(2) \qquad \mathbf{r}_m \leqq \sum_{i=1}^{m-1} \lambda_i \mathbf{r}_i, \ \sum_{i=1}^{m} \lambda_i = 1, \ \lambda_i \geqq 0, \quad (i=1, \ldots, m-1).$$

Let $\mathbf{u} \in X^{0\prime}$ and $\mathbf{w} \in Y^{0\prime}$. Then

$$(3) \qquad \begin{array}{l} \mathbf{r}_i \mathbf{w} \leqq v' \\ \mathbf{u}\mathbf{A}' \geqq v'\mathbf{1} \end{array} \qquad (i=1, \ldots, m-1).$$

We would like to prove that $\mathbf{u}' = (\mathbf{u}, 0)$ is an optimal strategy of player I. Since

$$\begin{array}{l} \mathbf{u}'\mathbf{A} \geqq v'\mathbf{1}, \\ \mathbf{r}_i \mathbf{w} \leqq v', \end{array} \qquad (i=1, \ldots, m-1)$$

it suffices to prove

$$\mathbf{r}_m \mathbf{w} \leqq v'$$

which readily follows from

$$(4) \qquad \mathbf{r}_m \mathbf{w} \leqq \sum_{i=1}^{m-1} \lambda_i \mathbf{r}_i \mathbf{w} \leqq v' \sum_{i=1}^{m-1} \lambda_i = v'.$$

Thus we have proved assertions (a) and (b).

If the dominance is strict, then (4) becomes

$$\mathbf{r}_m \mathbf{w} < v',$$

and by Theorem 6 of Chapter 9 the last component of any $\mathbf{x}^0 \in X^0$ is 0 which is precisely what assertion (c) says. ∎

An analogous theorem can be established for column dominance.

THEOREM 3. [79] Assume that a column of $\mathbf{A}$, say the last one, dominates some convex linear combination of the first $n-1$ columns. Drop the last column and let $\mathbf{A}'$ be the matrix game thus obtained, v' be the value, $X^{0\prime}$, $Y^{0\prime}$ be the set of optimal strategies, of the game $\mathbf{A}'$. Then

(a) $v = v'$,
(b) $X^{0'} = X^0$, $\bar{Y}^0 \subseteq Y^0$, where $\bar{Y}^0 = \{\mathbf{y}' | \mathbf{y}' = (\mathbf{y}, 0), \mathbf{y} \in Y^{0'}\}$,
(c) in case of strict dominance $\bar{Y}^0 = Y^0$.

(v, X^0, Y^0 is the same as in Theorem 2.) The proof of the theorem goes along the lines of Theorem 2.

As an example let us reduce matrix (1) using the dominance relations.

The third row dominates the first and second one while the fifth column is dominated by the first, the second and the sixth one. Dropping the corresponding rows and columns we get the matrix

$$\mathbf{A}^{(1)} = \begin{bmatrix} 3 & 5 & 4 \\ 9 & -3 & 3 \\ 6 & 1 & 4 \\ 5 & 0 & 4 \end{bmatrix}.$$

Now the third row dominates the fourth one. Thus we have

$$\mathbf{A}^{(2)} = \begin{bmatrix} 3 & 5 & 4 \\ 9 & -3 & 3 \\ 6 & 1 & 4 \end{bmatrix}.$$

The linear combination of the first two columns taken by the weights $\lambda_1 = \lambda_2 = \frac{1}{2}$ dominates the third one. Thus

$$\mathbf{A}^{(3)} = \begin{bmatrix} 3 & 5 \\ 9 & -3 \\ 6 & 1 \end{bmatrix}.$$

Since the average of the first two rows is exactly the third one any one of them (say the second one) can be dropped. Hence

$$\mathbf{A}^{(4)} = \begin{bmatrix} 3 & 5 \\ 6 & 1 \end{bmatrix}.$$

Easy calculation shows that the value of $\mathbf{A}^{(4)}$ is $v=\dfrac{27}{7}$ and the optimal strategies are as follows:

$$\bar{\mathbf{x}}^0=\left(\frac{5}{7},\frac{2}{7}\right),$$

$$\bar{\mathbf{y}}^0=\left(\frac{4}{7},\frac{3}{7}\right).$$

Out of these an equilibrium point of the original game can readily be obtained:

$$\mathbf{x}^0=\left(0,0,\frac{5}{7},0,\frac{2}{7},0\right),$$

$$\mathbf{y}^0=\left(0,0,\frac{4}{7},\frac{3}{7},0,0\right).$$

13.3. $2\times n$ GAMES

A very simple and straightforward method can be developed for the solution of matrix games having only two pure strategies for player I.

Let the pay-off matrix be

$$\text{(5)}\qquad \mathbf{A}=\begin{bmatrix} a_{11} & a_{12} & \dots & a_{1n} \\ a_{21} & a_{22} & \dots & a_{2n} \end{bmatrix}.$$

Any mixed strategy of player I can be given by a real number x, $(0\leqq x\leqq 1)$ representing the probability of his choosing the first pure strategy. Of course he applies his second pure strategy with probability $1-x$. If player II uses his jth pure strategy, then the expected pay-off to player I is

$$l_j(x)=xa_{1j}+(1-x)a_{2j}.$$

Let

$$\text{(6)}\qquad \Phi(x)=\min_j l_j(x)$$

and

$$(7) \qquad \Phi(x^0) = \max_{0 \leqq x \leqq 1} \Phi(x).$$

THEOREM 4. [79] The value of **A** is $\Phi(x^0)$, x^0 is an optimal strategy of player I. Player II's optimal strategy depends on x^0 in the following manner:

(a) if $x_0 = 0$, then the pure strategy belonging to the straight line $l_j(x)$ going through the point $(0, \Phi(0))$ and of minimal steepness is an optimal strategy of player II;

(b) if $x_0 = 1$, then the pure strategy belonging to the straight line $l_j(x)$ going through the point $(1, \Phi(1))$ and of maximal steepness is an optimal strategy of player II;

(c) if $0 < x_0 < 1$ and there is a straight line $l_{j_0}(x)$ going through the point $(x_0, \Phi(x_0))$ and of steepness 0, then the j_0th pure strategy is an optimal strategy of player II;

(d) if $0 < x_0 < 1$ and there is no straight line having the property specified in (c), then the mixed strategy $\mathbf{y}^0 = \lambda_1 \mathbf{e}_{j_1} + \lambda_2 \mathbf{e}_{j_2}$ is an optimal strategy of player II, where λ_1 and λ_2 are weights ($\lambda_1, \lambda_2 \geqq 0$, $\lambda_1 + \lambda_2 = 1$) for which the identity

$$(8) \qquad \lambda_1 l_{j_1}(x) + \lambda_2 l_{j_2}(x) \equiv \Phi(x^0)$$

holds. $l_{j_1}(x)$ and $l_{j_2}(x)$ are straight lines going through the point $(x^0, \Phi(x^0))$ and of minimal and maximal steepness resp.

Proof. If **y** is an arbitrary strategy of player II, then

$$(x^0, 1 - x^0)\mathbf{A}\mathbf{y} = \sum_{j=1}^{n} y_j l_j(x^0) \geqq y_j \left[\min_j l_j(x^0) \right] =$$

$$= \min_j l_j(x^0) = \Phi(x^0).$$

Let x be any strategy of player I and consider the cases (a), (b), (c) and (d).

(a) Let $l_{j_0}(x)$ be the straight line going through $(0, \Phi(0))$ and of minimum steepness. Since $\Phi(x)$ attains its maximum at 0 we have

$$l_{j_0}(x) \leqq \Phi(0), \qquad (0 \leqq x \leqq 1).$$

(b) By analogous arguments we can easily prove

$$l_{r_0}(x) \leqq \Phi(1), \qquad (0 \leqq x \leqq 1),$$

where $l_{r_0}(x)$ is the straight line going through the point $(1, \Phi(1))$ and of maximal steepness.

(c) The straight line $l_{j_0}(x)$ specified in the theorem is constant. Therefore

$$l_{j_0}(x) = \Phi(x^0), \qquad (0 \leqq x \leqq 1).$$

(d) Since x^0 is a maximumpoint there must be two straight lines with steepness of different sign and satisfying (8).

In all four cases $\Phi(x^0)$ is the value of the game and the equilibrium strategies are those specified in our theorem. ∎

Of course $m \times 2$ games can also be solved in the same way. All we have to do is solve the game with matrix $-\mathbf{A}^T$ and change the role of the players.

Theorem 4 gives rise to a graphical method since $\Phi(x)$ is a piecewise linear concave function the maximum of which can very easily be determined by simply drawing up the graph of $\Phi(x)$. The method is illustrated by an example.

The matrix game to be solved is given by

$$\mathbf{A} = \begin{bmatrix} 2 & 3 & 11 \\ 7 & 5 & 5 \end{bmatrix}.$$

Then

$$l_1(x) = 2x + 7(1-x) = 7 - 5x$$

$$l_2(x) = 3x + 5(1-x) = 5 - 2x$$

$$l_3(x) = 11x + 2(1-x) = 2 + 9x.$$

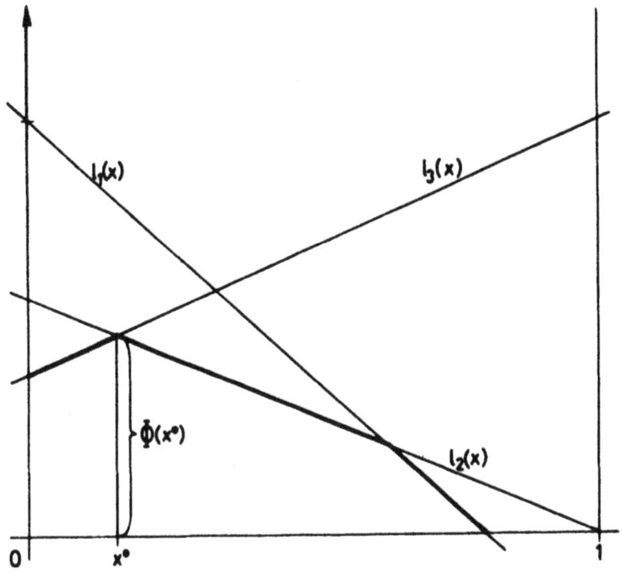

Figure 1

$\Phi(x)$ is shown in Figure 1. By elementary calculation we have

$$x^0=\frac{3}{11},\quad 1-x^0=\frac{8}{11};\quad \Phi(x^0)=\frac{49}{11}.$$

By solving

$$\lambda l_2(x)+(1-\lambda)l_3(x)\equiv 0$$

that is

$$-2\lambda+9(1-\lambda)=0$$

we get $\lambda=\frac{9}{11}$, $1-\lambda=\frac{2}{11}$. Hence

$$\mathbf{y}^0=\left(0,\frac{9}{11},\frac{2}{11}\right).$$

This graphical method (though it is much more complicated) can be used for solving $3\times n$ games, too, but of course no graphical method exists for $m>3$.

13.4. CONVEX (CONCAVE) MATRIX GAMES

A matrix $\mathbf{A}=[a_{ij}]$ is said to be *row-wise convex* (*concave*) if there exist convex (concave) functions $\varphi_i(x)$, $(i=1, \ldots, m)$ and real numbers r_j, $(j=1, \ldots, m)$ to satisfy $a_{ij}=\varphi_i(r_j)$, $(i=1, \ldots, m;$ $j=1, \ldots, n)$. The matrix $\mathbf{A}=[a_{ij}]$ is called *column-wise convex* (*concave*) if there exist convex (concave) functions $\Psi_j(x)$, $(j=1, \ldots, n)$ and real numbers q_i, $(i=1, \ldots, m)$ to satisfy $a_{ij}=\Psi_j(q_i)$ $(i=1, \ldots, m;$ $j=1, \ldots, n)$.

THEOREM 5. [147]

(i) If $\mathbf{A}$ is row-wise convex, then player II has an optimal strategy with at most two positive components.

(ii) If $\mathbf{A}$ is column-wise concave, then player I has an optimal strategy with no more than two positive components.

Before proving this theorem we establish a lemma.

LEMMA 1. Let Ξ be the set of random variables with possible values $r_1, \ldots, r_n$ and with expected value $\bar{r}$. If $r_{j_0} \leqq \bar{r} < r_{j_0+1}$ and f is a convex function, then there exist a $\xi_0 \in \Xi$ for which

$$\min_{\xi \in \Xi} M(f(\xi)) = M(f(\xi_0)).$$

In addition ξ_0 attains all possible values but r_{j_0} and r_{j_0+1} with probability 0. (Here M denotes the operator of taking expected values.)

Proof. Let $\xi \in \Xi$ and

$$M^+ = M(\xi \mid \xi > \bar{r}), \quad M^- = M(\xi \mid \xi \leqq \bar{r}).$$

Clearly $M^+ > r_{j_0+1}$ and $M^- \leqq r_{j_0}$. Furthermore

$$P(\xi > \bar{r})M^+ + P(\xi \leqq \bar{r})M^- = \bar{r}. \tag{9}$$

Since the function f is convex we have for any $\xi \in \Xi$

$$M(f(\xi)) \geqq f(M(\xi)).$$

Thus

$$(10) \quad M(f(\xi)) = P(\xi > \bar{r}) M(f(\xi) | \xi > \bar{r}) + \\ + P(\xi \leqq \bar{r}) M(f(\xi) | \xi \leqq \bar{r}) \geqq \\ \geqq P(\xi > \bar{r}) f(M^+) + P(\xi \leqq \bar{r}) f(M^-) .$$

Denoting the right-hand side of (10) by R the point $(\bar{r}, R)$ lies according to (9) and (10) on the line segment S with end-points $(M^-, f(M^-))$ and $(M^+, f(M^+))$. Since f is convex no point of the line segment with end-points $(r_{j_0}, f(r_{j_0}))$ and $(r_{j_0+1}, f(r_{j_0+1}))$ is above S. Thus there exists a $\xi_0 \in \Xi$ for which

$$(11) \quad M(f(\xi_0)) \leqq R ,$$

and all probabilities belonging to possible values of ξ_0 other than r_{j_0} and r_{j_0+1} are equal to 0. Relations (10) and (11) imply

$$\min_{\xi \in \Xi} M(f(\xi)) = M(f(\xi_0)) . \quad \blacksquare$$

Now we turn our attention to proving Theorem 5. It is enough to prove assertion (i), the case of (ii) is symmetric. Let $\mathbf{y}^0$ be an optimal strategy of player II and define

$$\bar{r} = \sum_{j=1}^{n} r_j y_j^0 .$$

By the optimality of $\mathbf{y}^0$ we have

$$\mathbf{A}\mathbf{y}^0 \leqq v\mathbf{1} .$$

Since $\mathbf{A}$ is row-wise convex this means

$$\sum_{j=1}^{n} f_i(r_j) y_j^0 \leqq v, \qquad (i = 1, \ldots, m) .$$

By Lemma 1, if $r_{j_0} \leqq \bar{r} < r_{j_0+1}$, then there exists a $\mathbf{y}^0$ with all components 0 except for the j_0th and j_0+1st. $\blacksquare$

At the same time Theorem 5 provides a method for solving game $\mathbf{A}$. All we have to do is solve the $n-1$ subgames defined by the matrices consisting of two adjacent columns of **A** and choose that one giving the smallest game value. The optimal strategy of player II in this subgame supplemented by 0's gives an overall optimal strategy for game **A**. For solving the subgames of size $m \times 2$ we can apply the graphical method of Section 13.3.

As an example let us consider the game defined by

$$\mathbf{A} = \begin{bmatrix} -1 & 0 & 1 \\ 1 & -1 & 1 \\ 1 & 0 & -1 \end{bmatrix}.$$

We form the submatrices

$$\mathbf{A}_1 = \begin{bmatrix} -1 & 0 \\ 1 & -1 \\ 1 & \boxed{0} \end{bmatrix}, \quad \mathbf{A}_2 = \begin{bmatrix} \boxed{0} & 1 \\ -1 & 1 \\ 0 & -1 \end{bmatrix},$$

where the elements squared are saddle points. We immediately get that the value of **A** is 0 and the vector

$$\mathbf{y}^0 = (0,1,0)$$

is an optimal strategy of player II. By Theorem 6 of Chapter 9,

$$\mathbf{x}^0 \mathbf{A} \mathbf{y}^0 = -x_2^0 = 0.$$

The remaining matrix is

$$\mathbf{B} = \begin{bmatrix} -1 & 0 & 1 \\ 1 & 0 & -1 \end{bmatrix},$$

from which $\mathbf{x}^0 = \left(\frac{1}{2}, 0, \frac{1}{2}\right)$ can easily be obtained.

14. Decomposition of matrix games

Let a matrix game be given by the hypermatrix

$$\mathbf{A}=\begin{bmatrix} \mathbf{A}_{11} & \mathbf{A}_{12} & \dots & \mathbf{A}_{1N} \\ \mathbf{A}_{21} & \mathbf{A}_{22} & \dots & \mathbf{A}_{2N} \\ \dots & \dots & \dots & \dots \\ \mathbf{A}_{M1} & \mathbf{A}_{M2} & \dots & \mathbf{A}_{MN} \end{bmatrix}.$$

Both from theoretical and computational aspects it is interesting to know what information can be obtained for the solution of game **A** by knowing the solutions of the component games $\mathbf{A}_{ij}$, $(i=1, \dots, M; j=1, \dots, N)$.

Let v denote the value of **A**, and v_{ij} the value of $\mathbf{A}_{ij}$, $(i=1, \dots, M; j=1, \dots, N)$. The set of optimal strategies for the hypergame is denoted by X^0 and Y^0, those for the component games are X^0_{ij}, Y^0_{ij}, $(i=1, \dots, M; j=1, \dots, N)$. Define a new game by the pay-off matrix $\mathbf{V}=[v_{ij}]$ whose optimal strategy sets are $\bar{X}$, $\bar{Y}$. The value of **V** is denoted by $\bar{v}$.

THEOREM 1. [119]

(i) If $\bigcap_j X^0_{ij}\neq\emptyset$, $(i=1, \dots, M)$, then $v\geq\bar{v}$. If $\mathbf{x}^0_i\in\bigcap_j X^0_{ij}$ and $\bar{\mathbf{x}}=(\bar{x}_1, \dots, \bar{x}_M)\in\bar{X}$, then the strategy

$$\text{(1)} \qquad \mathbf{x}^0=(\bar{x}_1\mathbf{x}^0_1, \dots, \bar{x}_M\mathbf{x}^0_M)$$

guarantees the pay-off $\bar{v}$ for player I.

(ii) If $\bigcap_i Y_{ij}^0 \neq \emptyset$, $(j=1, \ldots, N)$, then $v \leqq \bar{v}$. If $\mathbf{y}_j^0 \in \bigcap_i Y_{ij}^0$ and $\mathbf{y}=(\bar{y}_1, \ldots, \bar{y}_N) \in \bar{Y}$, then, by playing the strategy

$$(2) \qquad \mathbf{y}^0=(\bar{y}_1 \mathbf{y}_1^0, \ldots, \bar{y}_N \mathbf{y}_N^0),$$

player II does not lose more than $\bar{v}$.

Proof. We only prove assertion (i). Assertion (ii) can be verified in the same way.

Let $\mathbf{y}=(\mathbf{y}_1, \ldots, \mathbf{y}_N)$ be an arbitrary strategy of player II in game A. Obviously $\mathbf{y} \geqq \mathbf{0}$ and it has at least one positive component. Without loss of generality we may assume that the first component of $\mathbf{y}_1$ is positive. An arbitrarily small number ε can be chosen so that the vector

$$\mathbf{y}(\varepsilon)=\left(\mathbf{y}_1-\varepsilon \mathbf{e}_1, \mathbf{y}_2+\frac{\varepsilon}{N-1} \mathbf{e}_1, \ldots, \mathbf{y}_N+\frac{\varepsilon}{N-1} \mathbf{e}_1\right)$$

be nonnegative. Thus $\mathbf{y}(\varepsilon)$ is a strategy vector and can be written as

$$\mathbf{y}(\varepsilon)=(\alpha_1 \tilde{\mathbf{y}}_1, \ldots, \alpha_N \tilde{\mathbf{y}}_N)$$

where $\tilde{\mathbf{y}}_j$, $(j=1, \ldots, N)$ and $\boldsymbol{\alpha}=(\alpha_1, \ldots, \alpha_N)$ are also strategy vectors.

Then by (1) we have

$$\mathbf{x}^0 \mathbf{A} \mathbf{y}(\varepsilon)=\sum_{j=1}^N\left(\sum_{i=1}^M \bar{x}_i \mathbf{x}_i^0 \mathbf{A}_{ij} \alpha_j \tilde{y}_j\right) \geqq$$

$$\geqq \sum_{j=1}^N\left(\sum_{i=1}^M \bar{x}_i v_{ij} \alpha_j\right) \geqq \bar{v}.$$

Since ε is arbitrarily small, we obtain

$$\mathbf{x}^0 \mathbf{A} \mathbf{y} \geqq \bar{v}$$

for any strategy vector $\mathbf{y}$, which is exactly what assertion (i) of our theorem says. ∎

COROLLARY 1. If $\bigcap_j X_{ij}^0 \neq \emptyset$ and $\bigcap_i Y_{ij}^0 \neq \emptyset$, ($i=1, \ldots, M$; $j=1, \ldots, N$), then $v=\bar{v}$ and $(\mathbf{x}^0, \mathbf{y}^0)$ is an equilibrium point of $\mathbf{A}$.

COROLLARY 2. Let us consider the game

$$\mathbf{A}=\begin{bmatrix} \mathbf{A}_1 & \mathbf{0} & \ldots & \mathbf{0} \\ \mathbf{0} & \mathbf{A}_2 & \ldots & \mathbf{0} \\ \ldots & \ldots & \ldots & \ldots \\ \mathbf{0} & \mathbf{0} & \ldots & \mathbf{A}_N \end{bmatrix}.$$

Since any pair of strategies are optimal for the game $\mathbf{0}$, the assumptions of Corollary 1 are satisfied. Denote v_i the value of $\mathbf{A}_i$, ($i=1, \ldots, N$) and let $\mathbf{x}_i^0$, $\mathbf{y}_i^0$ be an arbitrary equilibrium point of $\mathbf{A}_i$, ($i=1, \ldots, N$). Now $\mathbf{V}$ is the diagonal matrix $\langle v_1, \ldots, v_N \rangle$ and the values of $\mathbf{A}$ and $\mathbf{V}$ coincide by Corollary 1. By Theorem 1 the strategies defined by (1) and (2) provide an equilibrium point of $\mathbf{A}$. The strategies $\bar{\mathbf{x}}$, $\bar{\mathbf{y}}$ are optimal strategies of the very simple game $\mathbf{V}$ which can readily be solved. It can be shown by direct calculation that if all $v_i \neq 0$ and are of the same sign, then the value of $\mathbf{V}$ is

$$\bar{v}=\frac{1}{\sum_{i=1}^{N} \frac{1}{v_i}},$$

and the strategy $\left(\frac{\bar{v}}{v_1}, \ldots, \frac{\bar{v}}{v_N}\right)$ is optimal for both players. If not all v_i are of the same sign, then the value of $\mathbf{A}$ is 0 and for arbitrary nonnegative numbers satisfying

$$h_i=0, \qquad \text{if} \quad v_i<0$$

$$r_i=0, \qquad \text{if} \quad v_i>0$$

$$h_0=\sum_{i=1}^{N} h_i>0, \quad r_0=\sum_{i=1}^{N} r_i>0$$

the strategies

$$\bar{\mathbf{x}} = \left(\frac{h_1}{h_0}, \ldots, \frac{h_N}{h_0}\right),$$

$$\bar{\mathbf{v}} = \left(\frac{r_1}{r_0}, \ldots, \frac{r_N}{r_0}\right)$$

are optimal for players I and II resp.

COROLLARY 3. Let a game be given by

$$\mathbf{A} = \begin{bmatrix} \mathbf{B}+c_{11}[1] & \mathbf{B}+c_{12}[1] & \ldots & \mathbf{B}+c_{1N}[1] \\ \mathbf{B}+c_{21}[1] & \mathbf{B}+c_{22}[1] & \ldots & \mathbf{B}+c_{2N}[1] \\ \ldots & \ldots & \ldots & \ldots \\ \mathbf{B}+c_{M1}[1] & \mathbf{B}+c_{M2}[1] & \ldots & \mathbf{B}+c_{MN}[1] \end{bmatrix}. \tag{3}$$

Then using the notation of Theorem 1 we have $\mathbf{A}_{ij}=\mathbf{B}+c_{ij}[\mathbf{1}]$. Since $\mathbf{A}_{ij}$ stems from $\mathbf{B}$ by adding a constant c_{ij}, $\mathbf{A}_{ij}$ and $\mathbf{B}$ are strategically equivalent and if the value of $\mathbf{B}$ is v, then that of $\mathbf{A}_{ij}$ is $v+c_{ij}$. Thus $\mathbf{V}=\mathbf{C}+v[1]$ and if v' denotes the value of $\mathbf{C}$, then the value of $\mathbf{V}$ is $v+v'$. Since the assumptions of Corollary 1 are satisfied the value of $\mathbf{A}$ is $v+v'$ and the strategies $\mathbf{x}^0$, $\mathbf{y}^0$ defined by (1) and (2) are optimal. By the special structure of $\mathbf{A}$ ($\mathbf{x}_1^0=\ldots=\mathbf{x}_M^0$, $\mathbf{y}_1^0=\ldots=\mathbf{y}_N^0$) is an equilibrium point of $\mathbf{B}$ and $(\bar{\mathbf{x}}, \bar{\mathbf{y}})$ is that of game $\mathbf{C}$.

We mention that by repeated application of construction (3) games played consecutively over time can be analyzed and overall optimal strategies can be gained via solving separate composite games.

To illustrate the decomposition principle laid down in Theorem 1 let us consider the matrix game

$$\mathbf{A}=\left[\begin{array}{cc|cc} -2 & 0 & 4 & 4 \\ 0 & -2 & 4 & 4 \\ \hline 2 & 2 & 0 & 2 \\ 2 & 2 & 2 & 0 \end{array}\right].$$

It is easy to see that the conditions of Corollary 1 are satisfied. (In general, if **A** can be partitioned in such a way that for each i there is at most one j and for each j there is at most one i so that not all components of $\mathbf{A}_{ij}$ are the same, then Corollary 1 applies.)

Easy calculation shows that

$$\mathbf{V}=\begin{bmatrix} -1 & 4 \\ 2 & 1 \end{bmatrix}, \quad \bar{v}=\frac{3}{2}.$$

Furthermore

$$X_{11}^0 \cap X_{12}^0 = X_{11}^0 = \left\{\left(\frac{1}{2}, \frac{1}{2}\right)\right\},$$

$$X_{21}^0 \cap X_{22}^0 = X_{22}^0 = \left\{\left(\frac{1}{2}, \frac{1}{2}\right)\right\},$$

$$Y_{11}^0 \cap Y_{12}^0 = Y_{11}^0 = \left\{\left(\frac{1}{2}, \frac{1}{2}\right)\right\},$$

$$Y_{21}^0 \cap Y_{22}^0 = Y_{22}^0 = \left\{\left(\frac{1}{2}, \frac{1}{2}\right)\right\},$$

$$\bar{X}=\left\{\left(\frac{1}{6}, \frac{5}{6}\right)\right\}, \quad \bar{Y}=\left\{\left(\frac{1}{2}, \frac{1}{2}\right)\right\}.$$

Thus by (1) and (2) we get

$$\mathbf{x}^0=\left(\frac{1}{6}\left(\frac{1}{2},\frac{1}{2}\right),\ \frac{5}{6}\left(\frac{1}{2},\frac{1}{2}\right)\right)=\left(\frac{1}{12},\frac{1}{12},\frac{5}{12},\frac{5}{12}\right),$$

$$\mathbf{y}^0=\left(\frac{1}{2}\left(\frac{1}{2},\frac{1}{2}\right),\ \frac{1}{2}\left(\frac{1}{2},\frac{1}{2}\right)\right)=\left(\frac{1}{4},\frac{1}{4},\frac{1}{4},\frac{1}{4}\right).$$

The value of $\mathbf{A}$ is $\frac{3}{2}$.

15. Examples of matrix games

15.1. EXAMPLE 1 [79]

Let us assume that the plan of a town can be viewed as a quadratic matrix. The rows and columns represent the streets, the elements of the matrix symbolize the buildings. In one of the buildings a bomb has been hidden. We have enough time to search through only one street. If the bomb has been placed in the street we are searching through, then we shall surely find it and get a pay-off equivalent to what the building to be blown up is worth. In case we cannot detect the bomb our pay-off is zero. What are the optimal strategies of both players, i.e., the "Seeker" and the "Hider", provided the value of each building is positive?

If **A** is the matrix containing the values of buildings, then player II chooses an element of **A** while player I picks a row or column and the pay-off matrix is the following:

Elements

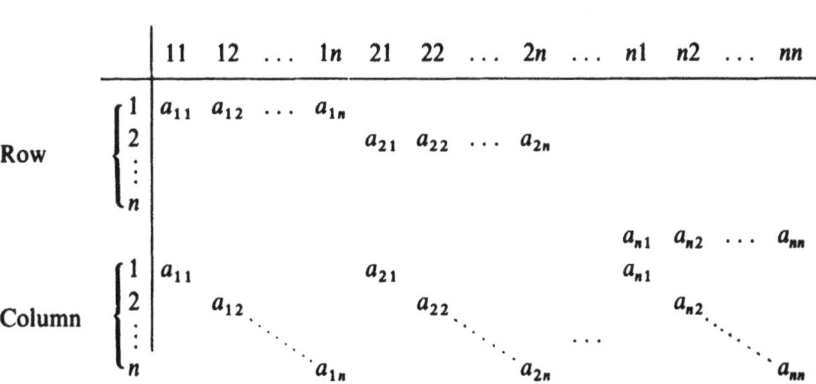

		11	12	...	1n	21	22	...	2n	...	n1	n2	...	nn
Row	1	a_{11}	a_{12}	...	a_{1n}									
	2					a_{21}	a_{22}	...	a_{2n}					
	⋮													
	n										a_{n1}	a_{n2}	...	a_{nn}
Column	1	a_{11}				a_{21}					a_{n1}			
	2		a_{12}				a_{22}			...		a_{n2}		
	⋮			⋱				⋱					⋱	
	n				a_{1n}				a_{2n}					a_{nn}

It is easy to show that the value of the game is positive. Therefore the optimal strategies can be determined from the optimal solutions of the linear program (see Section 12.1)

$$(1) \qquad y_{ij} \geqq 0$$

$$\sum_{j=1}^{n} a_{ij} y_{ij} + s_i = 1, \quad (i = 1, \ldots, n)$$

$$\sum_{i=1}^{n} a_{ij} y_{ij} + t_j = 1, \quad (j = 1, \ldots, n)$$

$$\sum_{i=1}^{n} \sum_{j=1}^{n} y_{ij} \to \max,$$

where s_i and t_j are slack variables. Introducing new variables

$$z_{ij} = a_{ij} y_{ij},$$

(1) takes the form

$$(2) \qquad z_{ij} \geqq 0$$

$$\sum_{j=1}^{n} z_{ij} + s_i = 1$$

$$\sum_{i=1}^{n} z_{ij} + t_j = 1$$

$$\sum_{i=1}^{n} \sum_{j=1}^{n} \left(\frac{1}{a_{ij}}\right) z_{ij} \to \min.$$

This is the well-known *assignment problem* (see [37]) solvable by numerous efficient methods. Having determined the optimal z_{ij}^0 values y_{ij}^0 can easily be calculated from which player II's optimal strategy can be obtained (see Theorems 1 and 2 of Chapter 12). Player I's optimal strategy comes from the optimal dual solution of (2) in a similar way.

15.2. EXAMPLE 2 [144]

In this example we put the famous von Neumann equilibrium model in a game theoretical framework.

Let us consider an economy producing m different products with n different activities (technologies). The technological relations in the economy are described by the nonnegative matrices $\mathbf{A}=(a_{ij})$ and $\mathbf{B}=(b_{ij})$. Here a_{ij} denotes the constant amount used by activity j from product i whereas unit application of activity j produces the amount b_{ij} of product i.

The model is closed, all goods consumed in any form being listed among the products (i.e., labour and capital consumption are also contained). The expansion of production factors left out of the model is assumed to be unlimited (e.g., land). Foreign trade and personal consumption are treated as activities, as well. Thereby their structure is assumed to be constant. The amount of products not consumed is used for the expansion of production. The activities are assumed to be continuous and realizable in one time-period.

Let $\mathbf{x}=(x_j)$, $(j=1, \ldots, n)$, $(\mathbf{x}\neq\mathbf{0})$ denote the level of activities and $\mathbf{p}=(p_i)$, $(i=1, \ldots, m)$, $(\mathbf{p}\neq\mathbf{0})$ the prices of products. The basic relations of the model are described by the inequality system

$$(3) \qquad \alpha\sum_j a_{ij}x_j \leqq \sum_j b_{ij}x_j$$

$$(4) \qquad \beta\sum_i a_{ij}p_i \geqq \sum_i b_{ij}p_i$$

where α denotes the expansion coefficient of the economy and β is the "profit factor" (1 + profit-rate).

(3) means that the upper bound of the consumption for each product is the amount produced times the expansion coefficient. (4) realizes the "non-profit principle", i.e., no activity is allowed to contain more profit than that determined by the profit factor.

We also stipulate that free goods are priced at 0, i.e.,

(5) $\alpha \sum_j a_{ij}x_j < \sum_j b_{ij}x_j \Rightarrow p_i = 0, \quad (i=1, \ldots, m)$

and non-profitable activities (or more exactly activities being less profitable than the average determined by β) must not be applied, i.e.,

(6) $\beta \sum_i a_{ij}p_i > \sum_i b_{ij}p_i \Rightarrow x_j = 0, \quad (j=1, \ldots, n).$

It can easily be seen that if **x** and **p** are solutions of the model (i.e., they satisfy relations (3)–(6) and $\mathbf{x} \geqq \mathbf{0}$, $\mathbf{p} \geqq \mathbf{0}$) then multiplying them by a positive number we also get a solution. Therefore we may assume that

(7) $\mathbf{1x} = \mathbf{1p} = 1.$

On the matrices **A** and **B** we make three additional fairly mild assumptions:

(i) for any j there is at least one i for which $a_{ij} > 0$,

(ii) for any i there is at least one j for which $b_{ij} > 0$,

(iii) $a_{ij} < 1$, $b_{ij} < 1$ for all i and j, which can always be assured by proper choice of scales.

Now we will show that finding a solution $\mathbf{x}^*$, $\mathbf{p}^*$, α^*, β^* to the system (3) through (7) is equivalent to finding an equilibrium point of a zero-sum two-person game.

Let

$$\Phi(\mathbf{x}, \mathbf{p}) = \frac{\mathbf{pBx}}{\mathbf{pAx}}.$$

($\mathbf{pAx} > 0$ for any probability vector **p**, and **x** by (i).) Let the feasible strategy sets of players I and II be L_x and L_p resp.

$$L_x = \{\mathbf{x} \mid \mathbf{x} \geqq \mathbf{0}, \quad \mathbf{1x} = 1\},$$

$$L_p = \{\mathbf{p} \mid \mathbf{p} \geqq \mathbf{0}, \quad \mathbf{1p} = 1\}.$$

The reader himself can easily verify that $\mathbf{x}^* \in L_x$, $\mathbf{y}^* \in L_p$, α^*, β^* is a solution to the von Neumann model if and only if $\mathbf{x}^*$, $\mathbf{y}^*$ is an equilibrium point of the two-person zerò-sum game

$$G=\{L_x, L_p, \Phi\}$$

and the value of the game is $v^*=\alpha^*=\beta^*$.

In the special case when all components of $\mathbf{B}$ are identical G is a matrix game. Generally G is not a matrix game but, as we will show, finding its equilibrium strategies can be reduced to the solution of matrix games.

Let $\alpha_0=0$, and consider the following sequence of matrix games:

$$\mathbf{C}_t=(c_{ij}^{(t)})$$

$$c_{ij}^{(t)}=b_{ij}-\alpha_t a_{ij}$$

$$\alpha_t=\alpha_0+\sum_{i=0}^{t-1} v_i, \qquad (t=1,2,\ldots)$$

where v_t is the value of the game given by $\mathbf{C}_t$ and $\mathbf{p}_t$, $\mathbf{x}_t$ are optimal strategies. $\mathbf{C}_t$ can also be written as

$$\mathbf{C}_t=\mathbf{C}_{t-1}-v_{t-1}\mathbf{A}, \qquad (t\geqq 1).$$

If $v_{t-1}=0$ for some $t\geqq 1$, then α_{t-1}, $\beta_{t-1}=\alpha_{t-1}$, $\mathbf{p}^{(t-1)}$, $\mathbf{x}^{(t-1)}$ is a solution to the von Neumann model which can be shown by simply substituting into the inequalities (3)–(7). It can also easily be seen that if $\sum_{i=0}^{t-1} v_i$ is convergent, then any cluster point of the sequences $\{\alpha_t\}$, $(\mathbf{p}^{(t)}\}$, $\{\mathbf{x}^{(t)}\}$ is a solution of the von Neumann model. (Note that the cluster point $\bar{\alpha}$, $\bar{\beta}=\bar{\alpha}$ is a limit point since $\{\alpha_t\}$ is monotone.) The existence of at least one cluster point is assured since L_p and L_x are closed and bounded. Thus it suffices to prove that $\sum_{i=0}^{t-1} v_i$ is convergent.

By condition (ii) $v_0>0$. Since v_t is the value of game $\mathbf{C}_t$ and $\mathbf{p}^{(t)}$, $\mathbf{x}^{(t)}$ is an equilibrium point we have

(8) $$v_t \leqq \sum_i \sum_j c_{ij}^{(t)} p_i^{(t-1)} x_j^{(t)} = \sum_i \sum_j c_{ij}^{(t-1)} p_i^{(t-1)} x_j^{(t)} - $$
$$- v_{t-1} \sum_i \sum_j a_{ij} p_i^{(t-1)} x_j^{(t)} \leqq$$
$$\leqq v_{t-1} \left(1 - \sum_i \sum_j a_{ij} p_i^{(t-1)} x_j^{(t)}\right).$$

Furthermore

(9) $$v_t \geqq \sum_i \sum_j c_{ij}^{(t)} p_i^{(t)} x_j^{(t-1)} = \sum_i \sum_j c_{ij}^{(t-1)} p_i^{(t)} x_j^{(t-1)} - $$
$$- v_{t-1} \sum_i \sum_j a_{ij} p_i^{(t)} x_j^{(t-1)} \geqq$$
$$\geqq v_{t-1} \left(1 - \sum_i \sum_j a_{ij} p_i^{(t)} x_j^{(t-1)}\right).$$

By condition (iii) inequalities (8) and (9) together with $v_0 > 0$ imply

$$0 < \frac{v_t}{v_{t-1}} < 1 .$$

In other words $\{v_t\}$ is a monotonically decreasing positive sequence.

Suppose now on the contrary that $\sum_{i=0}^{t-1} v_i$ is not convergent, i.e., $\lim_{t \to \infty} \alpha_t = \infty$. Choose an index t such that

(10) $$\alpha_t = \alpha_0 + \sum_{i=0}^{t-1} v_i > \frac{m}{a_{kl}}$$

where

$$a_{kl} = \min_{i,j} \{a_{ij} \mid a_{ij} > 0\} .$$

Let

$$A' = (a'_{ij})$$

$$a'_{ij} = \begin{cases} a_{kl}, & \text{if} \quad a_{ij} > 0 \\ 0, & \text{if} \quad a_{ij} = 0 . \end{cases}$$

Since v_t is the value of $\mathbf{C}_t$ we obtain for $i=1, \ldots, m$

$$(11) \quad v_t \leqq \sum_j c_{ij}^{(t)} x_j^{(t)} = \sum_j b_{ij} x_j^{(t)} - \alpha_t \sum_j a_{ij} x_j^{(t)} \leqq$$

$$\leqq \sum_j b_{ij} x_j^{(t)} - \alpha_t \sum_j a'_{ij} x_j^{(t)} .$$

By condition (i) any column of $\mathbf{A}'$ contains at least one positive element. Therefore there must be a row index u for which

$$(12) \quad \sum_j a'_{uj} x_j^{(t)} \geqq \frac{a_{kl}}{m} .$$

Thus by (11) we get

$$v_t \leqq \sum_j b_{uj} x_j^{(t)} - \alpha_t \sum_j a'_{uj} x_j^{(t)} .$$

Now condition (iii), (10) and (12) imply

$$v_t < 1 - \frac{m}{a_{kl}} \frac{a_{kl}}{m} = 0 ,$$

contradicting the positivity of v_t. Thus any cluster point of the sequence $\{\alpha_t\}$, $\{\mathbf{p}^{(t)}\}$, $\{\mathbf{x}^{(t)}\}$ is a solution of the von Neumann model. As we have seen such a cluster point (solution) can be approximated by solving a sequence of matrix games.

16. Games played over the unit square

Games played over the unit square can be considered a natural and relatively simple generalization of matrix games. In the case of matrix games each player has a finite number of pure strategies while the pure strategy set of both players engaged in a game over the unit square is the interval $[0, 1]$. The pay-off function is thus defined on the unit square. This is where the term "game played over the unit square" comes from. It is clear that if the set of pure strategies is an arbitrary closed interval $[a, b]$, then this interval can be transformed to $[0, 1]$ by a suitable continuous transformation.

Analogously to matrix games we are going to focus our attention on mixed extensions. To put it in precise terms we will study the following two-person zero-sum game:

$$\Gamma = \{X, Y, K\}$$

where X and Y are the family of distribution functions defined on $[0, 1]$ and $K(x, y)$ is the expected pay-off to player I:

$$K(x, y) = \int_0^1 \int_0^1 k(\xi, \eta)\, \mathrm{d}x(\xi)\, \mathrm{d}y(\eta)\,. \tag{1}$$

Here $k(\xi, \eta)$ is the pay-off belonging to the pure strategies $\xi, \eta \in [0, 1]$ and $x \in X$, $y \in Y$. The function k is called the *kernel* of the game Γ. (Of course $K(x, y)$ is well defined only for those kernels for which the Stieltjes integral in (1) exists.)

We can represent pure strategies by distribution functions having exactly one unit jump. In particular the distribution function

$$x_{\xi_0}(\xi)=\begin{cases}0, & \text{if } \xi \leqq \xi_0 \\ 1, & \text{if } \xi > \xi_0\end{cases}$$

represents the pure strategy ξ_0.

Without assuming more than integrability of the kernel k we cannot say much about the structure of the game $\Gamma=\{X, Y, K\}$. However we know (applying Theorem 6 of Chapter 3 to two-person zero-sum games) that if k is continuous, then there exists a pair of equilibrium strategies (distribution functions). In view of Theorem 2 of Chapter 7,

$$\min_{y \in Y} \max_{x \in X} K(x, y) = \max_{x \in X} \min_{y \in Y} K(x, y) = v\,. \tag{2}$$

(The symbols sup and inf could be changed to max and min because K is continuous and X, Y are compact.)

Equation (2) can be represented by an inequality system, too. A strategy pair (x^0, y^0) constitutes an equilibrium point and v is the value of the game if and only if

$$\begin{aligned} K(x^0, \eta) &\geqq v \quad \text{for all} \quad \eta \in [0, 1] \\ K(\xi, y^0) &\leqq v \quad \text{for all} \quad \xi \in [0, 1]\,, \end{aligned} \tag{3}$$

where $K(x^0, \eta) = K(x^0, y_\eta)$, $K(\xi, y^0) = K(x_\xi, y^0)$.

The proof of the above assertion can easily be carried out in the same way as it has been done in Theorem 3 of Chapter 9.

Unless otherwise stated kernel $k(\xi, \eta)$ will always be assumed to be continuous. Thus the game $\Gamma=\{X, Y, K\}$ is always solvable and relations (2), (3) hold.

Before dealing with some special games we give a few definitions and prove some simple theorems pertaining to general continuous games played over the unit square.

The *spectrum of a strategy* (distribution) $x(\xi)$ is defined to be the set of those points $\xi \in [0, 1]$ for which $x(a) < x(b)$ holds for arbitrary a, b satisfying $a < \xi < b$.

A pure strategy ξ is called *essential* if there is at least one equilibrium strategy x whose spectrum contains ξ.

A strategy x is said to be of *finite type* if its spectrum is a finite set. Obviously, in this case x can be written as a convex linear combination of finitely many pure strategies.

Strategies having as their spectrum the entire interval [0, 1] are called *completely mixed.*

The following theorem is an analogue of Theorem 6 of Chapter 9 and is very useful in the analysis of games played over the unit square.

THEOREM 1. [79] Let (x_0, y_0) be a pair of equilibrium strategies for the game $\Gamma = \{X, Y, K\}$ played over the unit square. If η_0 is in the spectrum of y_0, then $K(x_0, \eta_0) = v$ where v is the value of the game.

Proof. Since x_0 is an equilibrium strategy we have

$$K(x_0, \eta) \geqq v \quad \text{for all} \quad \eta \in [0, 1].$$

If this inequality were strict for $\eta = \eta_0$, then by the continuity of the kernel would be an interval [a, b] for which $a < \eta_0 < b$ and

$$(4) \qquad K(x_0, \eta) > v \quad \text{for all} \quad \eta \in [a, b].$$

Since η_0 is in the spectrum of y_0 we get $y_0(a) < y_0(b)$. Thus integrating both sides of (4) with respect to η weighted by the distribution y_0 we obtain

$$K(x_0, y_0) > v$$

contradicting the assumption that v is the value of the game. ∎

Of course, a similar theorem holds for player II.

A strategy x is said to be an *equalizer* if $K(x, \eta) = \text{constant}$ for all $\eta \in [0, 1]$.

As corollaries to Theorem 1 we get the following assertions:

COROLLARY 1. If player I has at least one completely mixed equilibrium strategy, then any equilibrium strategy of player II is an equalizer.

COROLLARY 2. If any pure strategy of player I is essential, then every equilibrium strategy of player II is an equalizer.

If the game $\Gamma = \{X, Y, K\}$ played over the unit square has a pair of pure equilibrium strategies (x_{ξ_0}, y_{η_0}), then wȩ call (ξ_0, η_0) a *saddle-point*. For a saddle point the following inequalities hold

$$k(\xi_0, \eta) \geqq k(\xi_0, \eta_0) \geqq k(\xi, \eta_0) \quad \text{for all} \quad \begin{cases} \xi \in [0, 1] \\ \eta \in [0, 1] \end{cases}.$$

One can easily prove that just as in case of finite games the set of equilibrium points is convex. (To obtain this result it is not necessary to assume the continuity of the kernel. It is enough to suppose that k is integrable.)

Further properties of the equilibrium points and methods, other than approximate, for finding them are only available for special games played over the unit square. Karlin's book [79] deals with this subject in great detail. In the following we are going to overview those special games where the structure of optimal strategies is more or less known. Nevertheless we will not go into details. For further studies the reader is referred to [79].

As a general method for solving continuous games played over the unit square we will give two approximative methods. One approximates the game by a matrix game, the other reduces it to a nonconvex programming problem.

17. Some special classes of games on the unit square

We deal briefly in this chapter with some special classes of games played on the unit square. The subject is covered in great detail in [79]. Therefore we only dwell upon those structural properties – mostly without proofs – that have some bearing on methods designed to solve them. To illustrate the general scope of these games we describe a few examples relating to conflict situations coming from different fields.

Separable games. A game on the unit square $\Gamma=\{X, Y, K\}$ is called *separable* if the kernel is separable i.e.

$$k(\xi,\eta)=\sum_{i=1}^{m}\sum_{j=1}^{n} a_{ij}r_i(\xi)s_j(\eta),$$

where $r_1, \ldots, r_m, s_1, \ldots, s_n$ is a continuous function on $[0, 1]$ and $\mathbf{A}=[a_{ij}]$ is a real matrix. As a special case we get *polynomial games*, where $k(\xi, \eta)$ is a polynomial of the variables ξ, η.

To every separable game $\Gamma=\{X, Y, K\}$ there belongs a game $H=\{U, W, L\}$. Here U is the set of m-vectors to which a distribution function $x(\xi)$ can be found to satisfy

$$u_i=\int_0^1 r_i(\xi)\,\mathrm{d}x(\xi), \quad (i=1, \ldots, m).$$

Similarly W is the set of n-vectors to which a distribution function $y(\eta)$ can be found satisfying

$$w_j=\int_0^1 s_j(\eta)\,\mathrm{d}y(\eta), \quad (j=1, \ldots, n).$$

The pay-off function L is defined by

$$L(\mathbf{u},\mathbf{w})=\sum_{i=1}^{m}\sum_{j=1}^{n}a_{ij}u_i w_j=$$

$$=\sum_{i=1}^{m}\sum_{j=1}^{n}a_{ij}\int_0^1 r_i(\xi)\,\mathrm{d}x(\xi)\int_0^1 s_j(\eta)\,\mathrm{d}y(\eta)=$$

$$=\int_0^1\int_0^1\left[\sum_{i=1}^{m}\sum_{j=1}^{n}a_{ij}r_i(\xi)s_j(\eta)\right]\mathrm{d}x(\xi)\,\mathrm{d}y(\eta)=$$

$$=\int_0^1\int_0^1 k(\xi,\eta)\,\mathrm{d}x(\xi)\,\mathrm{d}y(\eta)=K(x,y).$$

The relationship between Γ and H is demonstrated by the following theorem.

THEOREM 1. [108] Let $(\mathbf{u}_0, \mathbf{w}_0)$ be an equilibrium point of H. Then any pair of distribution functions x_0, y_0 satisfying

$$\text{(1)}\qquad \int_0^1 r_i(\xi)\,\mathrm{d}x_0(\xi)=u_i^0,\quad (i=1,\ldots,m),$$

$$\int_0^1 s_j(\eta)\,\mathrm{d}y_0(\eta)=w_j^0,\quad (j=1,\ldots,n)$$

is an equilibrium point of Γ.

As a consequence of this theorem we see that solving Γ amounts to solving the finite game H and then finding distribution functions satisfying the equations (1). The complexity of this task is greatly reduced by U and W being closed, bounded, convex sets.

The next theorem concerns the structure of optimal strategies.

THEOREM 2. [108] Let $\Gamma=\{X, Y, K\}$ be a separable game with kernel

$$k(\xi,\eta)=\sum_{i=1}^{m}\sum_{j=1}^{n}u_{ij}r_i(\xi)s_j(\eta).$$

Both players have optimal strategies with spectra consisting of at most $p=\min(m,n)$ points.

Games with convex and generalized convex kernels a game on the unit square $\Gamma=\{X, Y, K\}$ is called a *generalized convex game* if there exists a positive integer n such that either

$$(2) \qquad \frac{\partial^n k(\xi, \eta)}{\partial \eta^n} \geqq 0, \quad \text{for all} \quad \xi, \eta \in [0, 1]$$

or there is a positive integer m for which

$$(3) \qquad \frac{\partial^m k(\xi, \eta)}{\partial \xi^m} \leqq 0, \quad \text{for all} \quad \xi, \eta \in [0, 1]$$

where $k(\xi, \eta)$ is the kernel of the game.

Game Γ is called *convex* if $k(\xi, \eta)$ is a convex function of η for any fixed ξ. For differentiable kernels inequality (2) implies this property for $n=2$. In this case player II has a pure optimal strategy while player I has an optimal strategy with a spectrum of at most two points. This property enables us to readily solve these games. If in (2) $n=1$, then the situation is even simpler since in this case both players possess pure optimal strategies, i.e., the game has a saddle-point.

As an example for a convex game let us consider a military game of *attack and defense*. Player II defends two towns A and B against player I. We suppose that in the case of both towns if the attacking force exceeds the defence, then the loss to player II is proportional to the difference between the two forces. On the other hand if defence is stronger than attack, then no loss to player I is incurred. The loss in town B is λ times ($\lambda>0$) the loss in A. Both players are assumed to have equal forces, say 1.

We denote the pure strategies, by ξ and η which is the fraction of his force employed to attack and to defend resp. town A. Then the pay-off to player I looks like this

$$k(\xi, \eta)=\max\{\xi-\eta, \lambda(1-\xi-1+\eta)\}=$$
$$=\max\{\xi-\eta, \lambda(\eta-\xi)\}.$$

It is easy to see that $k(\xi, \eta)$ is convex for any fixed ξ and therefore the game is convex. It is not too difficult to determine the optimal strategies and the value of the game. Direct calculation shows that an optimal strategy of player II is the pure strategy

$$y^0(\eta) = y_{\frac{1}{1+\lambda}}(\eta),$$

while player I's optimal strategy is

$$x^0(\xi) = \frac{1}{1+\lambda} x_0(\xi) + \frac{\lambda}{1+\lambda} x_1(\xi).$$

The value of the game is

$$v = \frac{\lambda}{1+\lambda}.$$

In the general case ($n \geqq 3$ in (2)) there is no easy method for solving game Γ. But on the size of the spectrum we can establish the following result (for proof see [79]).

THEOREM 3. [79] In case of a generalized convex game satisfying (2) player I has an optimal strategy with a spectrum consisting of at most n points. The spectrum of player II's optimal strategy contains no more than $\frac{n}{2}$ points if inner points of the interval [0, 1] have a unit count and the boundary points 0 and 1 have half-counts.

Bell-shaped games. For the definition of a bell-shaped game we have to define *Polya-type functions.* A function of two variables $k(\xi, \eta) = \varphi(\xi - \eta)$ is called Polya-type if it meets the following requirements:

1. $\varphi(u)$ is defined and continuous for any real number u,
2. for any positive integer n and real numbers

$$\xi_1 < \ldots < \xi_n, \quad \eta_1 < \ldots < \eta_n$$

the determinant of matrix $\mathbf{F} = [\varphi(\xi_i - \eta_j)]$ is nonnegative,

3. for any sequence $\{\xi_i\}$, $(\xi_1 < \ldots < \xi_n)$ there exists a sequence $\{\eta_j\}$, $\{\eta_1 < \ldots < \eta_n)$ such that the determinant of $\mathbf{F}$ is positive. The

same must hold if $\{\eta_j\}$ is fixed first,

4. $$\int_{-\infty}^{\infty} \varphi(u)\,du < \infty.$$

A game on the unit square $\Gamma = \{X, Y, K\}$ is called *bell-shaped* if its kernel $k(\xi, \eta)$ is a positive, analytic Polya-type function.

The following theorems refer to the structure of optimal strategies to bell-shaped games. (For proofs see [79].)

THEOREM 4. [79] The value of any bell-shaped game is positive and all equilibrium strategies of both players have finite spectra.

THEOREM 5. [79] If player II has an optimal strategy with a spectrum consisting of n points, then the spectrum of any of player I's optimal strategies contains either n or $n-1$ points.

THEOREM 6. [79] If $\Gamma = \{X, Y, K\}$ is a bell-shaped game where $k(\xi, \eta) = \varphi(\xi - \eta)$, φ is differentiable and $\varphi'(0) = 0$, then

1. both players have unique optimal strategies,
2. the points 0 and 1 belong to the spectrum of player II's optimal strategy but do not belong to that of player I,
3. if $\{\xi_1, \ldots, \xi_n\}$ and $\{\eta_1, \ldots, \eta_s\}$, $(s = n$ or $s = n-1, \eta_1 = 0, \eta_s = 1)$ are the spectra of equilibrium strategies of player I and II resp., then $0 < \xi_1 < \eta_2$ and $\eta_{s-1} < \xi_n < 1$.

THEOREM 7. [79] Let $\Gamma = \{X, Y, K\}$ be a bell-shaped game with kernel $k(\xi, \eta) = \varphi(\xi - \eta)$. If $\varphi(u) = \varphi(-u)$ and $\varphi'(0) = 0$, then the optimal strategies of either player are symmetric to the point $\xi = \eta = \dfrac{1}{2}$. In other words if $\xi(\eta)$ belongs to the spectrum of an optimal strategy, then $1 - \xi(1 - \eta)$ also belongs to it and the weights attached to symmetric points are equal.

Bell-shaped games come up in practice mainly in situations where one of the players is required to estimate the value of a parameter controlled by the other player. The pay-off function (error-function) is assumed to be of Polya-type. As an example let us suppose that a submarine is hiding somewhere along a channel represented by the interval [0, 1]. A bomber not knowing exactly where the submarine is located discharges a bomb at a point of [0, 1]. If the submarine is at the point η and the bomb has reached water at ξ, then the damage incurred is assumed to be expressible by the error function $k(\xi, \eta)=e^{-\lambda(\xi-\eta)^2}$ where λ is a positive parameter. It is easy to see that the game $\Gamma(\lambda)=\{X, Y, K(\lambda)\}$ is bell-shaped and Theorems 4–7 can be applied to it. If $\lambda=3$, then easy calculation shows that the submarine must hide at the end points of the channel with probability $\frac{1}{2}$ while the bomber has to drop its charge at the points a and 1-a with probability $\frac{1}{2}$, ($a\approx 0.072$). The value of the game is $v=\frac{1}{2}(e^{-3a^2}+e^{-3(a-1)^2})$.

Games of timing. Games of timing are games on the unit interval where the pure strategies of the players represent a choice of a time to perform an action.

A typical example of such a game is the so-called *duel*. Let us suppose that each dueller is allowed to fire exactly once in the time interval [0, 1]. The later a dueller fires the more precise his shot will be, i.e., the probability of hitting his opponent will be greater. On the other hand if he keeps aiming too long, then he is taking the risk to be hit by the bullet fired at him by the other dueller. Under these circumstances there is a need for an "optimal strategy" which takes into account both tendencies.

Duels come up not only in military situations but also in business, politics and anywhere where choosing appropriate time for an action or course of actions is crucial.

As an example let us formulate the so-called "silent duel" as a game. Each dueller (player) fires exactly once in the time interval [0, 1]. The probability that player I hits his opponent is $P_1(\xi)$ if he fires at $\xi \in [0, 1]$. For player II this probability is $P_2(\eta)$. We assume that P_1 and P_2 are continuous monotonically increasing functions in [0, 1] and $P_1(0)=P_2(0)=0$, $P_1(1)=P_2(1)=1$. The pay-off to player I is 1 if he hits player II and he himself remains unhurt. If he is hit he gets -1 while the pay-off to both of them is 0 in the case of a tie (if both miss or simultaneously hit each other). Now the kernel of the game is the following:

$$k(\xi, \eta)=\begin{cases} P_1(\xi)-(1-P_1(\xi))P_2(\eta), & \text{if} \quad \xi<\eta \\ P_1(\xi)-P_2(\xi), & \text{if} \quad \xi=\eta \\ -P_2(\eta)+(1-P_2(\eta))P_1(\xi), & \text{if} \quad \xi>\eta\,. \end{cases}$$

From the mathematical point of view a special difficulty arises since the kernel k of a game of timing may not be continuous along the main diagonal of the unit square. (See, e.g., the silent duel described above.) Therefore the existence of equilibrium strategies is not always assured. Yet we will prove that under fairly general conditions both players have "ε optimal" strategies for any $\varepsilon>0$ and the game has a sup-inf value.

THEOREM 8. [79] Let $\Gamma=\{X, Y, K\}$ be a two-person, zero sum game played over the unit square with the kernel

$$k(\xi, \eta)=\begin{cases} M_1(\xi, \eta), & \text{if} \quad \xi<\eta \\ \varphi(\xi), & \text{if} \quad \xi=\eta \qquad \xi, \eta \in [0, 1]\,. \\ M_2(\xi, \eta), & \text{if} \quad \xi>\eta \end{cases}$$

The functions M_1, M_2 are defined and continuous on Δ_1 and Δ_2 resp., where $\Delta_1=\{(\xi, \eta) | 0 \leqq \xi \leqq \eta \leqq 1\}$, $\Delta_2=\{(\xi, \eta) | 0 \leqq \eta \leqq \xi \leqq 1\}$. Furthermore it is assumed that

$$(4) \qquad M_2(\xi, \xi) \leqq \varphi(\xi) \leqq M_1(\xi, \xi) \quad \text{for all} \quad \xi \in [0, 1]\,.$$

Then Γ has a sup-inf value and for any $\varepsilon>0$ there is an "ε optimal" strategy for each player.

Proof. M_1 and M_2 are uniformly continuous on Δ_1 and Δ_2 resp. Therefore for any $\varepsilon>0$ there is a $\delta=\delta(\varepsilon)>0$ such that

(5) $$|M_1(\xi,\eta)-M_1(\xi',\eta')|<\varepsilon, \quad \text{if} \quad |\xi-\xi'|<\delta, |\eta-\eta'|<\delta$$

and

(6) $$(\xi,\eta)\in\Delta_1, \quad (\xi',\eta')\in\Delta_1,$$

$$|M_2(\xi,\eta)-M_2(\xi',\eta')|<\varepsilon, \qquad \text{if} \quad |\xi-\xi'|<\delta, |\eta-\eta'|<\delta$$

and

$$(\xi,\eta)\in\Delta_2, \quad (\xi',\eta')\in\Delta_2.$$

Let $\varepsilon>0$ be arbitrary and divide the interval $[0, 1]$ by the grid points $0=\tau_1<\tau_2<\ldots<\tau_r=1$ so that $\tau_{i+1}-\tau_i<\delta(\varepsilon)$, $(i=1, \ldots, r-1)$, ($\delta(\varepsilon)$ is taken according to (5) and (6)).

Let Γ_ε be the matrix game where both players select one of the points $\tau_1, \ldots, \tau_r$ as their pure strategies and the pay-off matrix is: $\mathbf{A}=[a_{ij}]$ where

$$a_{ij}=k(\tau_i,\tau_j), \quad (i,j=1,\ldots,r).$$

By Theorem 1 of Chapter 9, Γ_ε is solvable and both players have mixed optimal strategies. Let $\mathbf{x}_\varepsilon=(x_1, \ldots, x_r)$ and $\mathbf{y}_\varepsilon=(y_1, \ldots, y_r)$ be optimal strategies and v_ε be the value of Γ_ε. Thus we have

$$\sum_{i=1}^{r} x_i k(\tau_i,\tau_j)\geqq v_\varepsilon, \quad (i=1,\ldots,r),$$

$$\sum_{j=1}^{r} y_j k(\tau_i,\tau_j)\leqq v_\varepsilon, \quad (j=1,\ldots,r).$$

Denote x_ε and y_ε distribution functions having jumps at $\tau_1, \ldots, \tau_r$ of magnitude $x_1, \ldots, x_r$ and $y_1, \ldots, y_r$ resp. Of course x_ε and y_ε are feasible strategies for Γ.

Let us consider the expected pay-off to player I in game Γ for arbitrary $\eta\in[0, 1]$ provided player I plays x_ε. If $\eta=\tau_j$, $(1\leqq j\leqq r)$, then

(7) $$K(x_\varepsilon,\eta)=K(x_\varepsilon,\tau_j)=\sum_{i=1}^{r} x_i k(\tau_i,\tau_j)\geqq v_\varepsilon.$$

If $\tau_j < \eta < \tau_{j+1}$, $(1 \leqq j \leqq r-1)$, then

$$K(x_\varepsilon, \eta) = \sum_{i=1}^{r} x_i k(\tau_i, \eta) = \sum_{i=1}^{j} x_i M_1(\tau_i, \eta) + \\ + \sum_{i=j+1}^{r} x_i M_2(\tau_i, \eta).$$

By the definition of $k(\xi, \eta)$ we have

$$K(x_\varepsilon, \tau_j) = \sum_{i=1}^{r} x_i k(\tau_i, \tau_j) = \\ = \sum_{i=1}^{j-1} x_i M_1(\tau_i, \tau_j) + x_j \varphi(\tau_j) + \\ + \sum_{i=j+1}^{r} x_i M_2(\tau_i, \tau_j).$$

Therefore

$$K(x_\varepsilon, \eta) - K(x_\varepsilon, \tau_j) = \sum_{i=1}^{j-1} x_i (M_1(\tau_i, \eta) - \\ - M_1(\tau_i, \tau_j)) + \sum_{i=j+1}^{r} x_i (M_2(\tau_i, \eta) - \\ - M_2(\tau_i, \tau_j)) + x_j (M_1(\tau_j, \eta) - \varphi(\tau_j)).$$

By (5), (6) and (7) we get

$$K(x_\varepsilon, \eta) - K(x_\varepsilon, \tau_j) \geqq \sum_{i=1}^{j-1} x_i (M_1(\tau_i, \eta) - \\ - M_1(\tau_i, \tau_j)) + \sum_{i=j+1}^{r} x_i (M_2(\tau_i, \eta) - \\ - M_2(\tau_i, \tau_j)) + x_j (M_1(\tau_j, \eta) - \varepsilon - \varphi(\tau_j)) \geqq -\varepsilon.$$

Hence

$$K(x_\varepsilon, \eta) \geqq K(x_\varepsilon, \tau_j) - \varepsilon. \tag{8}$$

Since $K(x_\varepsilon, \tau_j) \geqq v_\varepsilon$ we obtain

$$K(x_\varepsilon, \eta) \geqq v_\varepsilon - \varepsilon,$$

and therefore by (7) and (8) we get

$$(9) \qquad K(x_\varepsilon, y) \geqq v_\varepsilon - \varepsilon \quad \text{for all} \quad y \in Y,$$

which means that x_ε is "ε optimal".

In a similar way we can prove that y_ε is "ε optimal" for player II. From (9) it follows that

$$\sup_{x \in X} \inf_{y \in Y} K(x, y) \geqq v_\varepsilon - \varepsilon .$$

Since y_ε is "ε optimal" we get

$$K(x, y_\varepsilon) \leqq v_\varepsilon + \varepsilon \quad \text{for all} \quad x \in X,$$

and

$$\inf_{y \in Y} \sup_{x \in X} K(x, y) \leqq v_\varepsilon + \varepsilon .$$

Since, by Theorem 1 of Chapter 7

$$\inf_{y \in Y} \sup_{x \in X} K(x, y) \geqq \sup_{x \in X} \inf_{y \in Y} K(x, y),$$

we obtain

$$(10) \qquad v_\varepsilon - \varepsilon \leqq \sup_{x \in X} \inf_{y \in Y} K(x, y) \leqq \inf_{y \in Y} \sup_{x \in X} K(x, y) \leqq v_\varepsilon + \varepsilon .$$

Since v_ε is bounded, we can choose a subsequence of $\{v_\varepsilon\}$ converging to a real number v if $\varepsilon \to 0$. Obviously v is the sup-inf value of Γ. ∎

There are certain classes of games of timing where sup and inf in Theorem 8 can be replaced by max and min, i.e., there are optimal strategies for both players. In a lot of cases techniques taken from the theory of integral equations help to solve important special classes. In [79] a complete list of solutions to a number of different types of duels is given and proofs are also discussed in great detail.

18. Approximate solution of two-person zero-sum games played over the unit square

As we have already mentioned there is no general method for solving an arbitrary two-person zero-sum game played over the unit square. Apart from the special cases, of which we mentioned a few in the previous chapter, we have to make recourse to approximative methods. As a first step it seems acceptable to seek for "only" an "ε optimal" pair of strategies and an "approximate value" of the game $\Gamma=\{X, Y, K\}$ played over the unit square. To remind the reader of the definitions: we call a pair of strategies (x^0, y^0) "ε optimal" and a real number v_ε an "ε good" value of the game $\Gamma=\{X, Y, K\}$ if

$$(1)\qquad \begin{aligned} &K(x^0, y) \geqq v_\varepsilon - \varepsilon \quad \text{for all} \quad y \in Y, \\ &K(x, y^0) \leqq v_\varepsilon + \varepsilon \quad \text{for all} \quad x \in X, \\ &v_\varepsilon - v \leqq \varepsilon \end{aligned} \qquad (\varepsilon > 0)$$

hold, where v is the sup-inf value of Γ. (Of course, this definition applies for any two-person zero-sum game having a sup-inf value.)

At first glance it seems appealing to split the interval $[0, 1]$ into subintervals densely enough in order to obtain "ε optimal" strategies by solving the associated matrix game.

We are now going to estimate how close the grid points in $[0, 1]$ should be placed for a given ε. For the sake of simplicity we assume the interval $[0, 1]$ to have been divided uniformly. The game we are going to deal with is the game of timing defined in Theorem 8 in Chapter 17. It is clear that any zero-sum two-person game played on the unit square with a continuous kernel belongs to this class as a special case. In addition to the assumptions specified in Theorem 8 of Chapter 17, we suppose that there are constants K_1, K_2 to satisfy

$$|M_1(\xi, \eta)-M_1(\xi', \eta')| \leqq K_1 |(\xi, \eta)-(\xi', \eta')| \quad \text{for all} \quad (\xi, \eta)\in \Delta_1, (\xi', \eta')\in \Delta_1,$$

$$|M_2(\xi, \eta)-M_2(\xi', \eta')| \leqq K_2 |(\xi, \eta)-(\xi', \eta')| \quad \text{for all} \quad (\xi, \eta)\in \Delta_2, (\xi', \eta')\in \Delta_2 .$$

If, e.g., M_1 and M_2 are continuously differentiable functions on Δ_1 and Δ_2 resp., then the above condition is always met by taking the constants K_1, K_2 to satisfy the inequalities

$$K_1 \geqq \max_{(\xi, \eta)\in \Delta_1} |\nabla M_1(\xi, \eta)|,$$

$$K_2 \geqq \max_{(\xi, \eta)\in \Delta_2} |\nabla M_2(\xi, \eta)|,$$

where ∇M_1, ∇M_2 denote gradients of M_1, M_2 resp.

Let us now suppose that $\varepsilon > 0$ is given. If we take $n-1$ inner grid points equally spaced in the unit interval, then we get n subintervals of length $\frac{1}{n}$. Thus for any ξ, ξ' and η, η' we have

$$|\xi-\xi'| \leqq \frac{1}{n},$$

$$|\eta-\eta'| \leqq \frac{1}{n}.$$

Hence

$$|M_1(\xi, \eta)-M_1(\xi', \eta')| \leqq K_1 |(\xi, \eta)-(\xi', \eta')| \leqq$$

$$\leqq K_1 |\xi-\xi'| + K_1 |\eta-\eta'| \leqq \frac{2K_1}{n}.$$

Similarly

$$|M_2(\xi, \eta)-M_2(\xi', \eta')| \leqq \frac{2K_2}{n}.$$

Thus if $n \geqq \frac{2 \max [K_1, K_2]}{\varepsilon}$, then (5) and (6) of Chapter 17 hold. By the proof of Theorem 8 of Chapter 17 it follows that equilibrium strategies of the matrix game Γ_ε defined there are "ε optimal". We also get from (10) of Chapter 17 that the value v_ε of Γ_ε is an "ε good" value of Γ, since

$$v_\varepsilon - \varepsilon \leqq v \leqq v_\varepsilon + \varepsilon$$

implies

$$|v_\varepsilon - v| \leqq \varepsilon .$$

(Here v denotes the sup-inf value of Γ.)

As we have already seen, matrix games can be solved efficiently by a number of different methods. However, if we want good approximation and K is large, then n can also be very large, thereby increasing the computation time and storage requirements to a great extent. E.g., if we wished to approximate the separable game with the kernel

$$k(\xi, \eta) = \xi \sin \eta + \xi \cos \eta + 2\xi^2,$$

then taking $K_1 = K_2 = 7$ and $\varepsilon = 0{,}01$ we would have to work with a matrix game having 1400 rows and columns. It should be mentioned that by adjusting the density of the grid points to the shape of the kernel we can reduce this number considerably.

Because of the sensitivity of the above approximation to the parameters K_1, K_2, ε we would like to deal with another approximative method based on SUMT (Sequential unconstrained minimization technique). Although the programming problem thus obtained is usually of smaller size, it is nonlinear and generally nonconvex.

Our main purpose remains the same: to obtain a pair of "ε optimal" strategies and an "ε good" value to the game $\Gamma = \{X, Y, K\}$ played over the unit square.

The kernel $k(\xi, \eta)$ is assumed to satisfy a Lipschitz condition of order α, $(\alpha > 0)$

$$(2) \qquad |k(\xi, \eta)-k(\xi', \eta')| \leqq M |(\xi, \eta)-(\xi', \eta')|^{\alpha},$$

where ξ, ξ', η, η' are arbitrary points of $[0, 1]$ and M is a constant. Furthermore we suppose that game Γ is of *(m, n)-type*. This means that both players have equilibrium strategies having spectra consisting of at most m and n points resp. (Of the games discussed in Chapter 17, the separable and the generalized convex games are of (m, n)-type.)

Let us consider the following system of inequalities:

$$(3) \qquad 0 \leqq x_i \leqq 1, \quad 0 \leqq u_i \leqq 1, \qquad (i=1, \ldots, m)$$

$$0 \leqq y_j \leqq 1, \quad 0 \leqq z_j \leqq 1, \qquad (j=1, \ldots, n)$$

$$\sum_{i=1}^{m} u_i = 1$$

$$\sum_{j=1}^{n} z_j = 1$$

$$\sum_{i=1}^{m} k(x_i, \eta) u_i \geqq v \quad \text{for all} \quad \eta \in [0, 1],$$

$$\sum_{j=1}^{n} k(\xi, y_j) z_j \leqq v \quad \text{for all} \quad \xi \in [0, 1].$$

Since the game is of (m, n)-type, inequality system (3) has a solution. On the other hand, if $x_1^0, \ldots, x_m^0, y_1^0, \ldots, y_n^0, u_1^0, \ldots, u_m^0, z_1^0, \ldots, z_n^0, v^0$ is a solution to (3), then a pair of equilibrium strategies can very easily be obtained. E.g., if $x_{i_1}^0 = x_{i_2}^0 = \ldots x_{i_k}^0$, then player I plays $x_{i_1}^0$ with probability $\sum_{r=1}^{k} u_{i_r}$. This correspondence between solutions of (3) and strategies holds of course for player II, too, and obviously v^0 is the value of the game.

Instead of solving (3) exactly, we seek vectors $\mathbf{x}, \mathbf{y}, \mathbf{u}, \mathbf{z}$ and a scalar v which can violate the last two set of inequalities by at most $\varepsilon > 0$. ε is an arbitrarily small but fixed number. It is clear that to this approximate solution for (3) strategies can be associated in exactly the same way as it was done for exact solutions. Obviously the strategies obtained this way are "ε optimal" and v is an "ε good" value of the game Γ.

In order to get an approximate solution to (3), let us consider the following nonlinear programming problem:

$$\Phi(\mathbf{x}, \mathbf{u}, \mathbf{y}, \mathbf{z}, v, a) = \int_0^1 \exp\left\{a\left[v - \sum_{i=1}^{m} k(x_i, \eta)u_i\right]\right\} d\eta + \int_0^1 \exp\left\{a\left[\sum_{j=1}^{n} k(\xi, y_j)z_j - v\right]\right\} d\xi \to \min \tag{4}$$

$$0 \leqq x_i \leqq 1, \quad 0 \leqq u_i \leqq 1, \qquad (i=1, \ldots, m),$$

$$0 \leqq y_j \leqq 1, \quad 0 \leqq z_j \leqq 1, \qquad (j=1, \ldots, n),$$

$$\sum_{i=1}^{m} u_i = 1$$

$$\sum_{j=1}^{n} z_j = 1,$$

where a is a positive parameter.

THEOREM 1. [53] Given an $\varepsilon > 0$ there always can be found an $a(\varepsilon)$ such that, for any $a \geqq a(\varepsilon)$, each optimal solution of (4) provides a pair of "ε optimal" strategies and an "ε good" value of game Γ.

Proof. Let us subdivide the interval [0, 1] into subintervals I_d $(d=1, \ldots, q)$ so that $\bigcup_{d=1}^{q} I_d = [0, 1]$ and for any $t_{1d}, t_{2d} \in I_d$ the inequality $|t_{1d} - t_{2d}|^{\alpha} \leqq \frac{\varepsilon}{2M}$ holds, $(d=1, \ldots, q)$. (Here α and M are the same constants as in (2) and ε is a given positive number.) Let the length of the smallest subinterval be ρ. Choose a to satisfy

$$a \geqq \frac{2(\ln 2 - \ln \rho)}{\varepsilon}. \tag{5}$$

Let $\mathbf{x}^0, \mathbf{y}^0, \mathbf{u}^0, \mathbf{z}^0, v^0$ be one of the optimal solutions to (4). (It is easy to see that such an optimal solution always exists.) We are going to prove that

$$\sum_{j=1}^{n} k(\xi, y_j^0) z_j^0 \leqq v^0 + \varepsilon \quad \text{for all} \quad \xi \in [0, 1]. \tag{6}$$

Assume on the contrary that

$$\sum_{j=1}^{n} k(\xi_0, y_j^0) z_j^0 - v^0 > \varepsilon \tag{7}$$

for some $\xi_0 \in [0, 1]$.

Let $\xi_0 \in I_{d_0}$. Then

$$\sum_{j=1}^{n} k(\xi, y_j^0) z_j^0 - v^0 > \frac{\varepsilon}{2} \quad \text{for all} \quad \xi \in I_{d_0}, \tag{8}$$

since if there were a $\xi' \in I_{d_0}$ for which the inequality

$$\sum_{j=1}^{n} k(\xi', y_j^0) z_j^0 - v^0 \leqq \frac{\varepsilon}{2}$$

would hold, then (2) and the definition of I_d would imply

$$\begin{aligned}
\frac{\varepsilon}{2} &< \left| \sum_{j=1}^{n} k(\xi_0, y_j^0) z_j^0 - v^0 - \sum_{j=1}^{n} k(\xi', y_j^0) z_j^0 + v^0 \right| = \\
&= \left| \sum_{j=1}^{n} k(\xi_0, y_j^0) z_j^0 - \sum_{j=1}^{n} k(\xi', y_j^0) z_j^0 \right| \leqq \\
&\leqq \sum_{j=1}^{n} \left| k(\xi_0, y_j^0) - k(\xi', y_j^0) \right| z_j^0 \leqq \\
&\leqq \sum_{j=1}^{n} M |\xi_0 - \xi'|^\alpha z_j^0 = M |\xi_0 - \xi'|^\alpha \leqq \frac{\varepsilon}{2},
\end{aligned}$$

which is a contradiction. Therefore (8) holds. Hence

$$\int_{I_{d_0}} \exp\left\{\frac{a\varepsilon}{2}\right\} d\xi < \int_{I_{d_0}} \exp\left\{a\left[\sum_{j=1}^{n} k(\xi, y_j^0) z_j^0 - v^0\right]\right\} \leqq 2, \tag{9}$$

since the game Γ is of (m, n)-type and thus system (3) has a solution, say $(\bar{\mathbf{x}}, \bar{\mathbf{y}}, \bar{\mathbf{u}}, \bar{\mathbf{z}}, \bar{v})$ which in turn implies by the optimality of $(\mathbf{x}^0, \mathbf{y}^0, \mathbf{u}^0, \mathbf{z}^0, v^0)$

$$\Phi(\mathbf{x}^0, \mathbf{u}^0, \mathbf{y}^0, \mathbf{z}^0, \mathbf{v}^0, a) \leqq \Phi(\bar{\mathbf{x}}, \bar{\mathbf{u}}, \bar{\mathbf{y}}, \bar{v}, a) \leqq 2 .$$

On the other hand, using estimation (5) we get

$$\int_{I_{d_0}} \exp\left\{\frac{a\varepsilon}{2}\right\} d\xi \geqq \int_{I_{d_0}} \frac{2}{\rho} d\xi \geqq 2$$

contradicting (9).

Proving that

$$\sum_{i=1}^{m} k(x_i^0, \eta) u_i^0 \geqq v^0 - \varepsilon \quad \text{for all} \quad \eta \in [0, 1)$$

is done in the same way. Thus $(\mathbf{x}^0, \mathbf{y}^0, \mathbf{u}^0, \mathbf{z}^0, v^0)$ provides "ε optimal" strategies and an "ε good" value of game Γ. ∎

COROLLARIES

1. If $\varepsilon_k \to 0$, then any limit point of the sequence consisting of optimal solutions to (4) provides an equilibrium point and the value of game Γ.

2. Formula (5) can be used to estimate parameter a for a given ε.

It deserves mentioning that the number of variables in (4) depends only on m and n. If we want to increase accuracy, then only parameter a should be increased. In spite of all this, solving (4) for large enough m and n is not an easy job at all. This is primarily so because the objective function is not convex and local, but not global optima occur. However we have a good sufficient condition to check whether a particular local optimum is what we are looking for. All we have to do is substitute it into (3) and see that the constraints are not violated by more than ε. In this way local methods (such as gradient methods) can be applied using several different starting points. Of course there is no guarantee that this approach will always be successful.

It is worth noting that problem (4) is a convex program if the kernel of the game $k(\xi, \eta)$ is separable. We have seen in Chapter 17 that every separable game $\Gamma=\{X, Y, K\}$ with kernel

$$k(\xi, \eta)=\sum_{i=1}^{m} \sum_{j=1}^{n} a_{ij} r_i(\xi) s_j(\eta)$$

can be converted to a game $H=\{U, W, L\}$ where the strategy sets U, W are closed, bounded and convex and L is bilinear (see page 200) with matrix $\mathbf{A}=[a_{ij}]$. For finding an "ε-optimal" strategy pair to H we can apply the SUMT-method discussed above. This amounts to solving the following convex program:

$$\Psi(\mathbf{u}, \mathbf{w}, v)=\int_0^1 \exp\{a[v-\mathbf{u}\mathbf{A}\mathbf{s}(\eta)]\}\, d\eta+$$

$$+\int_0^1 \exp\{a[\mathbf{r}(\xi)\mathbf{A}\mathbf{w}-v]\}\, d\xi \to \min$$

$$\mathbf{u} \in U$$

$$\mathbf{w} \in W$$

where

$$\mathbf{s}(\eta)=(s_1(\eta), \ldots, s_n(\eta))$$

$$\mathbf{r}(\xi)=(r_1(\xi), \ldots, r_m(\xi))$$

and a is chosen according to (5).

To illustrate the method we give a numerical example. (We stress that the example is illustrative, in this special case other methods can be more effective.)

Solve the two-person zero-sum game $\Gamma=\{X, Y, K\}$ where the strategy-set of each player is the set of all distribution functions defined on the interval $\left[-\frac{\pi}{2}, \frac{\pi}{2}\right]$ and the kernel of the game is

$$k(\xi, \eta)=\sin(\xi-\eta)=\sin\xi \cos\eta - \cos\xi \sin\eta .$$

Since the game is symmetric we know that the value of the game is 0 and the set of optimal strategies for both players coincide. Using our standard notation we get

$$r_1(\xi) = \sin \xi \qquad s_1(\eta) = \cos \eta$$

$$r_2(\xi) = \cos \xi \qquad s_2(\eta) = \sin \eta$$

$$\mathbf{A} = \begin{bmatrix} 1 & 0 \\ 0 & -1 \end{bmatrix}.$$

It can easily be proved that the set U (and W) is the semicircle defined by the inequalities

$$0 \leqq u_1 \leqq 1$$

$$-\sqrt{1-u_1^2} \leqq u_2 \leqq \sqrt{1-u_1^2}\,.$$

Thus the programming problem to be solved takes the form:

$$\text{(10)} \qquad \Psi(u_1, u_2) = \int_{-\frac{\pi}{2}}^{\frac{\pi}{2}} \exp\{-a(u_1 \cos \eta - u_2 \sin \eta)\}\, d\eta \to \min$$

$$0 \leqq u_1 \leqq 1$$

$$-\sqrt{1-u_1^2} \leqq u_2 \leqq \sqrt{1-u_1^2}\,.$$

It is easy to verify that $a = 60$ is a proper choice for parameter a if the desired accuracy $\varepsilon = 0.1$.

For solution of (10) we get $u_1 = 1$, $u_2 = 0$. To find the equilibrium points of the original game Γ we have to solve the equation system for $p \geqq 0$

$$\text{(11)} \qquad 1 = p \sin x + (1-p) \sin y \qquad -\frac{\pi}{2} \leqq x \leqq \frac{\pi}{2},$$

$$0 = p \cos x + (1-p) \cos y \qquad -\frac{\pi}{2} \leqq y \leqq \frac{\pi}{2},$$

since the spectrum of both players consists of no more than two points. The unique solution of (11) is $p = 1$, $x = \frac{\pi}{2}$. This means that both players have to play the point $\frac{\pi}{2}$ with probability 1.

19. Two-person zero-sum games over metric spaces

For games defined on abstract spaces topological spaces provide the most general framework. Although metric spaces are less general, they can more easily be interpreted and in practical applications the generality offered by them is sufficient in most cases.

In this section we will consider two-person zero-sum games where the strategy sets are compact subsets of metric spaces and the pay-off function is continuous and convex-concave.[1] Since the strategy sets are not assumed to be subsets of a linear space a more general definition of convexity-concavity is needed.

DEFINITION. Let X and Y be arbitrary sets and define a real-valued function f on $X \times Y$. The function f is said to be *concave* on X if for any pair of elements $x_1, x_2 \in X$ and nonnegative real numbers $\xi_1, \xi_2, (\xi_1+\xi_2=1)$ an $x_0 \in X$ can be found to satisfy $f(x_0, y) \geqq \xi_1 f(x_1, y)+\xi_2 f(x_2, y)$ for all $y \in Y$.

Similarly f is defined to be *convex* on Y if for any $y_1, y_2 \in Y$ and nonnegative real numbers $\eta_1, \eta_2, (\eta_1+\eta_2=1)$ there can be found a $y_0 \in Y$ to satisfy $f(x, y_0) \leqq \eta_1 f(x, y_1)+\eta_2 f(x, y_2)$ for all $x \in X$.

If f is concave on X and convex on Y it is called *concave-convex*.

It can very easily be seen that the concave (convex) functions defined on convex subsets of linear spaces meet the requirements of the above definition.

[1] In Chapter 3 we already considered concave n-person games over linear metric spaces. However in this chapter no linear structure of the spaces involved is assumed.

Let us consider a two-person, zero-sum game $\Gamma=\{X, Y, f\}$ where X and Y are compact subsets of metric spaces and f is a bounded, continuous concave-convex function on $X\times Y$. In order to be able to show the existence of an equilibrium pair of strategies for Γ we have to prove a lemma.

LEMMA 1. If f is a concave-convex function on $X\times Y$ where X and Y are arbitrary sets, then for any finite sets $\{x_1, \ldots, x_m\}\subset X$, $\{y_1, \ldots, y_n\}\subset Y$ there can be found elements $x_0\in X$ and $y_0\in Y$ to satisfy the inequalities

$$\text{(1)} \qquad f(x_0, y_j)\geqq f(x_i, y_0), \quad (i=1, \ldots, m; \quad j=1, \ldots, n).$$

Proof. Let us consider the matrix game given by $\mathbf{A}=[a_{ij}]$, $a_{ij}=f(x_i, y_j)$. The matrix game defined by $\mathbf{A}^T$ is always solvable by Theorem 1 of Chapter 9. Let $\boldsymbol{\xi}$ and $\boldsymbol{\eta}$ be optimal strategies. By the definition of optimal strategies we have

$$\text{(2)} \qquad \max_{1\leqq i\leqq m} \sum_{j=1}^{n} \eta_j f(x_i, y_j)\leqq \min_{1\leqq j\leqq n} \sum_{i=1}^{m} \xi_i f(x_i, y_j).$$

Since f is concave-convex there exist $x_0\in X$, $y_0\in Y$ to satisfy

$$\text{(3)} \qquad f(x_i, y_0)\leqq \sum_{j=1}^{n} \eta_j f(x_i, y_j), \quad (i=1, \ldots, m)$$

$$\text{(4)} \qquad f(x_0, y_j)\geqq \sum_{i=1}^{m} \xi_i f(x_i, y_j), \quad (j=1, \ldots, n).$$

Thus (2), (3), (4) imply (1). ∎

THEOREM 1. [50] Game $\Gamma=\{X, Y, f\}$ is always solvable, i.e.,

$$\text{(5)} \qquad \max_{x\in X} \min_{y\in Y} f(x, y)= \min_{y\in Y} \max_{x\in X} f(x, y).$$

Proof. Both sides of (5) are well defined because a bounded, continuous function assumes its maximum and minimum on a compact set and the functions $\Phi(y)= \max f(x, y)$, $\Psi(x)= \min_{y\in Y} f(x, y)$ are continuous.

By Lemma 1, for any finite subset of X and Y the inequality

$$\text{(6)} \qquad \min_{y \in Y} \max_{1 \leq i \leq m} f(x_i, y) \leqq \max_{x \in X} \min_{1 \leq j \leq n} f(x, y_j)$$

holds. Therefore any real number α satisfies at least one of the following inequalities

$$\min_{y \in Y} \max_{1 \leq i \leq m} f(x_i, y) \leqq \alpha ,$$

$$\max_{y \in Y} \min_{1 \leq i \leq n} f(x_i, y) \geqq \alpha .$$

Let us now define the sets

$$L(x, \alpha) = \{y \in Y | f(x, y) \leqq \alpha\} ,$$

$$U(y, \alpha) = \{x \in X \,| f(x, y) \geqq \alpha\} .$$

By the continuity of f sets L and U are closed. Because of (6) at least one of the sets $\bigcap_{i=1}^{m} L(x_i, \alpha)$ and $\bigcap_{j=1}^{n} U(y_j, \alpha)$ is nonempty. L and U are closed therefore one of the sets $\bigcap_{x \in X} L(x, \alpha)$ and $\bigcap_{y \in Y} U(y, \alpha)$ is nonvoid by Cantor's intersection theorem. In other words, if there is no $y_0 \in Y$ for which $f(x, y_0) \leqq \alpha$ holds for any $x \in X$, then there exists an $x_0 \in X$ to satisfy $f(x_0, y) \geqq \alpha$ for any $y \in Y$. Thus any real number α satisfies one of the following inequalities

$$\min_{y \in Y} \max_{x \in X} f(x, y) \leqq \alpha ,$$

$$\max_{x \in X} \min_{y \in Y} f(x, y) \geqq \alpha .$$

Hence

$$\text{(7)} \qquad \min_{y \in Y} \max_{x \in X} f(x, y) \leqq \max_{x \in X} \min_{y \in Y} f(x, y) .$$

By Theorem 1 of Chapter 7, the opposite of (7) always holds implying the validity of (5). ∎

It is not too difficult to see that as special cases of Theorem 1 we get for $n = 2$ and zero-sum games

(i) Theorem 2 of Chapter 3, if X and Y are closed, bounded convex subsets of the Euclidean space, and no neighbourhood functions are considered,

(ii) Theorem 3 of Chapter 3, if X and Y are the sets of probability vectors of proper dimension,

(iii) Theorem 6 of Chapter 3, if X a Y is the family of all distribution functions defined on the interval $[0, 1]$.

In addition to these cases, Theorem 1 has a bearing on a large class of *games played over function spaces.* Without going thoroughly into the problem we describe a simple market situation which can be viewed as a two-person zero-sum game over function spaces.

Two companies A and B which are dominating the market sell a particular product. In the time interval $[0, 1]$ the marginal supply of A and B is given by the integrable functions $f(t)$ and $g(t)$ resp. Thus the amount offered for sale by A and B in the interval $[d, d+h]$ can be written as $\int_{d}^{d+h} f(t)\,dt$ and $\int_{d}^{d+h} g(t)\,dt$ resp. Denote by $p(t)$ and $q(t)$ the price of A's and B's products resp. at time t. The price must not exceed P. In the interval $[0, P]$ both companies ("players") choose a price independently, subject to the only restriction that by selling their whole supply $\int_0^1 f(t)\,dt$ and $\int_0^1 g(t)\,dt$ the total income must be at least K_a and K_b resp. We assume that selling everything at a price not greater than P is no problem.

The primary goal of the players is to squeeze the opponent out of the market by setting low prices. Too low prices cannot be maintained for a long time because of the lower bound on the income. Therefore the choice of a proper pricing policy is a real decision problem. The players should choose such price functions $p(t)$, $q(t)$ which assure the largest average market share. Denote $k(p(t), q(t))$ the relative market share of player A, $(0 \leqq k(p(t), q(t)) \leqq 1$, $0 \leqq t \leqq 1)$. Obviously the share of B is $1-k(p(t), q(t))$ and the game is constant-sum, i.e., essentially zero-sum. This game is in fact a model of a two-person competition where both companies apply "dumping-policy".

Let us now formulate the game in normal form: $\Gamma=\{X,Y,f\}$ where

$$X=\{p(t)\in F \mid 0\leqq p(t)\leqq P, \int_0^1 p(t)f(t)\,\mathrm{d}t\geqq K_a\}$$

$$Y=\{q(t)\in F \mid 0\leqq q(t)\leqq P, \int_0^1 q(t)g(t)\,\mathrm{d}t\geqq K_b\}$$

$$f[p(t), q(t)]=\int_0^1 k[p(t), q(t)]\,\mathrm{d}t,$$

where F is a properly chosen subset of the space of continuous functions on [0, 1].

By defining a suitable metric on F, e.g., if the distance between $r(t)$ and $s(t)$ is defined by

$$\rho(r,s)=\sup_{0\leqq t\leqq 1} |r(t)-s(t)|$$

and assuming $f(p,q)$ to be a continuous concave-convex function the assumptions of Theorem 1 are met provided F is defined in such a way as to make X and Y compact.

Unfortunately there is no general method for finding equilibrium points of games over function spaces. For some special cases (mainly military applications) there are methods, but these rely heavily on the special features of the problem (see [79]). As a general approach we may approximate the particular function space by a suitable subspace of finite dimension and then the available methods designed for finite dimensional spaces can be applied.

In the above example F can be approximated by the space of polynomials of degree not more than n. In this way the game reduces to a finite dimensional concave-convex game which can be handled computationally for not too large an n.

If the two-person zero-sum game over a function space is not concave-convex, then there is no guarantee that an equilibrium point exists. To introduce mixed strategies on function spaces involves special mathematical difficulties and is beyond the scope of our book.

20. Sequential games

When defining mixed extensions of games we usually think of the original game played several times. The players choose their pure strategies according to the distribution given by the mixed strategies. However, in each play we think in terms of the original game with the same strategy sets and pay-off functions. Sometimes the strategy choice in the kth play of the game has an impact on the strategy sets and the pay-off functions of the next play. Considering the whole sequence of plays as a "supergame" we can speak of a sequential game. For the sake of simplicity in our formal definition of a sequential game we will restrict ourselves to zero-sum two-person games.

A straightforward extension to the general case can be carried out very easily.

A sequence $\Gamma = \{\Gamma^{(1)}, \Gamma^{(2)}, \ldots, \Gamma^{(k)}, \Gamma^{(k+1)}, \ldots\}$ of two-person zero-sum games is called a *deterministic sequential two-person zero-sum game* if

$$\Gamma^{(k)} = \{\Sigma_1^{(k)}, \Sigma_2^{(k)}, f^{(k)}(\sigma_1, \sigma_2)\}, \quad (k = 1, 2, \ldots)$$

and

$$\Sigma_1^{(k+1)} = h_1^{(k)}(\bar{\sigma}_1^{(k)}, \bar{\sigma}_2^{(k)}), \Sigma_2^{(k+1)} = h_2^{(k)}(\bar{\sigma}_1^{(k)}, \bar{\sigma}_2^{(k)}),$$

$$f^{(k+1)} = g^k(\bar{\sigma}_1^{(k)}, \bar{\sigma}_2^{(k)})$$

provided the players used strategies $\bar{\sigma}_1^{(k)}, \bar{\sigma}_2^{(k)}$ in the subgame $\Gamma^{(k)}$, $(k = 1, 2, \ldots)$. (The functions $h_1^{(k)}$, $h_1^{(k)}$ and $g^{(k)}$ are set-valued and function-valued resp.)

If the functions $h_1^{(k)}$, $h_2^{(k)}$, $g^{(k)}$ determine the strategy sets and pay-off functions of $\Gamma^{(k+1)}$ only in a stochastic way, then we call Γ a *stochastic sequential two-person zero-sum game.*

In the following we are going to study two types of stochastic sequential games: (i) Shapley's stochastic game and (ii) recursive games. Both can be models of conflict situations where a "third party" (which can be conceived of as a random mechanism) has influence on the pay-off functions of the subgames.

20.1. SHAPLEY'S STOCHASTIC GAME

Assume we have N matrix games $\mathbf{A}^1, \mathbf{A}^2, \ldots, \mathbf{A}^N$, where $\mathbf{A}^k$ is of size m_k, n_k $(k=1, \ldots, N)$. The sequential game Γ is played as follows. When the game is in the kth position, i.e. the matrix game $\mathbf{A}^k$ is being played, and the players use their ith and jth pure strategies resp., then the probability of termination of the game is $s_{ij}^k > 0$ while the probability of transition to matrix game $\mathbf{A}^l$ is p_{ij}^{kl}. Of course $s_{ij}^k + \sum_{l=1}^{N} p_{ij}^{kl} = 1$. The pay-off to the players is the sum of the pay-offs obtained in the subgames. Depending on the starting position we get different games. Denoting the sequential game initialized from the kth position by $\Gamma^{(k)}$, the N-tuple of games $\Gamma = \{\Gamma^{(1)}, \ldots, \Gamma^{(k)}, \ldots, \Gamma^{(N)}\}$ is called *Shapley's stochastic game.*

When analyzing stochastic sequential games we are primarily interested in *stationary strategies.* An N-tuple of strategies $\mathbf{x} = (\mathbf{x}^1, \mathbf{x}^2, \ldots, \mathbf{x}^N)$ is called a stationary strategy of player I if he applies the mixed strategy $\mathbf{x}^k$ whenever subgame $\mathbf{A}^k$ is being played, $(k=1, 2, \ldots, N)$ no matter how the game got to the kth position. Player II's stationary strategies can similarly be defined. Let $\underline{v}^k$ be the greatest lower bound of the expected pay-off player I can guarantee himself and $\bar{v}^k$ the least upper bound player II can be sure of getting by proper choice of stationary strategies. In fact $\underline{v}^k$ and $\bar{v}^k$ are the sup-inf and inf-sup value of the game Γ^k resp., $(k=1, \ldots, N)$. The optimal (equilibrium) stationary strategies of the game Γ^k are those ensuring expected pay-offs $\underline{v}^k$ and

$\bar{v}^k$ resp. We say that Shapley's stochastic game Γ is *solvable* if Γ^k is solvable for each k, $(k=1, \ldots, N)$. Γ^k is said to be solvable if $\underline{v}^k = \bar{v}^k$ and both players have optimal stationary strategies. We will prove that Shapley's stochastic game is always solvable.

Let

$$s = \min_{k,i,j} s_{ij}^k ,$$

$$M = \max_{k,i,j} |a_{ij}^k|$$

where $\mathbf{A}^k = [a_{ij}^k]$, $(k=1, \ldots, N)$. Since s is positive, the game terminates after a finite number of steps. The expected pay-off of player I is bounded from above by

$$(1) \qquad M + (1-s)M + (1-s)^2 M + \ldots = \frac{M}{s} .$$

Given a matrix game $\mathbf{A}$ we will denote its value by $v(\mathbf{A})$, the optimal strategies by $X(\mathbf{A})$ and $Y(\mathbf{A})$. In the following we use the simple inequality

$$(2) \qquad |v(\mathbf{B}) - v(\mathbf{C})| \leqq \max_{i,j} |b_{ij} - c_{ij}|$$

where $\mathbf{B} = [b_{ij}]$, $\mathbf{C} = [c_{ij}]$. Let $\mathbf{A}^k(\boldsymbol{\alpha})$ denote a matrix depending on the N-vector of parameters $\boldsymbol{\alpha}$ the entries of which are

$$a_{ij}^k + \sum_{l=1}^{N} p_{ij}^{kl} \alpha^l, \quad (i=1, \ldots, m_k; j=1, \ldots, n_k) .$$

Let $\boldsymbol{\alpha}_{(0)}$ be arbitrary and define

$$\boldsymbol{\alpha}_{(t)} = (\alpha_{(t)}^1, \ldots, \alpha_{(t)}^k, \ldots, \alpha_{(t)}^N), \quad (t=1, 2, \ldots)$$

$$\alpha_{(t)}^k = v[\mathbf{A}^k(\boldsymbol{\alpha}_{t-1})] .$$

THEOREM 1. [159] The sequence $\{\alpha_{(t)}\}$ is convergent. The limit point $\lim_{t\to\infty} \boldsymbol{\alpha}_{(t)}$ is independent of $\boldsymbol{\alpha}_{(0)}$ and its components are the values of the games $\Gamma^{(k)}$, $(k=1, \ldots, N)$.

Proof. Consider transformation T

(3) $\quad T\boldsymbol{\alpha}=\boldsymbol{\beta}$

where $\beta^k = v[\mathbf{A}^k(\boldsymbol{\alpha})]$.

Define the norm of $\boldsymbol{\alpha}$ by

$$\|\boldsymbol{\alpha}\| = \max_{1 \leqq k \leqq N} |\alpha^k|.$$

Then using (2) we get

$$\|T\boldsymbol{\beta} - T\boldsymbol{\alpha}\| = \max_{1 \leqq k \leqq N} |v[\mathbf{A}^k(\boldsymbol{\beta})] - v[\mathbf{A}^k(\boldsymbol{\alpha})]| \leqq$$

$$\leqq \max_{k,i,j} \left| \sum_{l=1}^{N} p_{ij}^{kl} \beta^l - \sum_{l=1}^{N} p_{ij}^{kl} \alpha^l \right| \leqq$$

$$\leqq \max_{k,i,j} \left| \sum_{l=1}^{N} p_{ij}^{kl} \right| \max_{1 \leqq l \leqq N} |\beta^l - \alpha^l| = (1-s)\,\|\boldsymbol{\beta} - \boldsymbol{\alpha}\|.$$

Denoting $T^2 A = T(T\boldsymbol{\alpha})$ we have

$$\|T^2\boldsymbol{\alpha} - T\boldsymbol{\alpha}\| \leqq (1-s)\,\|T\boldsymbol{\alpha} - \boldsymbol{\alpha}\|.$$

Thus the sequence $\boldsymbol{\alpha}_{(0)}$, $T\boldsymbol{\alpha}_{(0)}$, $T^2\boldsymbol{\alpha}_{(0)}$, ... is convergent since $(1-s)<1$. Let $\boldsymbol{\varphi}$ be the limit point. It is easily seen that $\boldsymbol{\varphi} = T\boldsymbol{\varphi}$, i.e., $\boldsymbol{\varphi}$ is a fixed point of the transformation T. This fixed point is unique since if there were another fixed point, say $\boldsymbol{\psi} = T\boldsymbol{\psi}$, then

$$\|\boldsymbol{\psi} - \boldsymbol{\varphi}\| = \|T\boldsymbol{\psi} - T\boldsymbol{\varphi}\| \leqq (1-s)\,\|\boldsymbol{\psi} - \boldsymbol{\varphi}\|,$$

implying $\|\boldsymbol{\psi} - \boldsymbol{\varphi}\| = 0$ and $\boldsymbol{\varphi} = \boldsymbol{\psi}$. The fixed point $\boldsymbol{\varphi}$ is independent of the initial vector $\boldsymbol{\alpha}_{(0)}$.

Let $(\mathbf{x}_0, \mathbf{y}_0)$ be a pair of stationary strategies for which $\mathbf{x}_0^l \in X[\mathbf{A}^l(\boldsymbol{\varphi})]$, $\mathbf{y}_0^l \in Y[\mathbf{A}^l(\boldsymbol{\varphi})]$, $(l=1, \ldots, N)$. We will now prove that $(\mathbf{x}_0, \mathbf{y}_0)$ is a pair of optimal stationary strategies for any game Γ^k and the value of Γ^k is φ^k, $(k=1, \ldots, N)$.

Let us consider a "finite truncation" of the game Γ^k, i.e., we assume that at step t when the game is in the position h player I gets $a_{ij}^h + \sum_{l=1}^{N} p_{ij}^{hl} \varphi^l$ instead of a_{ij}^h and the game terminates.

In the "finite" game Γ_t^k thus obtained $\mathbf{x}_0$ is an optimal strategy of player I, and φ^k is the value of the game. To prove this assertion we proceed by induction. For $t=1$ the assertion is obviously valid since $\boldsymbol{\varphi}$ is a fixed point of transformation T and $\mathbf{x}_0$ is optimal for $\Gamma_1^{(k)}$, $(k=1, \ldots, N)$ by definition. (Apparently $\Gamma_1^{(k)}$ is an ordinary matrix game.) Let us now assume that $\mathbf{x}_0$ is optimal in the game $\Gamma_t^{(k)}, (t \geqq 1)$ and the value of the game is φ^k. In game $\Gamma_{t+1}^{(k)}$ after step t if the game goes to the lth position, player I gets at least φ^l since $\boldsymbol{\varphi}$ is a fixed point of the transformation T. Thus truncating the game by omitting the last step and giving φ^l to player I after step t we arrive at game $\Gamma_t^{(k)}$. But by induction in game $\Gamma_t^{(k)}$ player I can assure himself at least φ^k by playing $\mathbf{x}_0$. Thus $\mathbf{x}_0$ is a maximin strategy.

It can be proved similarly that $\mathbf{y}_0$ is a minimax strategy and therefore φ^k is the value of game $\Gamma_t^{(k)}$ for any $t \geqq 1$.

In our original game $\Gamma^{(k)}$, by playing $\mathbf{x}_0$, player I's expected pay-off after step t is at least

$$\varphi^k-(1-s)^{t-1} \max_{h,i,j} \sum_{l=1}^{N} p_{ij}^{hl} \varphi^l \tag{4}$$

if before step t we were in the hth position. (4) is bounded from above by

$$\varphi^k-(1-s)^t \max_{1 \leqq l \leqq N} \varphi^l .$$

Thus player I's expected pay-off in game Γ^k is at least

$$\varphi^k-(1-s)^t \max_{1 \leqq l \leqq N} \varphi^l-(1-s)^t \frac{M}{s} .$$

Letting $t \to \infty$ the stationary strategy $\mathbf{x}_0$ ensures at least φ_k for player I.

In the same way we can prove that strategy $\mathbf{y}_0$ holds down the loss of player II at a level not exceeding φ^k. ∎

The proof of Theorem 1 has been constructive thereby providing a method for finding the fixed point of transformation T. Of course, this method is not necessarily finite.

20.2. RECURSIVE GAMES

Recursive games differ from Shapley's stochastic game in two main points.

1. Both players are paid off only at the end of the game.
2. The probability of termination s_{ij}^k is not necessarily positive.

Thus – using the same notation as before – if the game is in the kth position where the players apply their ith and jth pure strategy resp., then the game ends with probability s_{ij}^k and player I receives a_{ij}^k from player II. The probability that the game gets into the lth position is p_{ij}^{kl} and there is no pay-off.

If the game goes on infinitely long, the pay-off is 0.

We state the following existence theorem without proof. For details the interested reader is referred to [129].

THEOREM 2. [129] A recursive matrix game $\Gamma = \{\Gamma^{(1)}, \ldots, \Gamma^{(N)}\}$ has a value for each possible starting matrix and there exist ε-optimal stationary strategies for each player. Furthermore, the same strategy works for any starting matrix.

Good examples of recursive games are the so-called *games of survival*. Let us suppose that two players begin to play a game with initial sums R_1 and R_2. At each run of the game they play a matrix game $\mathbf{A} = [a_{ij}]$ and player II pays a_{ij} to player I. The matrix game is repeated as long as one of the players runs out of money.

This game can be thought of as a recursive game if $\mathbf{A}$, R_1, R_2 are integers ($R_1 \geqq 1$, $R_2 \geqq 1$). Let us consider the recursive matrix game which has $R-1$ ($R = R_1 + R_2$) component games Γ^k, ($k = 1, \ldots, R-1$), where k denotes the actual amount of money at player I's disposal. If in the kth position Γ^k the players pick their ith and jth strategy resp., then the game proceeds as follows:

(a) go to $\Gamma^{k+a_{ij}}$, if $1 \leqq k + a_{ij} \leqq R-1$,

(b) the game ends and player I gets 0, if $k + a_{ij} \leqq 0$,

(c) the game ends and player I gets 1, if $k + a_{ij} \geqq R$. Note that this game is player II's survival game, player I is going to ruin him.

More can be found about games of survival in [98], [115].

21. Games against nature

The theory of games is concerned with analyzing conflict situations where the participants of the game are conscious, rational entities seeking to reach (or approach) some objective. Very often one of the participants (players) cannot be regarded as a conscious individual having his own preferences and objectives. Thus the other players cannot count on his rational behaviour.

The participant representing the forces uncontrollable by the players usually is referred to as "nature". Therefore games of this type are called *games against nature*.

In the strict sense of the word games against nature do not fit into the subject of game theory. However, because of their practical importance they deserve a few words.

We are going to deal with only "two-person games" where player I is called the Decision Maker (DM) and player II is "nature". Both players are assumed to have finitely many pure strategies and the m by n "pay-off" matrix $\mathbf{A}=[a_{ij}]$ is known. Of course the pay-off matrix is meaningful only for player I and a_{ij} is assumed to represent the gain obtained by player I if he applies his ith strategy while nature is in state j. (Nature's strategies should rather be called "states" instead of strategies.)

It is our basic assumption that nothing is known about the probability distribution governing nature's "selection" of states. Statistical decision theory deals with decision making problems where these probabilities (objective or subjective alike) come into the picture. The reader can find many a good book on statistical decision theory (e.g., [98], [155], [199]).

How should a rational DM choose his strategy in a game against nature? This question cannot be answered definitely. It depends on the criteria the DM applies, on his attitude towards risk, on his ideas about gain and loss, etc. We mention four well-known and simple concepts for strategy selection.

1. *Laplace's criterion.* According to Laplace all states of nature should be regarded as equally probable because nothing is known about the real probabilities. Thus the DM has to choose strategy i^* if

$$\sum_{j=1}^{n} \frac{1}{n} a_{i^*j} \geqq \frac{1}{n} \sum_{j=1}^{n} a_{ij}, \quad (i=1, \ldots, m).$$

Among the severe conceptual deficiencies of the idea we only mention its sensitivity to the definition of possible states.

2. *Wald's (maximin) criterion.* This criterion is that of a pessimist. Nature is supposed to be acting against the DM who adopts a *maximin strategy*, i.e., an optimal strategy of player I in a matrix game **A** against nature as player II.

There are situations where using this criterion is justified (e.g., choosing medication for someone whose disease is not exactly diagnosed) but in most cases the DM is willing to take some risk in the hope of increasing his gain.

3. *Maximax criterion.* As opposed to Wald's this criterion corresponds to the philosophy of an optimist whose risk aversion is very low. The DM picks that row (strategy) of the matrix **A**, where the maximal element is maximal, therefore the name maximax.

4. *Hurwicz's criterion.* This approach is a kind of combination of the maximin and maximax criterion. Hurwicz defines a number α between 0 and 1 called "the index of optimism" which is supposed to measure the attitude of the DM toward risk. If $\alpha = 1$, then the DM is most optimistic, if $\alpha = 0$ then he is most pessimistic, i.e., he applies Wald's criterion. Let s_i and S_i be the least and greatest elements resp. in row i of **A** and define

$$h_i = \alpha S_i + (1-\alpha) s_i, \quad (i=1, \ldots, m).$$

Then the DM chooses the strategy giving the greatest h_i. Thus, e.g., in the following game

		Nature 1	2
DM	1	1	2
	2	0	1000

the DM has to choose strategy 1 by the maximin criterion but if $\alpha > \frac{1}{999}$, then Hurwicz tells him to choose strategy 2.

Hurwicz's criterion can be criticized first of all in that the choice of α is very subjective. In addition it does not meet other "rationality" requirements which a decision making criterion is supposed to satisfy (e.g., if $\alpha > 0$, then the convex combination of two optimal decisions is not necessarily optimal).

5. *Savage's criterion.* Savage proposes to set up a "regret matrix", $\mathbf{R} = [r_{ij}]$. r_{ij} is the loss incurred when the DM chooses his ith strategy and nature is in state j. Then, in the matrix game $\mathbf{R}$ the DM should apply the maximin (Wald's) criterion.

In our previous example the regret matrix is the following

		Nature 1	2
DM	1	0	998
	2	1	0

DM's (mixed) maximin strategy is $\left(\frac{1}{999}, \frac{998}{999}\right)$. The weak point of this criterion is the concept of "regret". In reality decision makers do not generally think in terms of "opportunity losses"; rather they think in terms of "real" gains and losses.

Since none of these five criteria is satisfactory in all respects, the question arises whether there exist criteria meeting a system of requirements concerning the behaviour of a "rational" decision maker. This axiomatic approach of decision theory is used by Luce and Raiffa. For details the interested reader should consult their book [98].

To illustrate situations which can be looked upon as games against nature let us consider the following example.

A distributor is commissioned by a customer to supply an equipment consisting of three main parts of high value.

In their contract it was specified that

(a) if the equipment supplied is faultless, then amount α is to be paid to the distributor,

(b) if the equipment is defective, then a forfeit β is due for the customer.

The distributor can examine the equipment when taking it over from the factory. However this quality control costs money. In particular, examining a main part costs γ. The distributor has no prior knowledge of the probabilities according to which the main parts can be defective. Therefore his decision problem can very well be regarded as a game against nature. The distributor takes into account four pure strategies:

1. to accept the equipment from the factory without quality control,

2. to pick a main part at random (choosing any of the main parts with probability $\frac{1}{3}$), accepting the equipment if this is faultless and sending it back if not,

3. to pick two main parts at random, accepting the equipment if both are faultless and sending it back if not, (of course if the main part examined first is defective, then there is no need to examine the second one),

4. to examine all three main parts and accepting the equipment only if each is faultless.

Nature has also four states (strategies): 0, 1, 2 or 3 if three main parts are defective. The pay-off matrix **A** can easily be determined

$$\mathbf{A}=\begin{bmatrix} \alpha & -\beta & -\beta & -\beta \\ \alpha-\gamma & -\frac{2}{3}\beta-\gamma & -\frac{1}{3}\beta-\gamma & -\gamma \\ \alpha-2\gamma & -\frac{1}{3}\beta-\frac{5}{3}\gamma & -\frac{4}{3}\gamma & -\gamma \\ \alpha-3\gamma & -2\gamma & -\frac{4}{3}\gamma & -\gamma \end{bmatrix}.$$

For example, if nature is in state 1 and the DM plays his third strategy, then the probability that he discovers the faulty part when examining the first main part is $\frac{1}{3}$. The probability of finding the defective one at the second examination is $\frac{2}{3}\cdot\frac{1}{2}=\frac{1}{3}$, thus the probability that the faulty part remains undetected is also $\frac{1}{3}$. The costs of the distributor in each case are $-\gamma$, -2γ and $-\beta-2\gamma$ resp. Thus the "pay-off" to the DM (the expected cost) is

$$\frac{1}{3}(-\gamma)+\frac{1}{3}(-2\gamma)+\frac{1}{3}(-\beta-2\gamma)=$$

$$=-\frac{1}{3}\beta-\frac{5}{3}\gamma.$$

The optimal strategies of the DM depend on the parameters α, β, γ and of course on the criterion he applies. Using Wald's criterion we get a matrix game. This game does not necessarily have an equilibrium point (a saddle point) in pure strategies for any value of the parameters α, β, γ. We are going to determine the ranges of the parameters where the matrix $\mathbf{A}$ has a saddle point.

(a) The first row has three minimal elements a_{12}, a_{13}, a_{14}.

(i) The element a_{12} provides a saddle point if it is maximal in its column, i.e.,

$$-\beta \geqq -\frac{2}{3}\beta-\gamma,$$

$$-\beta \geqq -\frac{1}{3}\beta-\frac{5}{3}\gamma,$$

$$-\beta \geqq -2\gamma,$$

which means $\beta \leqq 2\gamma$.

(ii) a_{13} is a saddle point if

$$-\beta \geqq -\frac{1}{3}\beta-\gamma,$$

$$-\beta \geqq -\frac{4}{3}\gamma$$

that is, $\beta \leqq \frac{4}{3}\gamma$.

(iii) a_{14} is obviously a saddle point if and only if $\beta \leqq \gamma$.

(b) The least element in the second row is a_{22}. This can be a saddle point only if

$$-\frac{2}{3}\beta-\gamma \geqq -\beta,$$

$$-\frac{2}{3}\beta-\gamma \geqq -\frac{1}{3}\beta-\frac{5}{3}\gamma,$$

$$-\frac{2}{3}\beta-\gamma \geqq -2\gamma.$$

But this inequality system is inconsistent for positive β and γ.

(c) a_{31} cannot be a saddle point since it is not maximal in its column. Thus in the third row a_{32} is the only candidate for a saddle point. But the inequality system

$$-\frac{1}{3}\beta-\frac{5}{3}\gamma \geqq -\beta,$$

$$-\frac{1}{3}\beta-\frac{5}{3}\gamma\geqq-\frac{2}{3}\beta-\gamma,$$

$$-\frac{1}{3}\beta-\frac{5}{3}\gamma\geqq-2\gamma,$$

$$-\frac{1}{3}\beta-\frac{5}{3}\gamma\leqq\alpha-2\gamma,$$

does not have a feasible solution for positive α, β, γ.

(d) Similarly in the fourth row only a_{42} can be a saddle point. The condition for this is

$$-2\gamma\geqq-\frac{1}{3}\beta-\frac{5}{3}\gamma,$$

$$-2\gamma\geqq-\frac{2}{3}\beta-\gamma,$$

$$-2\gamma\geqq-\beta,$$
$$-2\gamma\leqq\alpha-3\gamma,$$

which can be reduced to $\beta\leqq2\gamma$ and $\gamma\leqq\alpha$.

Summing up all cases:

(i) If $\beta>2\gamma$ and $\gamma>\alpha$, then there is no saddle point.

(ii) If $2\gamma\geqq\beta>\frac{4}{3}\gamma$, then there is a unique saddle point.

(iii) If $\beta\leqq\frac{4}{3}\gamma$, then there are two saddle points.

22. Cooperative games in characteristic function form

Throughout the previous chapters the players have been assumed to choose their strategies independently. Their performance in the game has been measured solely by the pay-off they got. However, in a number of games the "benefit" gained can be increased considerably if certain players more or less give up their independence. In other words, some of the players may form "coalitions" in order to increase their pay-off. Of course forming a coalition can generally be profitable only if the coalition assures more than the sum of the pay-offs the members can achieve playing alone. Players gathered in a coalition will not choose their strategies independently any more. They strive at maximizing the coalition's overall "profit" and possibly their own share.

Maintaining our hypothesis of rational players in the theory of cooperative games, too, our investigations will center around three questions:

1. Which of the theoretically possible coalitions are likely to realize?

2. How stable is the coalitional structure (system of coalitions) thus realized?

3. How is the "profit" gained by a coalition to be distributed among the members?

These three questions obviously are interrelated and will be studied later in detail.

In the noncooperative n-person game the behaviour (strategy) of $n-1$ players determined what the remaining nth player should do, provided he is "rational" and seeks to maximize his own pay-off.

In cooperative games change in a single player's behaviour (e.g., leaving a coalition and joining another one) may lead to a completely different coalitional structure.

Now we turn our attention to the mathematical formulation of the intuitive concepts discussed above. To this end we need a number of new definitions.

Let $N=\{1, \ldots, n\}$ be the *set of players* in an n-person game. A nonempty subset of N is called a *coalition*. We allow the "extreme" cases, i.e., coalitions consisting of a single player and the coalition of all the n players, the "*grand-coalition*".

If the game under consideration is (given in normal form) $\Gamma=\{\Sigma_1, \ldots, \Sigma_n; K_1, \ldots, K_n\}$, and $M \subset N$, $(M=\{i_1, \ldots, i_k\})$ is a coalition, then the strategy set of M is the Cartesian product $\Sigma_{i_1} \times \ldots \times \Sigma_{i_k}$ and the pay-off function is the sum of the individual functions $\sum_{\nu=1}^{k} K_{i_\nu}$.

Assume that a coalition M is formed. It is now expected that the remaining players also gather in a coalition $N-M$ since this is the most effective way to counterweigh coalition M. In this manner a two-person game played by coalitions M and $N-M$ arises. Of course other more general situations can also be visualized if the formation of more than two coalitions is permitted. Then we speak about a *coalition structure*

$$\mathfrak{C}=\{M_1, \ldots, M_m\}, \quad M_i \subset N, \quad M_i \cap M_k = \emptyset, \quad (i \neq k), \quad (i=1, \ldots, m)$$

where M_i denotes the set of players in the ith coalition.

We denote by $\{i\}$ the set (coalition) consisting of the single player i.

Consider now the case where the players split into coalitions M and $N-M$. We assign to M the maximin value of the two-person game played between M and $N-M$

(1) $$v(M) = \max_{\sigma_M} \min_{\sigma_{N-M}} K_M(\sigma_M, \sigma_{N-M}),$$

where $\sigma_M(\sigma_{N-M})$ varies over $\Sigma_{i_1} \times \ldots \times \Sigma_{i_k}, (\Sigma_{j_1} \times \ldots \times \Sigma_{j_{n-k}})$ and

$$K_M(\sigma_M, \sigma_{N-M}) = \sum_{\nu=1}^{k} K_{i_\nu}(\sigma_1, \ldots, \sigma_n)$$

and $M = \{i_1, \ldots, i_k\}$.

The function v which assigns to each coalition $M \subseteqq N$ the maximum value defined above is called the *characteristic function*[1] of the game Γ. It is convenient to define

(2) $$v(\emptyset) = 0.$$

Thus the characteristic function is defined for all subsets M of N, and $v(M)$ can be interpreted as the amount of pay-off the members of M can obtain, no matter what the remaining players do.

An immediate consequence of the above definition of the characteristic function is the inequality

(3) $$v(M \cup M') \geqq v(M) + v(M'),$$

$$M, M' \subset N, \quad M \cap M' = \emptyset$$

called the *superadditivity property.*

By induction we get the superadditivity property for arbitrarily many disjoint coalitions

$$v\left(\bigcup_k M_k\right) \geqq \sum_k v(M_k)$$

if $\bigcup_k M_k \subseteqq N, \quad M_k \cap M_l = \emptyset, \quad (k \neq l)$.

An n-person game is given in *characteristic function form* by a real-valued function v, defined on the subsets of N, satisfying conditions (2) and (3). As it appears from this definition the maximin origin of numbers $v(M)$ is of no importance any more. Putting it in a

[1] The fundamental concept of characteristic function was first introduced by von Neumann and Morgenstern [126].

precise mathematical framework, an n-person cooperative game Γ, in characteristic function form, is defined to be a pair (N, v), where $N=\{1, \ldots, n\}$ is the set of players and $v: 2^N \rightarrow \mathbb{R}$ is a function satisfying $v(\emptyset)=0$. Γ is said to be *superadditive* if in addition v fulfils condition (2). Games derived from the normal form by (1) are always superadditive but more general games in characteristic function form may exist which do not necessarily possess the superadditivity property. Even in this case $v(M)$ is interpreted as the value (utility) coalition M can achieve whatever the other players do.

A game $\Gamma=\{N, v\}$ is said to be *monotonic* if $S \supset T$ implies $v(S) \geqq v(T)$. It is clear that any superadditive game is monotonic but the converse is not true.

A game Γ given in characteristic function form is called *constant-sum* if

$$(4) \qquad v(M)+v(N-M)=v(N)$$

holds for any coalition $M \subset N$.

It is obvious that constant-sum games in characteristic function form are not necessarily constant-sum if given in normal form and vice versa.

We have already seen that n-person games in normal form always give rise to cooperative games in characteristic function form. The converse of this assertion generally is not true. However if the cooperative game Γ in characteristic function form is superadditive, then Γ can always be thought of as a game originated from an n-person non-cooperative game via the maximin principle (1).

THEOREM 1. [126] Let v be a real-valued function defined on subsets of $N=\{1, \ldots, n\}$ having the following properties:

(a) $v(\emptyset)=0$,

(b) for any two disjoint subsets $M, M' \subseteqq M$ the inequality

$$v(M)+v(M') \leqq v(M \cup M')$$

holds.

Then, there always exists an n-person game, whose characteristic function is v as defined by (1).

If, in addition

$$v(M)+v(N-M)=v(N)$$

for any $M\subset N$, then there exists an n-person constant-sum game with characteristic function v.

Proof. Our proof will be constructive. Let v be a function satisfying conditions (a) and (b). Consider the n-person game $\Gamma=\{\Sigma_1,\ldots,\Sigma_n;K_1,\ldots,K_n\}$, where Σ_i consists of those subsets $M\subseteqq N$, for which $i\in M$. The pay-off functions are defined as

$$(5)\qquad K_i(M_1,\ldots,M_n)=\begin{cases}\dfrac{1}{|M_i|}v(M_i), & \text{if } M_j=M_i \\ & \text{for all } \; j\in M_i \\ v(\{i\}) \quad \text{otherwise,} & \\ & (i=1,\ldots,n),\end{cases}$$

where $M_i\in\Sigma_i$ and $|M_i|$ denotes the number of players in M_i.

For the sake of notational simplicity we call player i a special player – or briefly *S-player* – with respect to the n-tuple of strategies $(M_1,\ldots,M_n)$ if $M_j=M_i$ for any $j\in M_i$.

Let $\bar{v}$ denote the characteristic function of the game Γ defined above. We will prove that $\bar{v}(M)=v(M)$ for any subset $M\subseteqq N$. In the trivial case $M=\emptyset$ we obviously have $\bar{v}(M)=v(M)=0$. Let $M\neq\emptyset$ and assume that $M_j=M$ for any $j\in M$, i.e., each player $j\in M$ has chosen strategy M. Consider now the two-person game

$$\Gamma_M=\left\{\mathop{\times}_{j\in M}\Sigma_j,\ \mathop{\times}_{k\in N-M}\Sigma_k;\ \sum_{j\in M}K_j,\ \sum_{k\in N-M}K_k\right\}.$$

Putting together the strategies $M_j=M,(j\in M)$ we get a strategy M^* of players gathered in M. Let L be an arbitrary strategy of the coalition $N-M$. Then by (5) we obtain

$$\sum_{j\in M}K_j(M^*,L)=\sum_{j\in M}\frac{1}{|M|}v(M)=v(M).$$

By definition of the characteristic function the inequality

(6) $\bar{v}(M) \geqq v(M)$

holds. For the players in $N-M$ we assume again that $M_k = N-M$ for any $k \in N-M$; i.e., each player $k \in N-M$ has picked strategy $N-M$. Let this be L^*. (Symbols M, $N-M$, M_j are used for denoting coalitions and strategies as well, depending on the context.) Let M' be an arbitrary strategy of the coalition M in game Γ_M and $j \in M$ an S-player with respect to the n-tuple of strategies (M', L^*). Consider an arbitrary $r \in M_j$. Then $r \in M$, because otherwise $r \in N-M$ would hold, implying $M_r = N-M$ by the definition of L^*. Thus $M_j = M_r = N-M$ contradicting the assumption $j \in M_j = M$. Consequently, if $j \in M$ is an S-player with respect to (M', L^*), then $M_j \subseteqq M$. It is also clear that if $j_1, j_2 \in M$ are S-players, then M_{j_1} and M_{j_2} are either identical or disjoint subsets of N.

According to the above analysis M can be partitioned as

$$M = M_{j_1} \cup M_{j_2} \cup \ldots \cup M_{j_\alpha} \cup \{r_1\} \cup \ldots \cup \{r_\beta\}$$

where $j_1, \ldots, j_\alpha$ are the S-players with respect to (M', L^*) and $r_1, \ldots, t_\beta$ are the remaining ones. By (5) the pay-off for coalition M in Γ_M is given by

$$\sum_{j \in M} K_j(M', L^*) = v(M_{j_1}) + \ldots + v(M_{j_\alpha}) +$$

$$+ v(\{r_1\}) + \ldots + v(\{r_\beta\}) .$$

By condition (b) the right-hand side of the above equality does not exceed $v(M)$, implying $\bar{v}(M) \leqq v(M)$. Taking into account (6) we have $\bar{v}(M) = v(M)$ for arbitrary $M \subseteqq N$.

In order to prove the second assertion of our theorem we define new pay-off functions:

$$K_i'(M_1, \ldots, M_n) = K_i(M_1, \ldots, M_n) +$$

$$+ \frac{1}{n}\left(v(N) - \sum_{i=1}^{n} K_i(M_1, \ldots, M_n)\right), \quad (i = 1, \ldots, n),$$

where functions K_i are defined by (5). Clearly $\sum_{i=1}^{n} K'_i = v(N)$, thus $\Gamma' = \{\Sigma_1, \ldots, \Sigma_n; K'_1, \ldots, K'_n\}$ is a constant-sum game. We will again show that the characteristic function $\bar{v}'$ of Γ' coincides with v. The equality $\bar{v}'(\emptyset) = v(\emptyset) = 0$ is trivial in this case, too.

Let $M \subsetneqq N$ be arbitrary and suppose that $M_j = M$ for any $j \in M$, i.e., all members of M are S-players. We can define Γ'_M analogously to Γ_M as

$$\Gamma'_M = \left\{ \mathop{\times}_{j \in M} \Sigma_j, \mathop{\times}_{k \in N-M} \Sigma_k; \sum_{j \in M} K'_j, \sum_{k \in N-M} K'_k \right\}.$$

Just as for Γ_M we get

$$\text{(7)} \qquad \sum_{j \in M} K'_j(M^*, L) = v(M) + \frac{|M|}{n} \left(v(N) - \sum_{i=1}^{n} K_i(M^*, L) \right),$$

where $M^* = (M, \ldots, M)$ is the first player's strategy, while L is an arbitrary strategy of player II (coalition $N - M$). By the same argumentation applied in the first part of the proof we can write M as a union of disjoint subsets

$$M = M_{k_1} \cup \ldots \cup M_{k_\varrho} \cup \{r_1\} \cup \ldots \cup \{r_\nu\}$$

where $k_1, \ldots, k_\rho$ are S-players with respect to the n-tuple of strategies (M^*, L), and $r_1, \ldots, r_\nu$ are the remaining ones.

Then by (5) and condition (b) we get

$$\sum_{i=1}^{n} K_i(M^*, L) = v(M_{k_1}) + \ldots + v(M_{k_\varrho}) +$$
$$+ v(\{r_1\}) + \ldots + v(\{r_\nu\}) \leqq v(M).$$

Using (7) we have the inequality

$$\sum_{j \in M} K'_j(M^*, L) \geqq v(M)$$

for arbitrary L implying

$$\text{(8)} \qquad \bar{v}'(M) \geqq v(M).$$

Since Γ'_M is constant-sum therefore $\bar{v}'(M)+\bar{v}'(N-M)=\bar{v}'(N)==v(N)$. Characteristic function v has been assumed to satisfy $v(M)+v(N-M)=v(N)$. Thus we get $\bar{v}'(M)+\bar{v}'(N-M)\leqq v(M)++v(N-M)$ which together with (8) means that $\bar{v}'(M)=v(M)$ for any $M\subseteqq N$. ∎

We have just seen that a characteristic function represents a whole class of games which may have different normal forms but the same characteristic function. In the following we mostly define games as $\Gamma=\{N, v\}$ where N is the player set and v is the characteristic function. Sometimes we simply speak of game v if it does not cause any confusion.

In a cooperative game, as we have already mentioned in the introduction, the question of distributing the total pay-off $v(N)$ plays a crucial role. This gives rise to the definition of an *imputation*. An imputation is a real n-vector $(x_1, \ldots, x_n)$ satisfying conditions

(i) $\sum_{i\in N} x_i=v(N)$

(ii) $x_i\geqq v(\{i\}), \quad (i=1, \ldots, N)$.

Condition (i) means that the total amount available if all N players cooperate is to be divided. Only distributions, assuring each player at least as much as he can get without cooperating with anyone are allowed by Condition (ii).

Adding up the inequalities in (ii) we get

$$(9) \qquad v(N)\geqq \sum_{i\in N} v(\{i\}).$$

Games for which (9) holds we call *rational games*. All superadditive games are clearly rational and the set of all imputations $X(\Gamma)$ is never empty if Γ is rational. If equality holds in (9) then none of the players is motivated to form a coalition. Thus games satisfying

$$v(N)=\sum_{i\in N} v(\{i\})$$

are called *inessential* and are not dealt with in our book. Games for which

$$v(N) > \sum_{i \in N} v(\{i\})$$

holds are called *essential* and are of particular interest both from the point of view of coalition formation and pay-off distribution.

A simple criterion for a game Γ to be inessential is provided by the following theorem.

THEOREM 2. [126] A game $\Gamma = \{N, v\}$ is inessential if and only if $v(M) + v(M') = v(M \cup M')$ for any disjoint pair of coalitions $M, M' \subseteqq N$.

Proof. The reader is invited to prove this theorem. ∎

The definition of an imputation indicates that there are numerous possible distributions of the overall pay-off $v(N)$. In order to select some "reasonable" imputations we need the important notion of domination. Let $M \subseteqq N$ be an arbitrary coalition. Imputation $\mathbf{x} = (x_1, \ldots, x_n)$ is said to *dominate imputation* $\mathbf{y} = (y_1, \ldots, y_n)$ through coalition M (notation $\mathbf{x} \overset{M}{\succ} \mathbf{y}$) if conditions

1. $x_i > y_i$,
2. $\sum_{i \in M} x_i \leqq v(M)$

are satisfied.

Condition 1 states that each player in M is better off with imputation $\mathbf{x}$ than with $\mathbf{y}$. Condition 2 ensures that coalition M has the "power" to give its members x_i, $(i \in M)$. It is trivial that no dominance can occur through $M = N$.

An imputation $\mathbf{x}$ is said *to dominate* $\mathbf{y}$ (notation $\mathbf{x} \succ \mathbf{y}$) if there is a coalition $M \subset N$, for which $\mathbf{x} \overset{M}{\succ} \mathbf{y}$.

THEOREM 3. [126] If $\mathbf{y} \overset{M}{\succ} \mathbf{x}$, then $1 < |M| < |N|$.

Proof. If $M = \{j\}$ for some j, then $v(\{j\}) \geqq y_j > x_j \geqq v(\{j\})$ which is impossible. On the other hand, if $M = N$, then $v(N) = \sum_{i \in N} y_i > \sum_{i \in N} x_i = v(N)$, also a contradiction. ∎

THEOREM 4. [126] Let $\mathbf{x}=(x_1, \ldots, x_n)$ be an imputation of the superadditive game $\Gamma=\{N, v\}$ and $M \subseteqq N$ an arbitrary coalition. The following two assertions are equivalent:

(a) $\sum_{i \in M} x_i < v(M)$,

(b) there exists an imputation $\mathbf{y}$, which dominates $\mathbf{x}$ through M.

Proof. If $\mathbf{y} \overset{M}{\succ} \mathbf{x}$, then $v(M) \geqq \sum_{i \in M} y_i > \sum_{i \in M} x_i$ by definition of dominance. To prove that (a) implies (b) let us assume that $\delta = v(M) - \sum_{i \in M} x_i > 0$. By superadditivity we have

$$v(N) - v(M) \geqq v(N-M) \geqq \sum_{i \in N-M} v(\{j\})$$

or

$$\delta' = v(N) - v(M) - \sum_{j \in N-M} v(\{j\}) \geqq 0 .$$

Let

$$y_i = \begin{cases} x_i + \dfrac{\delta}{|M|}, & \text{if } \; i \in M , \\[2ex] v(\{i\}) + \dfrac{\delta'}{|N-M|}, & \text{if } \; i \in N-M . \end{cases}$$

Now we have $\sum_{i \in M} y_i = v(M) + v(N) - v(M) = v(N)$ and $y_i > x_i$, if $i \in M$, which means $\mathbf{y} \overset{M}{\succ} \mathbf{x}$. ∎

Let now $X(\Gamma)$ denote the set of all imputations of the game $\Gamma=\{N, v\}$. Two games $\Gamma=\{N, v\}$, $\Gamma'=\{N, v'\}$ are said to be *isomorphic* if there exists a one-to-one mapping $\mathbf{f}: X(\Gamma) \leftrightarrow X(\Gamma')$ which preserves dominance, i.e., $\mathbf{x} \overset{M}{\succ} \mathbf{y} \leftrightarrow \mathbf{f}(\mathbf{x}) \overset{M}{\succ} \mathbf{f}(\mathbf{y})$ for any $M \subseteqq N$.

Games $\Gamma=\{N, v\}$ and $\Gamma'=\{N, v'\}$ are called *M-equivalent* (or strategically equivalent) if there exist numbers $\alpha > 0$ and $\beta_1, \ldots, \beta_n$ such that

$$v(M) = \alpha v'(M) + \sum_{i \in M} \beta_i \tag{10}$$

for any $M \subseteq N$. It is easy to see that the equivalence defined above is reflexive, symmetric and transitive.

THEOREM 5. [126] If games $\Gamma = \{N, v\}$, $\Gamma' = \{N, v'\}$ are M-equivalent, then they are isomorphic.

Proof. Let Γ and Γ' be M-equivalent, i.e., they satisfy (10). Denote $\boldsymbol{\beta} = (\beta_1, \ldots, \beta_n)$. Then the mapping $\mathbf{f}(\mathbf{x}) = \alpha \mathbf{x} + \boldsymbol{\beta}$ is a one-to-one mapping between $X(\Gamma)$ and $X(\Gamma')$ which preserves dominance since $\mathbf{x} \overset{M}{\succ} \mathbf{y}$ implies $\mathbf{f}(\mathbf{x}) \overset{M}{\succ} \mathbf{f}(\mathbf{y})$ as can be shown by simple substitution. ∎

Much harder is to prove the converse of this theorem. The interested reader can find a proof in [85]. Thus we state without proof

THEOREM 6. [85] Two games are isomorphic if and only if they are M-equivalent.

A game $\Gamma = \{N, v\}$ is said to be (0, 1) *normalized* if

$$v(\{i\}) = 0 \quad \text{for all} \quad i \in N$$

and

$$v(N) = 1\,.$$

Clearly, all (0, 1) normalized games are essential.

If $\Gamma = \{N, v\}$ is an essential game, then it is easy to show by elementary algebra that the equality system

$$\alpha v(\{i\}) + \beta_i = 0\,,$$

$$\alpha v(N) + \sum_{i=1}^{n} \beta_i = 1\,,$$

has exactly one solution for $\alpha, \beta_1, \ldots, \beta_n$ and $\alpha > 0$. Thus we have:

THEOREM 7. [126] In the class of games M-equivalent to an essential game Γ there exists exactly one (0, 1) normalized game.

In this way a (0, 1) normalized game represents an equivalence class of essential games. Of course, (a, b) normalized games can also be defined similarly, but for our purposes (0, 1) normalization will always suffice.

Let N be fixed. Then a (0, 1) normalized game is given by a (characteristic) function v satisfying

1. $v(\emptyset)=0$,
2. $v(\{i\})=0$ for all $i \in N$,
3. $v(N)=1$,
4. $v(M \cup M') \geqq v(M)+v(M')$, for all $M \cap M'=\emptyset$, $M, M' \subseteqq N$.

It is clear that if functions v and v' satisfy these four conditions, then their convex linear combinations also do so which means that the set of possible characteristic functions is convex. This property carries over to the case when only constant-sum games are considered, i.e., if, in 4. equality holds for $M'=N-M$.

23. Solution concepts for n-person cooperative games

In this chapter we will dwell upon the most important solution concepts for n-person rational games. Our main concern is which of the imputations can be considered to represent some kind of "equilibrium" or "stability". Since these notions are rather vague and can be interpreted differently, a number of different solution concepts have been proposed in the literature. Though each of them is theoretically well founded and based on certain assumptions as to the rational behaviour of players in a game, none of them is accepted unanimously as "the solution" of a cooperative n-person game. That is why we would like to give a survey of various solution concepts instead of analyzing one of them in detail.

23.1. THE VON NEUMANN–MORGENSTERN SOLUTION

Historically the first solution concept for a cooperative game $\Gamma=\{N, v\}$ was proposed by Neumann and Morgenstern in their fundamental work [126].

A subset V of all imputations $X(\Gamma)$ of the game $\Gamma=\{N, v\}$ is said to be a *stable-set* or *von Neumann–Morgenstern* (briefly $N-M$) *solution* if

1. $\mathbf{x} \not\succ \mathbf{y}$ for any $\mathbf{x}, \mathbf{y} \in V$ (internal stability),

2. for any imputation $\mathbf{x} \notin V$ there exists a $\mathbf{y} \in V$ for which $\mathbf{y} \succ \mathbf{x}$ (external stability).

The stable-set V was interpreted by Neumann and Morgenstern as a "standard of behaviour". If such a "standard of behaviour" is generally accepted, then there is no tendency for an imputation outside V "to get into" V because there is at least one coalition which objects to it (external stability) and an imputation in V will not "leave" V since there is no coalition in V through which it would be dominated.

The main difficulty with the $N-M$-solution is that neither existence nor uniqueness of the stable-sets can be established. A 10-person game which has no $N-M$ solution has been constructed by Lucas [95]. On the other hand, most of the games possess a lot of stable sets, which makes it difficult to accept one of them as a "solution". These problems motivated further research to look for other "solution concepts".

A very detailed analysis of stable sets in case of various special games can be found in [126].

23.2. THE CORE

The *core* $C(\Gamma)$ of a rational game $\Gamma=\{N, v\}$ is defined to be the set of imputations $\mathbf{x}=(x_1, \ldots, x_n)$ satisfying the following system of inequalities

$$(1) \qquad \sum_{i\in M} x_i \geqq v(M) \quad \text{for all} \quad M\subset N,$$

$$\sum_{i\in N} x_i = v(N).$$

The concept of the core is in close relationship with that of efficiency in multiobjective programming and Pareto-optimality in economics as demonstrated by Theorem 1. To state this theorem we need a definition. A game $\Gamma=\{N, v\}$ is called *weakly superadditive* if

$$v(N)\geqq v(M)+\sum_{i\in N-M} v(\{i\})$$

holds for any coalition M. Clearly, superadditive games are weakly superadditive.

THEOREM 1. [178] The core of a weakly superadditive game is the set of undominated imputations.

Proof. We have only to prove that, for any coalition M,

$$\sum_{i\in M} x_i < v(M)$$

if and only if there exists an imputation $\mathbf{y}$ which dominates imputation $\mathbf{x}=(x_1, \ldots, x_n)$ through M.

But this assertion can be proved in exactly the same way as was done when proving Theorem 4 of Chapter 22. ∎

Thus the core represents a certain kind of stability: once a distribution of "money" (imputation) belonging to the core is agreed upon, then it is of no coalition's interest to change it.

The definition of the core can also be given in terms of *additive set-functions*: $C(\Gamma)$ is the set of all additive set-functions x such that

$$x(N)=v(N),$$

$$x(S) \geqq v(S) \quad \text{for all} \quad S \subseteqq \mathrm{N}.$$

It should also be noted that $C(\Gamma)$ is a convex polyhedron. In case of an inessential game this polyhedron is a single point.

THEOREM 2. [178] The core of an inessential game $\Gamma=\{N, v\}$ consists of the imputation

$$(x_1, \ldots, x_n)=(v(\{1\}), \ldots, v(\{n\})).$$

Proof. Since Γ is inessential, therefore $\sum_{i\in N} v(\{i\})=v(N)$ which implies that $x_i=v(\{i\})$, $(i=1, \ldots, N)$ is the only solution of (1). ∎

THEOREM 3. [178] If $\Gamma=\{N, v\}$ is a two-person game, then $C(\Gamma)$ is the set of all imputations, i.e., the set of vectors $\mathbf{x}=(x_1, x_2)$ satisfying

$$x_1+x_2=v(\{1,2\}),$$

$$x_1 \geqq v(\{1\}),$$

$$x_2 \geqq v(\{2\}).$$

Proof. The assertion of the theorem readily follows from the definition of the core. ∎

THEOREM 4. [131] If $\Gamma = \{N, v\}$ is an essential constant-sum game, then $C(\Gamma) = \emptyset$.

Proof. Suppose $C(\Gamma) \neq \emptyset$ and let $\mathbf{x} \in C(\Gamma)$ be arbitrary. Then

$$\sum_{j \in N - \{i\}} x_j \geqq v(N - \{i\})$$

holds for all $i \in N$. Since Γ is constant-sum, therefore

$$v(N - \{i\}) = v(N) - v(\{i\}).$$

Thus

$$\sum_{j \in N - \{i\}} x_j + v(\{i\}) \geqq v(N), \quad \text{for all} \quad i \in N.$$

By property (i) of an imputation we get $x_i \leqq v(\{i\})$. Since Γ is essential, therefore

$$\sum_{j \in N} x_i \leqq \sum_{i \in N} v(\{i\}) < v(N)$$

which is a contradiction. ∎

As we have seen, for an important class of games the core is empty and of course cannot be interpreted as a solution. On the other hand for many games of economic origin (e.g., market games) it does exist and constitutes an important feature of the economic situation under analysis.

A necessary and sufficient condition can be given for a game Γ to have a nonempty core in terms of balanced sets. Let $S_1, \ldots, S_p$ be distinct, nonempty, proper subsets of a fixed finite set N. The set

$$\mathfrak{S} = \{S_1, \ldots, S_p\}$$

is said to be *balanced* if there exist positive coefficients $\gamma_1, \ldots, \gamma_p$ such that

$$\sum_{j \mid i \in S_j} \gamma_j = 1$$

holds for all $i \in N$. (The summation is taken over all j for which S_j contains i.) The numbers $\gamma_1, \ldots, \gamma_p$ are called *weights* and are said to balance $\mathfrak{S}$. If they are all equal to 1, then $\mathfrak{S}$ is a partition of N. Thus balanced sets may be regarded as generalized partitions.

Let us call a linear form $L(v) = \sum_j \alpha_j v(S_j)$ *balanced* if it vanishes identically for additive functions v. Similarly, we call an inequality (or equation) between two linear forms *balanced* if their difference is balanced. Thus

$$\sum_{j=1}^{p} \gamma_j v(S_j) \leqq \sum_{k=1}^{q} \delta_k v(T_k)$$

is balanced if for each $i \in N$, the sum of the γ_j such that $i \in S_j$ is equal to the sum of the δ_k such that $i \in T_k$.

Let now $\Gamma = \{N, v\}$ be a game (not necessarily superadditive). $\Gamma = \{N, v\}$ is said to be *balanced* if

$$(2) \qquad \sum_{S \in \mathfrak{B}} \gamma_S v(S) \leqq v(N)$$

holds for every balanced set $\mathfrak{B}$ with weights $\{\gamma_S\}$. We will call Γ *weak* if it has a nonempty core. It can easily be seen that the set of all weak games describes a closed convex polyhedral cone W of full dimension in the linear space of dimension $2^n - 1$ with coordinates $v(S)$, $S \neq \emptyset$, $S \subseteqq N$.

We will now show that the set of weak games can be defined by a system of simultaneous inequalities. To obtain this result we have to prove the following lemma.

LEMMA 1. Any linear inequality $L(v) \leqq M(v)$ satisfied by all weak games $\{N, v\}$ is necessarily balanced.

Proof. The characteristic function of a game is a set function on the subsets of N. If a is an arbitrary additive function, then the definition of the core implies that the additive set function x is in the core of $\{N, v\}$ if and only if $x+a$ is in the core of $\{N, v+a\}$. Thus $\{N, v\}$ is weak if and only if $v+a$ is weak for all additive a. Since any unbalanced inequality $L(v) \leqq M(v)$ can be made false by adding a suitable additive function to v we get the assertion of our lemma. ∎

Since a closed convex cone is the intersection of the closed half-spaces that contain it, Lemma 1 implies that the set of weak games can be described by a system of inequalities.

THEOREM 5. [162] A game has a nonempty core if and only if it is balanced.

Proof. Let $\Gamma = \{N, v\}$ be weak, i.e., Γ has a nonempty core. Then reducing the value of $v(S)$ for any $S \neq N$ will preserve weakness. This means that $v(S)$ must occur on the "less" side of any defining inequality in which it occurs at all (assuming positive coefficients). Hence, all defining inequalities can be put in the form of (2). By Lemma 1, they must be balanced inequalities.

It remains to be shown that all balanced inequalities of form (2) are satisfied by all weak games. Let $\Gamma = \{N, v\}$ be weak, and let the additive function x be in its core. By the definition of a balanced inequality, we have

$$\gamma_1 x(S_1) + \ldots + \gamma_p x(S_p) = x(N).$$

Since x is in the core $x(S_j) \geqq v(S_j)$, $(j = 1, \ldots, p)$ and $x(N) = v(N)$. By the positivity of the γ_j we get (2) for any balanced set $S \in \mathfrak{B}$. ∎

23.3. THE STRONG ε-CORE

Consider a rational game $\Gamma = \{N, v\}$ and let $\mathbf{x} = (x_1, \ldots, x_n)$ be an arbitrary imputation. (We denote the set of all imputations by $X(\Gamma)$.) Define for any coalition S

$$x(S)=\begin{cases}\sum_{i\in S} x_i & \text{if } S\neq\emptyset\\ 0 & \text{if } S=\emptyset\end{cases}$$

and

$$(3)\qquad e(S,\mathbf{x})=v(S)-x(S).$$

The expression $e(S,\mathbf{x})$ is called the *excess* of S with respect to imputation $\mathbf{x}$ (in game Γ). It represents the gain (loss—if it is negative) that is caused to the coalition S if its members depart from an agreement represented by $\mathbf{x}$ and form their own coalition.

We can also define the core in terms of the excess: the core $C(\Gamma)$ is the set of all pay-off vectors that give rise only to non-positive excesses:

$$C(\Gamma)=\{\mathbf{x}\mid \mathbf{x}\in X(\Gamma) \text{ and } e(S,\mathbf{x})\leqq 0 \text{ for all } S\subseteqq N\}.$$

Let now ε be a real number. A generalization of the core is the *strong ε-core*[1] $C_\varepsilon(\Gamma)$ of the game Γ which is defined as the set of all imputations which give rise only to excesses not greater than ε, for coalitions other than $\emptyset$ and N:

$$(4)\qquad C_\varepsilon(\Gamma)=\{\mathbf{x}\mid \mathbf{x}\in \mathbb{R}^n, x(N)=v(N), e(S,\mathbf{x})\leqq\varepsilon \text{ for any } S\neq N, S\neq\emptyset\}.$$

The strong ε-core can be interpreted as a set of "generalized imputations" (Property (ii) of an imputation is not required here). which cannot be blocked by any "proper" coalition if blocking requires a cost of ε in order to form. (We allow the possibility of ε being negative.) Clearly, $C_0(\Gamma)=C(\Gamma)$; i.e., the core is the strong 0-core. Also, $C_{\varepsilon_1}(\Gamma)\supset C_{\varepsilon_2}(\Gamma)$ if $\varepsilon_1>\varepsilon_2$. By proper choice of ε

$$(5)\qquad \varepsilon\geqq \max_{S\mid S\neq N,\emptyset}\Big[v(S)-\sum_{i\in S} v(\{i\})\Big]$$

we have $C_\varepsilon(\Gamma)\supset X(\Gamma)$, which means that for sufficiently large ε the strong ε-core is never empty. On the other hand, if $n\geqq 2$ and ε is

[1] C_ε was first studied in [106].

small enough, then $C_\varepsilon(\Gamma)=\emptyset$. The bounds depend on the characteristic function.

Both the core and the strong ε-core are compact convex polyhedra bounded by at most 2^n-2 hyperplanes.

$$(6)\qquad H_S=\{\mathbf{x}\mid x(S)=v(S)\}\,,\quad S\subset N,\quad S\neq\emptyset, N,$$

and

$$(7)\qquad H_S^\varepsilon=\{\mathbf{x}\mid x(S)=v(S)-\varepsilon\}\,,\quad S\subset N\,,\quad S\neq\emptyset\,,N\,,$$

respectively.

Figure 1 shows typical core and strong ε-core of a 3-person game. Here $v(\{1,2,3\})=60$, $v(\{1,2\})=20$, $v(\{1,3\})=10$, $v(\{2,3\})=50$, $v(\{i\})=0$ for $i=1,2,3$, $\varepsilon=15$.

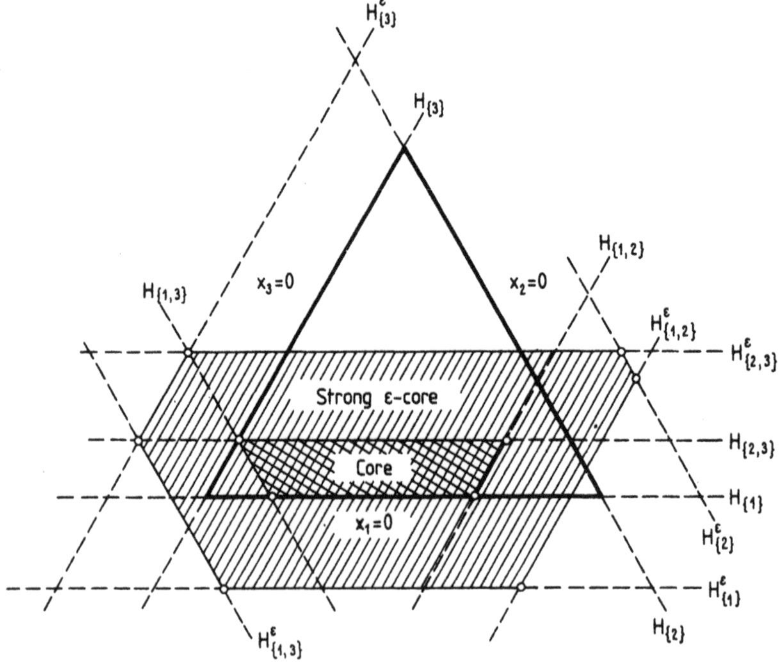

Figure 1

Note that the strong ε-core indeed extends beyond $X(\Gamma)$. Observe also that although H_S and H_S^ε are paralleled and equidistant (this is not true, in general, if $n>3$), the shapes of the various $C_\varepsilon(\Gamma)$'s may be different. In our example $C(\Gamma)$ is a quadrilateral, whereas $C_{15}(\Gamma)$ is a hexagon. For ε large enough, however, all the hyperplanes (7) touch the strong ε-core. This is a direct consequence of the following theorem which adds to the geometric characterization of the strong ε-core.

THEOREM 6. [106] Let S be a fixed coalition of the n-person game $\Gamma=\{N, v\}$. For sufficiently large ε, $H_S^\varepsilon(\Gamma)\cap C_\varepsilon(\Gamma)\neq\emptyset$.

Proof. There exists a vector $\mathbf{x}^0=(x_i^0)_{i\in S}$ and a nonnegative real number δ, satisfying:

$$x^0(S)=v(S)+\delta\,,$$

$$x^0(T)\geqq v(T) \quad \text{for all} \quad T\subset S\,.$$

Choose ε large enough, so that

$$\varepsilon\geqq |S|\delta$$

$$\frac{v(N)-v(S)+\varepsilon}{|N-S|}\geqq \max_{T\subset S}\{v(T),\quad v(T)-v(S\cap T)\}\,,$$

holds for all $T\subset N$.

Define

$$x_i=\begin{cases} x_i^0-t & \text{for } i\in S \\ \dfrac{v(N)-v(S)+\varepsilon}{|N-S|} & \text{for } i\in N-S \end{cases} \qquad (i=1,\ldots,n)$$

where $t=\dfrac{\varepsilon+\delta}{|S|}$.

We will show that $\mathbf{x}=(x_1,\ldots,x_n)\in H_S^\varepsilon(\Gamma)\cap C_\varepsilon(\Gamma)$. Indeed,

$$x(S)=x^0(S)-t\,|S|\,=v(S)+\delta-(\varepsilon+\delta)=v(S)-\varepsilon\,,$$

$$x(N)=x(S)+x(N-S)=v(S)-\varepsilon+v(N)-v(S)+\varepsilon=$$
$$=v(N),$$

which means that $\mathbf{x}\in H_S^\varepsilon(\Gamma)$.

If $T\subset S$, $T\neq\emptyset$, then

$$x(T)=x^0(T)-|T|t\geqq v(T)-\frac{|T|(\varepsilon+\delta)}{|S|}\geqq v(T)-$$
$$-\frac{(|S|-1)\left(\varepsilon+\dfrac{\varepsilon}{|S|}\right)}{|S|}=v(T)-\frac{|S|^2-1}{|S|^2}\varepsilon\geqq v(T)-\varepsilon.$$

If $T\subseteqq N-S$, $T\neq\emptyset$, then

$$x(T)\geqq\frac{v(N)-v(S)+\varepsilon}{|N-S|}\geqq v(T)\geqq v(T)-\varepsilon.$$

Finally, if $T\cap S\neq\emptyset$ and $T-S\neq\emptyset$, then

$$x(T)=x(T-S)+x(T\cap S)\geqq$$
$$\geqq\frac{v(N)-v(S)+\varepsilon}{|N-S|}+v(S\cap T)-\varepsilon\geqq$$
$$\geqq v(T)-v(S\cap T)+v(S\cap T)-\varepsilon=v(T)-\varepsilon.$$

Thus $\mathbf{x}\in C_\varepsilon(\Gamma)$ which concludes the proof. ∎

COROLLARY. For any n-person game $\Gamma=\{N,v\}$ there exists a real number ε_0 such that for all $\varepsilon\geqq\varepsilon_0$, the strong ε-core $C_\varepsilon(\Gamma)$ has a boundary which consists of 2^n-2 non-empty subsets of the hyperplanes H_S^ε.

23.4. THE KERNEL

Let $\Gamma=\{N,v\}$ be a rational n-person game. For $i,j\in N$, $i\neq j$ we denote by T_{ij} the set of coalitions containing i but not containing j:

$$T_{ij}=\{S\mid S\subset N,\quad i\in S,\quad j\notin S\}.$$

For each imputation $\mathbf{x} \in X(\Gamma)$ we define the maximum surplus of i over j to be:

$$s_{ij}(\mathbf{x}) = \max_{S \in T_{ij}} e(S, \mathbf{x}).$$

We say that i *outweighs* j with respect to $\mathbf{x}$, if

$$s_{ij}(\mathbf{x}) > s_{ji}(\mathbf{x})$$

and

$$x_j > v(\{j\}).$$

We say that i and j are *in equilibrium* with respect to $\mathbf{x}$ if neither of them outweighs the other. The *kernel* $K(\Gamma)$ is the set of imputations with respect to which every two players are in equilibrium, i.e., $\mathbf{x} \in K(\Gamma)$ if $\mathbf{x} \in X(\Gamma)$ and

$$(8) \qquad (s_{ij}(\mathbf{x}) - s_{ji}(\mathbf{x}))(x_j - v(\{j\})) \leqq 0 \quad \text{for all} \quad i, j; i \neq j.$$

The interpretations of the kernel as a fair division scheme are less convincing than that of the core or the stable set. The quantity $s_{ij}(\mathbf{x})$, which measures player i's "strength" against player j, can be interpreted as the maximum gain (and, if negative, in absolute value — the minimal loss) player i would obtain by "bribing" some members other than player j to depart from $\mathbf{x}$, giving each of them a very small bonus. Player i can be thought of as being in a "strong bargaining position" against player j if his maximum gain is greater than that of player j's, i.e., if i outweighs j. If no player outweighs the other with respect to $\mathbf{x}$, i.e., $\mathbf{x}$ belongs to the kernel, then $\mathbf{x}$ can be considered as a "fair division" of $v(N)$ since there is no player in a "strong bargaining position". The weak point of this interpretation is that interpersonal comparison of "utilities" $s_{ij}(\mathbf{x})$ and $s_{ji}(\mathbf{x})$ is assumed.

The kernel is accepted as a solution concept mainly because of its attractive mathematical properties, first of all its existence and its close relation to the core.

THEOREM 7. [106] Any rational n-person game Γ has a nonempty kernel. If Γ has a non-empty core, then the intersection of the core and the kernel is also nonempty.

Proof. The proof of this theorem is too lengthy and involved to be given here. It can be found in [106].

Observe that both the kernel and the strong ε-core are relative invariants under strategic equivalence, i.e., if a positive constant α and an n-tuple of real constants $\boldsymbol{\beta}=(\beta_1, \ldots, \beta_n)$ exist such that for each coalition S $v(S)=\alpha w(S)+\sum_{i\in S}\beta_i$, then the kernel (the strong ε-core) of the game $\{N, v\}$ is obtained from the kernel (the strong ε-core) of the game $\{N, w\}$ by the linear transformation $\mathbf{x}\to\alpha\mathbf{x}+\beta$. Therefore, inasmuch as relations between these solution concepts are discussed there will be no loss of generality in assuming that the underlying game $\{N, v\}$ is 0-normalized, i.e.,

$$(9) \qquad v(\{i\})=0, \quad (i=1, \ldots, n).$$

We will assume (9) whenever convenient. The following two theorems provide conditions under which (8) can be considerably simplified, either for the entire kernel or for a subset of it.

THEOREM 8. [106] Let $\mathbf{x}$ belong to the core $C(\Gamma)$ of a rational n-person game Γ. Then $\mathbf{x}\in K(\Gamma)$ if and only if

$$(10) \qquad s_{ij}(\mathbf{x})=s_{ji}(\mathbf{x}) \quad \text{for all} \quad i, j;\ i\neq j.$$

Proof. Clearly (10) implies (8), hence every imputation in $X(\Gamma)$ satisfying (10) lies in the kernel (even if $\mathbf{x}\notin C(\Gamma)$). Conversely, suppose $\mathbf{x}\in K(\Gamma)\cap C(\Gamma)$ and let $i, j\in N$, $i\neq j$. It is sufficient to show that $s_{ij}(\mathbf{x})\leqq s_{ji}(\mathbf{x})$. Indeed, if $s_{ij}(\mathbf{x})>s_{ji}(\mathbf{x})$, then, by (8), $x_j=v(\{j\})$, because for each imputation $\mathbf{x}$ we have $x_j\geqq v(\{j\})$ by definition. Since $\{j\}\in T_{ji}$, it follows from the definition of T_{ji} and $s_{ji}(\mathbf{x})$ that $s_{ji}(\mathbf{x})\geqq v(\{j\})-x_j=0$. Thus $s_{ij}(\mathbf{x})>0$. By the definition of $C(\Gamma)$ $\mathbf{x}$ does not belong to the core, which is a contradiction. ∎

A game $\Gamma=\{N, v\}$ is said to be monotonic in the 0-normalization, if the 0-normalized game Γ^* is monotonic.

We state without proof a theorem analogous to Theorem 8 for monotonic games. The proof can be found in [106].

THEOREM 9. If $\Gamma=\{N, v\}$ is monotonic in the 0-normalization, then an imputation $\mathbf{x} \in X(\Gamma)$ belongs to the kernel $K(\Gamma)$ if and only if

$$(11) \quad s_{ij}(\mathbf{x})=s_{ji}(\mathbf{x}) \quad \text{for all} \quad i, j;\ i\neq j. \qquad \blacksquare$$

Other structural and geometric properties of the kernel can be found in [38], [105], [106], [197].

23.5. THE NUCLEOLUS

One of the newest solution concepts to n-person cooperative games, the nucleolus was introduced by Schmeidler [156]. Let $\Gamma=\{N, v\}$ be a rational n-person game. For each $\mathbf{x} \in X(\Gamma)$, let $\Theta(\mathbf{x})$ be a 2^n-tuple whose components are the numbers $e(S, \mathbf{x})$, $S \subseteqq N$, arranged in a non-increasing order, i.e.,

$$(12) \quad \Theta_i(\mathbf{x}) \geqq \Theta_j(\mathbf{x}), \quad 1 \leqq i \leqq j \leqq 2^n.$$

The *nucleolus* $N(\Gamma)$ of Γ is the set of imputations for which $\Theta(\mathbf{x})$ is minimal in the lexicographic order, i.e.,

$$(13) \quad N(\Gamma)=\{\mathbf{x} \mid \mathbf{x} \in X(\Gamma) \quad \text{and} \quad \Theta(\mathbf{x}) \precsim \Theta(\mathbf{y}) \quad \text{whenever } \mathbf{y} \in X(\Gamma)\},$$

where $\Theta(\mathbf{x}) \prec \Theta(\mathbf{y})$ if and only if there exists an index i_0 (possibly $i_0=0$), such that

$$(14) \quad \Theta_\nu(\mathbf{x})=\Theta_\nu(\mathbf{y}), \quad \nu=1, \ldots, i_0$$

and

$$\Theta_{i_0+1}(\mathbf{x}) < \Theta_{i_0+1}(\mathbf{y}).$$

(The symbol $\precsim$ means that $\succ$ does not hold.)

Intuitively, the nucleolus represents a kind of "fair" or "even" division of the common gain $v(N)$, inasmuch as it minimizes the largest deviation from the amount a coalition can assure regardless of the behaviour of other players.

We will now introduce the *lexicographic core* of Γ. We start by denoting $\Sigma_0=\{\emptyset, N\}=\Sigma^0$, $X_0=X(\Gamma)$ and defining

$$(15)\qquad \varepsilon_1=\min_{\mathbf{x}\in X_0}\ \max_{S\mid S\notin\Sigma^0} e(S,\mathbf{x})$$

$$(16)\qquad X_1=\{\mathbf{x}\mid \mathbf{x}\in X_0, e(S,\mathbf{x})\leqq\varepsilon_1 \text{ for all } S\notin\Sigma^0\}.$$

Clearly, $X_1=C_{\varepsilon_1}(X)\cap X(\Gamma)$ and ε_1 is minimal under the requirement that $X_1\neq\emptyset$. We now define Σ_1 to be the set of coalitions $S\notin\Sigma_0$, such that $H_S^{\varepsilon_1}$ (see (7)) contains X_1, i.e.,

$$(17)\qquad \Sigma_1=\{S\mid e(S,\mathbf{x})=\varepsilon_1 \text{ for all } \mathbf{x}\in X_1, S\notin\Sigma^0\}.$$

Suppose $\varepsilon_1,\varepsilon_2,\ldots,\varepsilon_{i-1},X_1,X_2,\ldots,X_{i-1},\Sigma_1,\Sigma_2,\ldots,\Sigma_{i-1}$ have been defined, ($i\geqq 2$), and denote $\Sigma^{i-1}=\Sigma_0\cup\Sigma_1\cup\ldots\cup\Sigma_{i-1}$. We now define

$$(18)\qquad \varepsilon_i=\min_{\mathbf{x}\in X_{i-1}}\ \max_{S\notin\Sigma^{i-1}} e(S,\mathbf{x}),$$

$$(19)\qquad X_i=\{\mathbf{x}\mid \mathbf{x}\in X_{i-1}, e(S,\mathbf{x})\leqq\varepsilon_i \quad \text{for all } S\notin\Sigma^{i-1}\},$$

$$(20)\qquad \Sigma_i=\{S\mid e(S,\mathbf{x})=\varepsilon_i \quad \text{for all } \mathbf{x}\in X_i,\ S\notin\Sigma^{i-1}\}.$$

This sequence terminates at stage l, when $\Sigma^l=2^N$. Then X_l is called the *lexicographic core* of Γ.

To illustrate this rather complicated notion we take two examples. Consider first a 3-person 0-normalized game, where $v(\{1,2\})=v(\{1,3\})=30$, $v(\{2,3\})=80$, $v(\{1,2,3\})=100$.

It is easy to verify that $\varepsilon_1=-10$, $\varepsilon_2=-25$, $\varepsilon_3=-45$, that $\Sigma_1=\{\{1\},\{2,3\}\}$, $\Sigma_2=\{\{1,2\},\{1,3\}\}$, $\Sigma_3=\{\{2\},\{3\}\}$ and that $X_1=\{\mathbf{x}\mid x_1=10,\ x_2\leqq 60,\ x_3\leqq 60,\ x_2+x_3=90\}$, $X_2=X_3=$ $=\{(10,45,45)\}$.

Figure 2 shows the sets X_i as well as a few hyperplanes $H_S^{\varepsilon_i}$. In addition, the core is also indicated in this figure.

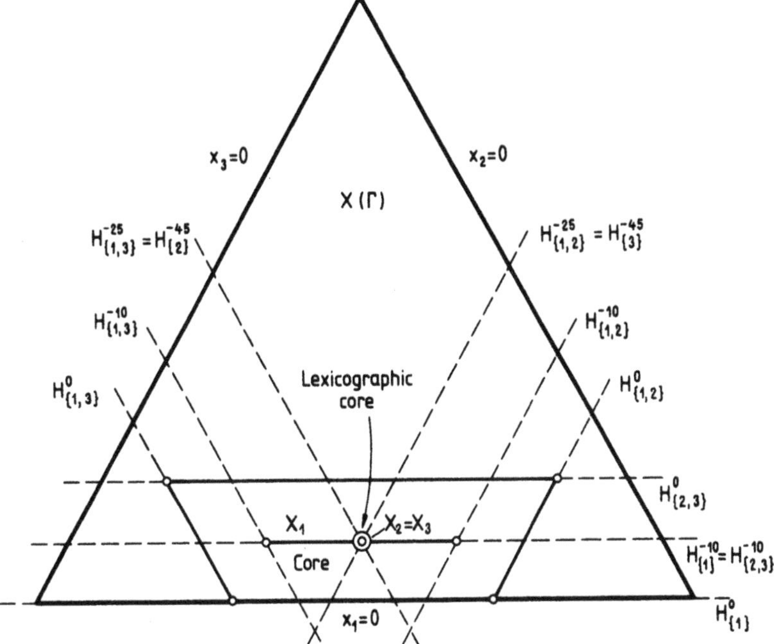

Figure 2

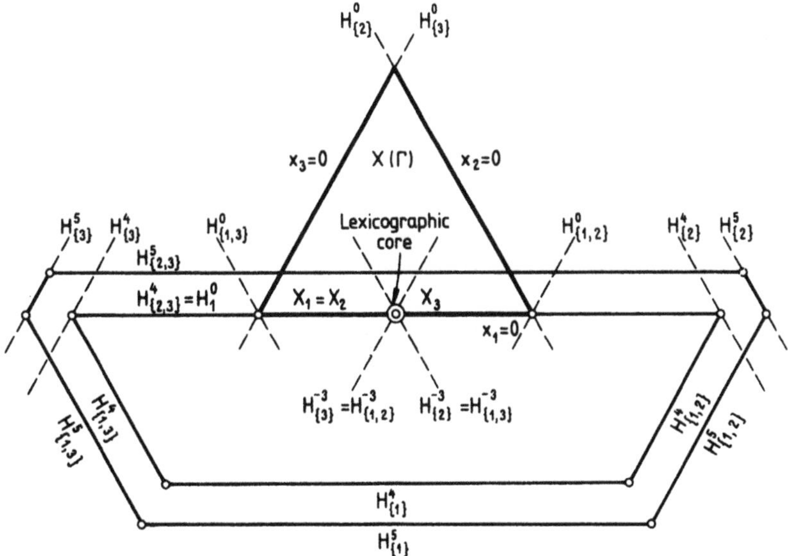

Figure 3

As a second example let us consider the following coreless 3-person, 0-normalized game: $v(\{1, 2\})=v(\{1, 3\})=0$, $v(\{2, 3\})=10$, $v(\{1, 2, 3\})=6$. This time $\varepsilon_1=4$, $\varepsilon_2=0$, $\varepsilon_3=-3$, $\Sigma_1=\{\{2, 3\}\}$, $\Sigma_2=\{\{1\}\}$, $\Sigma_3=\{\{2\}, \{3\}, \{1, 2\}, \{1, 3\}\}$, $X_1=X_2=\{\mathbf{x}|\mathbf{x}\in X(\Gamma), x_1=0\}$, $X_3=\{(0, 3, 3)\}$.

Figure 3 shows the process of reaching X_3. The strong 5-core is also indicated in this figure.

Reaching the lexicographic core can also be described in geometric terms: One starts with an arbitrary strong ε-core that intersects X and pushes "inside" all the hyperplanes H_S^{ε}, $S\neq\emptyset, N$.

The push is performed at equal l_1-distances and it is stopped either when any further push renders the interior empty (see Figure 2), or when any further push makes the interior disjoint from $X(\Gamma)$ (see Figure 3). Thus, the amount of pushing depends both on the shape of the strong ε-core and its location in the space of imputations. The push brings us to the set X_1.

By (17), $H_S^{\varepsilon_1}$, $S\neq\emptyset, N$ contains X_1 if and only if $S\in\Sigma_1$, any further push of such a hyperplane will make X_1 empty. We therefore continue to push only hyperplanes $H_S^{\varepsilon_1}$ where $S\notin\Sigma^1$ These we push at equal maximum l_2-distances, so that the interior of the strong ε-core modified in this fashion is neither empty nor disjoint from $X(\Gamma)$. This brings us to X_2. The process continues in the same manner until all the hyperplanes $H_S^{\varepsilon_i}$, $S\in\Sigma_i$, contain a set X_l, which is the lexicographic core.

In order to establish the existence and uniqueness of the lexicographic core we need two lemmas.

LEMMA 2. Under notation (15)–(20), and for $i=1, \ldots, l$,

(i) ε_i are well defined,

(ii) X_i are non-empty compact and convex sets,

(iii) $X_0\supset X_1\supset\ldots\supset X_l$,

(iv) $\Sigma_i\neq\emptyset$,

(v) $l<\infty$

(vi) $\varepsilon_1>\varepsilon_2>\ldots>\varepsilon_l$.

Proof. Relation (19) proves (iii). Claim (i) implies claim (ii) and this claim implies claim (i) when $i+1$ replaces i, $(i<l)$. Since ε_1 is well defined, (i) and (ii) are thus proved by induction.

Assume now that $\Sigma_i=\emptyset$ and $\Sigma^{i-1}\neq 2^N$.[1] This means that for each S in $2^N-\Sigma^{i-1}$ there exists an imputation $\mathbf{x}^{(S)}\in X_i$ such that $e(S,\mathbf{x}^{(S)})<\varepsilon_i$. Let m be the number of such coalitions, then $m\geqq 1$, and by the convexity of X_i,

$$\text{(21)}\qquad \tilde{\mathbf{x}}=\frac{1}{m}\sum_{R\notin\Sigma^{i-1}}\mathbf{x}^{(R)}\in X_i\subset X_{i-1}\,.$$

Clearly, for $S\notin\Sigma^{i-1}$,

$$\text{(22)}\qquad e(S,\tilde{\mathbf{x}})=v(S)-\frac{1}{m}\sum_{R\notin\Sigma^{i-1}}X^{(R)}(S)=$$

$$=\frac{1}{m}\sum_{R\notin\Sigma^{i-1}}e(S,x^{(R)})<\varepsilon_i\,,$$

contrary to (18). This contradiction proves (iv). Assertion (v) follows from assertion (iv), since the number of coalitions is finite.

To prove claim (vi), observe that for each coalition R in $2^N-\Sigma^i$, there has to exist a point $\mathbf{x}^{(R)}\in X_i$ for which $e(R,\mathbf{x}^{(R)})<\varepsilon_i$. Thus, if $2^N-\Sigma^i$ contains t coalitions, then

$$\text{(23)}\qquad \hat{\mathbf{x}}=\frac{1}{t}\sum_{R\in\Sigma^i}\mathbf{x}^R$$

belongs to X_i and satisfies $e(T,\hat{\mathbf{x}})<\varepsilon_i$ whenever $T\in\Sigma^i$. Therefore $\varepsilon_{i+1}<\varepsilon_i$ for $i<l$. ∎

LEMMA 3. An imputation $\mathbf{x}\in X(\Gamma)$ belongs to X_i, $(i=1,\ldots,l)$ if and only if

$$\text{(24)}\qquad \theta^*(\mathbf{x})=(\varepsilon_1,\ldots,\varepsilon_1,\varepsilon_2,\ldots,\varepsilon_2,\ldots,\varepsilon_i\ldots\varepsilon_i,\; e(R_1,\mathbf{x}),\ldots,e(R_p,\mathbf{x}))$$

[1] 2^N denotes the set of all subsets of $N=\{1,\ldots,n\}$.

and

(25) $e(R_\nu, \mathbf{x}) \leqq \varepsilon_i$, $(\nu = 1, \ldots, p)$,

where the coalitions which take the excesses ε_μ are precisely the coalitions of Σ_μ, $(\mu = 1, \ldots, i)$. Here $\theta^*(\mathbf{x})$ is obtained from $\theta(\mathbf{x})$ by deleting the components $e(N, \mathbf{x})$ and $e(\emptyset, \mathbf{x})$, while p is the number of coalitions not contained in $\bigcup_{\mu=1}^{i} \Sigma_\mu$.

Proof. The assertion of this lemma readily follows from relations (15)–(20) and Lemma 2. ∎

THEOREM 10. [106] The lexicographic core consists of a unique point.

Proof. By construction, the lexicographic core is non-empty. By Lemma 3, the excess of each coalition is constant for all points in the lexicographic core. In particular, the excess of the single-person coalitions is also constant. Consequently, the coordinates of the points in the lexicographic core are constant and this can happen only if it consists of a unique point. ∎

THEOREM 11. [106] The lexicographic core coincides with the nucleolus.

Proof. We first prove that if $\mathbf{x} \in X_i$ and $\mathbf{y} \in X(\Gamma) - X_i$, then $\Theta(\mathbf{x}) \prec \Theta(\mathbf{y})$, $(i = 1, \ldots, l)$. Verification of this assertion for $i = 1$ is straightforward (see (15) and (16)). Suppose the assertion is true for $i = 1, \ldots, k-1$, $(2 \leqq k \leqq l)$. By claim (iii) of Lemma 2, we have only to examine the case $\mathbf{x} \in X_k$, $\mathbf{y} \in X_{k-1} - X_k$. It follows from Lemma 3 and (19), (for $i = k$) that $\Theta^*(\mathbf{x}) \prec \Theta^*(\mathbf{y})$, consequently $\Theta(\mathbf{x}) \prec \Theta(\mathbf{y})$. Thus $\Theta(\mathbf{x}_l) \prec \Theta(\mathbf{y})$ for any $\mathbf{y} \in X(\Gamma)$, $\mathbf{x}_l \neq \mathbf{y}$ where $X_l = \{x_l\}$ is the (unique) lexicographic core. But, this is the definition of the nucleolus. ∎

Assume now that the polyhedral set $X(\Gamma)$ is bounded, i.e., $X(\Gamma)$ is a polytope. The nucleolus can then be obtained as an optimal solution of a properly chosen linear programming problem:

(26) $$z \to \min$$

$$z \geqq \sum_{j=1}^{2^n} t^{\pi(j)}(v(S_j) - x(S_j)) \quad \text{for all} \quad \pi \in \Pi$$

$$\mathbf{x} \in X(\Gamma),$$

where Π denotes the group of permutations on $\{1, \ldots, 2^n\}$, and t is a positive parameter. Our aim is to prove that for large enough t the nucleolus $N(\Gamma)$ is the unique solution of (26). Before doing so we have to introduce a new function and to prove a lemma.

For $\mathbf{y} \in \mathbb{R}^m$, define

$$f_t(\mathbf{y}) = \sum_{j=1}^{m} t^{m+1-j} y_j, \quad \mathbf{y} = (y_1, \ldots, y_m).$$

LEMMA 4. Let C be the union of a finite number of polytopes in $\mathbb{R}^m$, and let $\mathbf{y}^*$ be the lexicographical minimum of C. Then there exists a number t_0, such that for all $t > t_0$, $\mathbf{y}^*$ is the unique point of C at which the minimum of f_t is attained.

Proof. Denote the extreme points of the polytopes, the union of which is C, by $\mathbf{y}_1, \ldots, \mathbf{y}_k$. If $\mathbf{y}_i \neq \mathbf{y}^*$, then $\mathbf{y}_i$ is lexicographically larger than $\mathbf{y}^*$, and a number t_i exists such that for all $t > t_i$, $f_t(\mathbf{y}_i) > f_t(\mathbf{y}^*)$. Now if $\mathbf{y}$ is any point of C which differs from $\mathbf{y}^*$, then

$$\mathbf{y} = \sum_{i=1}^{k} r_i \mathbf{y}_i,$$

where $r_i \geqq 0$, $\sum_{i=1}^{k} r_i = 1$ and for some $1 \leqq j \leqq k$, $r_j > 0$ and $\mathbf{y}_j \neq \mathbf{y}^*$. From the linearity of f_t, it follows that $f_t(\mathbf{y}) > f_t(\mathbf{y}^*)$ for all $t > t_0 = \max_{\substack{i \\ \mathbf{y}_i \neq \mathbf{y}^*}} t_i$. ∎

Define the vector of excesses for all coalitions of $\Gamma = \{N, v\}$

$$\mathbf{e}(\mathbf{x}) = (v(S_1) - x(S_1), \ldots, v(S_{2^n}) - x(S_{2^n})).$$

Also define $q: E^{2^n} \to E^{2^n}$ as the function that rearranges the components of each vector in order of decreasing magnitude.

THEOREM 12. [82] There exists a positive number t_0 such that the nucleolus $N(\Gamma)$ of the game $\Gamma=\{N, v\}$ is the unique solution of (26) for any $t>t_0$.

Proof. We note that for $t>1$

$$\max_{\pi\in\Pi}\sum_{j=1}^{2^n} t^{\pi(j)}(v(S_j)-x(S_j))=f_t qe(\mathbf{x})$$

(where $f_t qe(\mathbf{x})$ is short for $f_t(q(e(\mathbf{x})))$ and $m=2^n$). Thus we have to establish the existence of a number t_0, such that, for $t>t_0$, $N(\Gamma)$ is the unique point of $X(\Gamma)$ at which the minimum of $f_t qe$ is attained.

The transformation $e(\mathbf{x}): X(\Gamma)\to\mathbb{R}^{2^n}$ is affine. This implies that $e(X(\Gamma))=\{\mathbf{y}\,|\,\mathbf{y}=e(\mathbf{x}),\ \mathbf{x}\in X(\Gamma)\}$ is a polytope.

Let

$$e_\pi=\{\mathbf{y}\in\mathbb{R}^{2^n}\,|\,\mathbf{y}_{\pi(1)}\geqq\mathbf{y}_{\pi(2)}\geqq\cdots\geqq\mathbf{y}_{\pi(2^n)}\},$$

where $\pi\in\Pi$. Clearly, $e(X(\Gamma))\cap E_\pi$ is a bounded set which is the intersection of a finite number of closed half-spaces, hence it is a polytope. From the linearity of q on E_π, it follows that $q(e(X(\Gamma))\cap E_\pi)$ is also a polytope. Since

$$qe(X(\Gamma))=\bigcup_{\pi\in\Pi} q(e(X(\Gamma))\cap E_\pi),$$

$qe(X(\Gamma))$ is the union of a finite number of polytopes.

By the definition of the nucleolus, $qe(N(\Gamma))$ is lexicographically least on this set. We now apply Lemma 4 to find a number t_0 such that, for all $t>t_0$, $qe(N(\Gamma))$ is the unique point of $qe(X(\Gamma))$ at which the minimum of f_t is attained. From the uniqueness of the nucleolus it follows that $N(\Gamma)$ is the unique point of $X(\Gamma)$ at which the minimum of $f_t qe$ is attained, i.e., $N(\Gamma)$ is the unique solution of (26). ∎

It can also be shown [82] that the number t in Theorem 12 may be chosen independently of the characteristic function v, i.e., there exists a number t_0 such that, for all $t>t_0$ the nucleolus is the unique solution of (26) inasmuch as $X(\Gamma)$ is bounded and t_0 depends only on the number of players n.

The linear program (26) becomes cumbersome as n grows since both the number of columns and that of the rows grows very fast. Thus (26) provides an efficient method for computing $N(\Gamma)$ only for small n.

23.6. THE SHAPLEY-VALUE

Just as the nucleolus, the Shapley-value is also a unique n-vector which is supposed to "measure" the "value" of each player in a game $\Gamma=\{N, v\}$. Shapley approaches his value axiomatically.

Denote by $V\subset\mathbb{R}^{2^n}$ the set of all n-person games given in characteristic function form:

$$V=\{\mathbf{v}\in\mathbb{R}^{2^n} \mid v_1=0\}\,.$$

The first component of each $v\in V$ is zero since $v(\emptyset)=0$ for any game. The other coalitions also correspond to fixed components of $\mathbf{v}$.

Let $\mathbf{\Phi}$ be a function $\mathbf{\Phi}: V\to\mathbb{R}^n$ which we interpret as follows: $\Phi_i(\mathbf{v})$ is the *value* of the ith player in the game $\mathbf{v}$. ($\mathbf{v}$ stands for the game $\Gamma=\{\mathrm{N}, \mathrm{v}\}$, where v assigns the components of $\mathbf{v}$ to each coalition.)

Shapley proposes three axioms which the function Φ ought to satisfy. In order to state them it is necessary first to define a few concepts.

(i) S is called a *carrier* for $\mathbf{v}$ if

$$v(T)=v(T\cap S) \quad \text{for all} \quad T\subseteqq N\,.$$

Intuitively, this means that any player who does not belong to a carrier is a "dummy" – a player who cannot contribute anything to any coalition.

(ii) If $\pi: N\to N$ is a permutation of N, then the game $\pi(\mathbf{v})$ is defined by

$$\pi(v(T))=v(\pi(T)) \quad \text{for all} \quad T\subseteqq N\,.$$

Actually, the game $\pi(\mathbf{v})$ is the game $\mathbf{v}$, with the roles of the players interchanged by the permutation.

(iii) Given any two games $\Gamma_1 = \{N, v_1\}$, and $\Gamma_2 = \{N, v_2\}$, the game $\Gamma = \{N, v_1 + v_2\}$ is defined by

$$(v_1 + v_2)(T) = v_1(T) + v_2(T) \quad \text{for all} \quad T \subseteqq N .$$

which means that Γ is determined by the vector $\mathbf{v}_1 + \mathbf{v}_2 \in V$ if Γ_1 and Γ_2 were determined by $\mathbf{v}_1 \in V$ and $\mathbf{v}_2 \in V$ resp.

Now Shapley's axioms are:

S1. If S is any carrier for $\Gamma = \{N, v\}$, then

$$\sum_{i \in S} \Phi_i(\mathbf{v}) = v(S) .$$

Intuitively this axiom means that 0 value is assigned to dummies.

S2. For any permutation π and $i \in N$,

$$\Phi_{\pi(i)}(\pi(\mathbf{v})) = \Phi_i(\mathbf{v}) .$$

This means that changing the roles of the players by permuting them does not affect their values.

S3. If $\mathbf{v}_1$ and $\mathbf{v}_2$ are any games, then

$$\mathbf{\Phi}(\mathbf{v}_1 + \mathbf{v}_2) = \mathbf{\Phi}(\mathbf{v}_1) + \mathbf{\Phi}(\mathbf{v}_2) ,$$

i.e., the value function is assumed to be additive.

THEOREM 13. [43] There is a unique function $\mathbf{\Phi}$, defined on V, which satisfies the axioms S1, S2, S3.

Proof.[1] For each coalition S define the game with characteristic function $v_{S,c}$ by

$$v_{S,c}(T) = \begin{cases} 0 & \text{if} \quad S \nsubseteq T \\ c & \text{if} \quad S \subseteqq T , \end{cases}$$

where c is a constant. Then it is clear that S and its supersets are all carriers for $v_{S,c}$. Therefore, by S1,

[1] This proof is due to Dubey [43].

$$\sum_{i \in S} \Phi_i(\mathbf{v}_{S,c}) = c\,,$$

and

$$\sum_{i \in S \cup \{j\}} \Phi_i(\mathbf{v}_{S,c}) = c\,,$$

whenever $j \notin S$. This implies that $\Phi_j(\mathbf{v}_{S,c}) = 0$, whenever $j \notin S$. Also if π is a permutation of N which interchanges i and j (for any $i \in S$ and $j \in S$) and leaves the other players fixed, then it is clear that $\pi(\mathbf{v}_{S,c}) = \mathbf{v}_{S,c}$ and thus, by S2,

$$\Phi_i(\mathbf{v}_{S,c}) = \Phi_j(\mathbf{v}_{S,c}) \quad \text{for any } i \in S \text{ and } j \in S\,.$$

Consequently $\mathbf{\Phi}(\mathbf{v}_{S,c})$ is unique, if $\mathbf{\Phi}$ exists, and is given by

$$\Phi_i(\mathbf{v}_{S,c}) = \begin{cases} \dfrac{c}{|S|}, & \text{if } \; i \in S, \\ 0, & \text{if } \; i \notin S\,. \end{cases}$$

Now, consider the games $\mathbf{v}'_{S,c} \in V$, ($c \in \mathbb{R}$, $S \neq \emptyset$) defined by

$$v'_{S,c}(T) = \begin{cases} c & \text{if } \; T = S, \\ 0 & \text{if } \; T \neq S. \end{cases}$$

Any game $\mathbf{v}$ can be written as a finite sum of games of the type $\mathbf{v}'_{S,c}$. If we can show that each $\mathbf{\Phi}(\mathbf{v}'_{S,c})$ is unique, then the uniqueness of $\mathbf{\Phi}$ follows, using S3.

Assume that $\mathbf{\Phi}(\mathbf{v}'_{S,c})$ is unique for $|S| = k+1, \ldots, n$. (This is obviously true for $|S| = n$ because $\mathbf{v}'_{N,c} = \mathbf{v}_{N,c}$.) We will then show that $\mathbf{\Phi}(\mathbf{v}'_{S,c})$ is unique for $|S| = k$. Let $S_1, \ldots, S_l$ be all of the proper supersets of S. Note that $|S_i| > k$ for $i = 1, \ldots, l$, thus $\mathbf{\Phi}(\mathbf{v}'_{S_i,c})$ is unique by the inductive assumption. But

$$\mathbf{v}_{S,c} = \mathbf{v}'_{S,c} + \mathbf{v}'_{S_1,c} + \ldots + \mathbf{v}'_{S_l,c}\,.$$

Therefore, by S3,

$$\mathbf{\Phi}(\mathbf{v}_{S,c}) = \mathbf{\Phi}(\mathbf{v}'_{S,c}) + \mathbf{\Phi}(\mathbf{v}'_{S_1,c}) + \ldots + \mathbf{\Phi}(\mathbf{v}'_{S_l,c})\,. \tag{27}$$

$\mathbf{\Phi}(\mathbf{v}'_{S,c})$ is unique since all other terms in (27) have been shown unique. This concludes the proof that $\mathbf{\Phi}$, if it exists, is unique. ▮

Now we construct $\mathbf{\Phi}$. Suppose

$$(28)\qquad \Phi_i(v'_{S,c})=\begin{cases}\dfrac{(s-1)!(n-s)!}{n!}\cdot c\,, & \text{if } i\in S,\\[2ex] -\left(\dfrac{s}{n-s}\right)\dfrac{(s-1)!(n-s)!}{n!}\cdot c\,, & \text{if } i\notin S,\end{cases}$$

for $s=|S|=k+1,\ \ldots,\ n$. This is obviously true for $|S|=n$, since $\mathbf{v}'_{N,c}=\mathbf{v}_{N,c}$. It follows, using (27), that (28) holds for $|S|=k$ which makes (28) valid for any S.

It is now straightforward to obtain $\mathbf{\Phi}(\mathbf{v})$ for any $\mathbf{v}$. Since

$$\mathbf{v}=\sum_{\emptyset\neq S\subseteq N}\mathbf{v}'_{S,v(S)}$$

we have, by S3.

$$\mathbf{\Phi}(\mathbf{v})=\sum_{\emptyset\neq S\subseteq N}\mathbf{\Phi}(\mathbf{v}'_{S,v(S)})\,.$$

From the right-hand side, when simplified, we get

$$(29)\qquad \Phi_i(\mathbf{v})=\sum_{\{i\in T\subseteq N\}}\frac{(t-1)!(n-t)!}{n!}\Big[\mathbf{v}(T)-\mathbf{v}(T-\{i\})\Big],\qquad (i=1,\ldots,n)\,.$$

It is easy to verify that $\mathbf{\Phi}$, defined as above, satisfies the axioms where $t=|T|$, S1, S2, S3. ▮

Apart from the axiomatic treatment, the Shapley-value, as defined by (29) can be given another heuristic explanation. Assume the players agree to gather at a specified place and time. Because of random fluctuation, they will arrive at different times. We assume, however, that all possible orders of arrival are of the same probability: $\dfrac{1}{n!}$. We suppose that if player i arrives and finds there the members of coalition $T-\{i\}$, then he obtains the pay-off $v(T)-v(T-\{i\})$ which can be considered as his contribution to the coalition. Then the Shapley value, $\Phi_i(\mathbf{v})$ is the expected pay-off to player i under this randomization scheme.

Formally

$$\Phi_i(\mathbf{v})=\frac{1}{n!}\sum_{w\in\Omega}[v(P_{w,i}\cup\{i\})-v(P_{w,i})] \tag{30}$$

where Ω is the set of all orderings of N and $P_{w,i}$ is the set of predecessors of player i in the ordering w.

We remark that the Shapley-value can be derived in other ways, too, [31]. The Shapley-value behaves "nicely" for M-equivalent games, as stated by the next theorem.

THEOREM 14. [178] If $\Gamma_1=\{N, u\}$ and $\Gamma_2=\{N, v\}$ are M-equivalent and

$$v(M)=\alpha u(M)+\sum_{i\in M}\beta_i,$$

for any coalition $M\subseteqq N$, $\alpha>0$ and $\boldsymbol{\beta}=(\beta_1, \ldots, \beta_n)$ arbitrary, then

$$\boldsymbol{\Phi}(\mathbf{v})=\alpha\boldsymbol{\Phi}(\mathbf{u})+\boldsymbol{\beta}.$$

Proof. Using the explicit formula (29) we obtain

$$\begin{aligned}\Phi_i(\mathbf{v})=&\sum_{i\in M\subseteqq N}\frac{(|M|-1)!(n-|M|)!}{n!}[\alpha u(M)-\\ &-\alpha u(M-\{i\})+\beta_i]=\\ &=\alpha\Phi_i(\mathbf{u})+\beta_i\sum_{i\in M\subseteqq N}\frac{(|M|-1)!(n-|M|)!}{n!}=\\ &=\alpha\Phi_i(\mathbf{u})+\beta_i\sum_{m=1}^{n}\frac{(m-1)!(n-m)!}{n!}\binom{n-1}{m-1}=\\ &=\alpha\Phi_i(\mathbf{u})+\beta_i, \qquad (i=1, \ldots, n).\ \blacksquare\end{aligned}$$

As an example, we give the Shapley-value of a two-person game with characteristic function v:

$$\Phi_1(\mathbf{v}) = \frac{v(\{1\}) + v(\{1,2\}) - v(\{2\})}{2},$$

$$\Phi_2(\mathbf{v}) = \frac{v(\{2\}) + v(\{1,2\}) - v(\{1\})}{2}.$$

In geometric terms, the point $(\Phi_1(\mathbf{v}), \Phi_2(\mathbf{v}))$ bisects the line-segment determined by $x_1 + x_2 = v(\{1,2\})$, $x_1 \geqq v(\{1\})$, $x_2 \geqq v(\{2\})$.

For a three-person (0, 1) normalized game we can get the Shapley-values, by elementary calculation:

$$\Phi_1(\mathbf{v}) = \frac{v(\{1,2\}) + v(\{1,3\}) - 2v(\{2,3\}) + 2}{2}$$

$$\Phi_2(\mathbf{v}) = \frac{v(\{1,2\}) + v(\{2,3\}) - 2v(\{1,3\}) + 2}{6}$$

$$\Phi_3(\mathbf{v}) = \frac{v(\{1,3\}) + v(\{2,3\}) - 2v(\{1,2\}) + 2}{6}.$$

As another type of example we will consider a simple symmetric market game [164]. The model can be formulated in terms of gloves. Each one of the n players starts with one glove – either right- – or left-handed, and the players may trade them, or buy or sell them for money. At the end of the game, an assembled pair is worth \$ 1 to whoever holds it.

The characteristic function of the game, which states the dollar potential of each coalition S, is given by the equation

$$(31) \qquad v(S) = \min\{|S \cap R|, |S \cap L|\}.$$

We denoted by R and L the original sets of owners of right- and left-handed gloves, resp., and the notation $|A|$ means the number of elements of the set A.

To calculate the Shapley-value of this game we will make use of the "random order" version of the definition (see (30)). Let $r=|R|$ and $l=|L|$ and suppose $r \geqq l$.

Let $\Phi_{\text{right}}(r, l)$ denote the sum of the values to the r members of R. If we consider separately those orderings that end with a member of R $\left(\text{probability } \frac{r}{r+l}\right)$, and those orderings that end with a member of L $\left(\text{probability } \frac{l}{r+l}\right)$, we obtain the following difference equation:

$$\Phi_{\text{right}}(r, l) = \frac{r}{r+l}\,[\Phi_{\text{right}}(r-1, l) + v(R \cup L) - v(R' \cup L)] + \frac{l}{r+l}\,[\Phi_{\text{right}}(r, l-1)], \tag{32}$$

where R' is any set satisfying $R' \subset R$, $|R'| = r-1$.

The solution of this difference equation is

$$\Phi_{\text{right}}(r, l) = \frac{r}{2} - \frac{r-l}{2} \sum_{k=1}^{l} \frac{r!\,l!}{(r+k)!\,(l-k)!}$$

which can be verified by substituting it into (32). This amount is divided equally among the members of R. By axiom S1. we get

$$\Phi_{\text{right}} + \Phi_{\text{left}} = v(R \cup L) = l$$

and thus the values to members of L are easily determined. So the Shapley-values are as follows:

$$\Phi_i(\mathbf{v}) = \frac{1}{2} - \frac{r-l}{2r} \sum_{k=0}^{l} \frac{r!\,l!}{(r+k)!\,(l-k)!} \quad \text{for} \quad i \in R \tag{33}$$

$$\Phi_j(\mathbf{v}) = \frac{1}{2} + \frac{r-l}{2} \sum_{k=1}^{l} \frac{r!\,l!}{(r+k)!\,(l-k)!} \quad \text{for} \quad j \in L.$$

The case $r \leqq l$ is symmetrical. The table below gives an idea how these values look like for small numbers of traders.

Value to a member of R

r \ l	0	1	2	3	4	5	6	7	8
1	0	0.500	0.667	0.750	0.800	0.833	0.857	0.875	0.889
2	0	0.167	0.500	0.650	0.733	0.786	0.822	0.847	0.867
3	0	0.083	0.233	0.500	0.638	0.720	0.774	0.811	0.838
4	0	0.050	0.133	0.272	0.500	0.629	0.710	0.764	0.802
5	0	0.033	0.086	0.168	0.297	0.500	0.622	0.701	0.755
6	0	0.024	0.060	0.113	0.194	0.315	0.500	0.616	0.693
7	0	0.018	0.044	0.081	0.135	0.214	0.330	0.500	0.610
8	0	0.014	0.033	0.061	0.099	0.153	0.230	0.341	0.500

The Shapley-value solution definitely favours the "short" side of the market. For example, if $l<r$, the members of L, with less than half the population, get more than half the total profit, which is $v(R \cup L)=l$. On the other hand, the "long" side of the market is not totally defeated, since it gives some credit for the bargaining position of the group in oversupply.

It is worth mentioning that the core is a unique imputation where any trader in R gets \$ 1 and any player in L gets nothing if $r<l$.

24. Stability of pay-off configurations

The concepts which we have previously studied are mainly concerned with the distribution of the pay-off which the grand coalition can achieve. In this chapter we will study the stability of this distribution provided a coalition structure (not necessarily the grand coalition) has formed. By a *coalition structure* for an n-person cooperative game $\Gamma = \{N, v\}$ we mean a partition

$$\mathfrak{C} = \{M_1, \ldots, M_m\},$$

$$M_i \subseteqq N, \quad M_i \neq \emptyset, \qquad (i = 1, \ldots, m)$$

of the player set N such that $\bigcup_i M_i = N$ and the M_i are disjoint.

Suppose that structure $\mathfrak{C}$ is reached. Now the question arises: How will the amount $v(M_k)$ realizable by coalition M_k, $(k = 1, \ldots, m)$ be divided among its members taking into consideration the "power" of the player in the bargaining process which may actually take place during a play of the game.

For the time being we only assume the bargaining process to consist of "threats" and "counter-threats". A more realistic but mathematically less appealing model of the bargaining process will be discussed in the next chapter.

Now we define a *pay-off configuration* to be a pair

$$(\mathbf{x}, \mathfrak{C}) = (x_1, \ldots, x_n, M_1, \ldots, M_m)$$

where $\mathfrak{C}$ is a coalition structure and $\mathbf{x}$ is an n-vector satisfying

$$(1) \qquad \sum_{i \in M_k} x_i = v(M_k), \qquad (k=1, \ldots, m).$$

In a sense $(\mathbf{x}, \mathfrak{C})$ can be considered stable if $\mathbf{x}$ belongs to the core since in this case there is no coalition strong enough to disrupt the coalition structure. But the core is often empty, therefore we approach the question of stability from a different angle.

We consider objections which can be raised against a pay-off configuration. The concept of *Ψ-stability* is based on the idea of allowing only certain objections to a particular outcome. A pay-off configuration is said to be stable if the set of "allowable" objections is empty.

In mathematical terms, a function Ψ is given which assigns to each partition of N a collection of subsets of N (the allowable coalitions). Heuristically, $S \in \Psi(\mathfrak{C})$ if and only if the disrupting coalition S is allowed to form from the present coalition structure $\mathfrak{C}$. We call the pay-off configuration $(\mathbf{x}, \mathfrak{C})$ stable if

(i) $\sum_{i \in T} x_i \geqq v(T) \quad \text{for all} \quad T \in \Psi(\mathfrak{C})$,

(ii) $x_i > v(\{i\}), \quad \text{if} \quad \{i\} \notin \mathfrak{C}$.

Condition (i) states that none of the "allowable" coalitions has the power to disrupt $(\mathbf{x}, \mathfrak{C})$. Condition (ii) states that only those players can form a "one-member coalition" who cannot achieve more than $v(\{i\})$ in any "non-trivial" coalition.

The principal difficulty with Ψ-stability is deciding on a "reasonable" choice of function Ψ. If Ψ is too restrictive, then there will be many stable pay-off configurations. On the other hand, if Ψ is not restrictive enough, then there may not exist any stable pay-off configuration at all.

One "reasonable" possibility for Ψ would be what is called *k-stability*, where k is a positive integer. In this case, Ψ is defined by saying that $T \in \Psi(\mathfrak{C})$ if and only if there is a $C \in \mathfrak{C}$ such that $(T-C) \cup (C-T)$ has at most k elements.

This means that Ψ allows only those coalitions which differ from any coalition of $\mathfrak{C}$ by no more than k members. The "rational" behind this definition is based on the observation that the more players have to be "moved" the more difficult to disrupt a coalition structure could be.

If a game has at least one k-stable pay-off configuration, then it is called *k-stable*. Let k be an integer with $0 \leqq k \leqq n-2$ and let $\mathfrak{C}$ be a coalition structure, then a subset $C \subset N$ is called a *k-critical* coalition of $\mathfrak{C}$ if there exists a coalition $T \in \mathfrak{C}$ such that $|(C-T) \cup (T-C)| \leqq k$. It is clear that if $T \in \mathfrak{C}$, then T is a k-critical coalition of $\mathfrak{C}$ for every k.

In the following we will establish a few results concerning k-stability for two special classes of games: symmetric and quota games. From now on, unless otherwise stated, we will deal with superadditive $0-1$ normalized games.

A game $\Gamma = \{N, v\}$ is said to be *symmetric* if the characteristic function depends only on the size of a coalition, i.e.

$$v(T) = v(T') \quad \text{if} \quad |T| = |T'| .$$

We will simply write $v(T) = v(|T|)$.

A game $\Gamma = \{N, v\}$ is called a *quota game* if there exists a real n-tuple $\mathbf{q} = (q_1, \ldots, q_n)$, called the *quota*, such that

$$\text{(a)} \qquad \sum_{i=1}^{n} q_i = 1 ,$$

$$\text{(b)} \qquad v(\{i, j\}) = q_i + q_j , \quad i, j \in N , \quad i \neq j .$$

In the following we will use the notation $x(T) = \sum_{i \in T} x_i$ if T is a nonempty coalition and $(x_1, \ldots, x_n)$ is an n-vector.

THEOREM 1. [97] A symmetric game $\Gamma = \{N, v\}$ is k-stable if and only if $v(i) \leqq \frac{i}{n}$ for $i = 0, 1, \ldots, k+1$.

Proof. It is clear that $\left(\frac{1}{n}, \ldots, \frac{1}{n}; \{1\}, \{2\}, \ldots, \{n\}\right)$ is k-stable if the condition of the theorem is met.

Conversely, suppose $(\mathbf{x}, \tau)$ is k-stable and that $v(k+1) > \frac{k+1}{n}$. (Note that k-stability implies i-stability for $i \leqq k-1$, therefore for our indirect proof it is sufficient to assume that $v(i) \leqq \frac{i}{n}$ does not hold for $i = k+1$). Consider any positive integer a such that $a(k+1) \leqq n$. Since we may partition any coalition of $a(k+1)$ elements into a disjoint coalition of $k+1$ elements, we obtain, by superadditivity

$$(2) \qquad v[a(k+1)] \geqq av(k+1) > \frac{a(k+1)}{n}.$$

For any $T_i \in \tau$ it is clear that since $0 < |T_i| < n$ and $0 \leqq k \leqq n-2$ we may write

$$|T_i| = a_i(k+1) + b_i$$

where a_i, $(a_i > 0)$, b_i are integers such that

$$0 < a_i(k+1) < n \quad \text{and} \quad -k \leqq b_i \leqq k.$$

We consider three cases:

1. $b_i = 0$. By definition of k-stability using (2) we have

$$x(T_i) \geqq v(|T_i|) = v[a_i(k+1)] > a_i \frac{k+1}{n} = \frac{|T_i|}{n}.$$

2. $b_i < 0$. We first show that it is always possible to find a set B_i such that

$$(3) \qquad B_i \subset N - T_i, \quad |B_i| = |b_i|$$

and

$$(4) \qquad x(B_i) \leqq \frac{[1 - x(T_i)]\,|b_i|}{n - |T_i|}.$$

If this were not the case, then we would have to assume that for the $\binom{n-|T_i|}{|b_i|}$ coalitions B_i satisfying (3)

$$(5) \qquad x(B_i) > \frac{[1-x(T_i)]\,|b_i|}{n-|T_i|}.$$

Observe that each $j \in N - T_i$ appears in exactly $\binom{n-|T_i|-1}{|b_i|-1}$ of these sets, and so if we sum over all of them we obtain

$$\sum_{B_i} x(B_i) = \binom{n-|T_i|-1}{|b_i|-1} x(N-T_i) =$$

$$= \binom{n'-|T_i|-1}{|b_i|-1} [1-x(T_i)] >$$

$$> \binom{n-|T_i|}{|b_i|} \frac{[1-x(T_i)]\,|b_i|}{n-|T_i|} =$$

$$= \binom{n-|T_i|-1}{|b_i|-1} [1-x(T_i)],$$

which contradiction establishes the existence of a B_i meeting the conditions (3) and (4).

Since $|B_i| = |b_i| \leqq k$, $T_i \cup B_i$ is a k-critical coalition of τ and so

$$x(T_i) + \frac{[1-x(T_i)]\,|b_i|}{n-|T_i|} \geqq x(T_i) + x(B_i) \geqq$$

$$\geqq v(|T_i| + |b_i|) = v[a_i(k+1)] > a_i \frac{(k+1)}{n}.$$

Thus

$$\frac{x(T_i)(n-|T_i|-|b_i|) + |b_i|}{n-|T_i|} > a_i \frac{(k+1)}{n},$$

or

$$x(T_i) > \frac{(n-|T_i|)a_i(k+1)-n|b_i|}{n(n-|T_i|-|b_i|)} =$$

$$= \frac{(n-|T_i|)(|T_i|+|b_i|)-n|b_i|}{n(n-|T_i|-|b_i|)} = \frac{|T_i|}{n}.$$

3. $b_i > 0$. We first show that it is always possible to find a set B_i such that

$$(6) \qquad B_i \subset T_i, \quad |B_i| = b_i$$

and

$$(7) \qquad x(B_i) \geqq \frac{x(T_i)b_i}{|T_i|}.$$

If this were not the case, then we may sum over all $\binom{|T_i|}{b_i}$ sets B_i satisfying (6) and we get

$$\sum_{B_i} x(B_i) = \binom{|T_i|-1}{b_i-1} x(T_i) < \binom{|T_i|}{b_i} \frac{x(T_i)b_i}{|T_i|} =$$

$$= \binom{|T_i|-1}{b_i-1} x(T_i),$$

which is a contradiction. Observe that for any B_i meeting the conditions (6) and (7) the set $T_i - B_i$ is a k-critical coalition of τ, hence

$$x(T_i) - \frac{x(T_i)b_i}{|T_i|} \geqq x(T_i) - x(B_i) \geqq$$

$$\geqq v[a_i(k+1)] > \frac{a_i(k+1)}{n} = \frac{|T_i| - b_i}{n}.$$

Thus,

$$x(T_i) > \frac{|T_i|}{n}.$$

We have therefore shown that for any $T_i \in \tau$ the inequality $x(T_i) > \frac{|T_i|}{n}$ holds. Thus

$$1=x(N)=\sum_{T_i} x(T_i)>\sum_{T_i}\frac{|T_i|}{n}=1,$$

which is impossible and so $(\mathbf{x},\tau)$ is not k-stable which is a contradiction. ∎

In order to be able to characterize k-stable quota games it is useful to define what we mean by a *weak player*. Player i of a game $\Gamma=\{N,v\}$ with quota $\mathbf{q}$ is called weak if $q_i<0$. Since $v(\{i,j\})\geqq 0$, there is at most one weak player, and when n is odd there is no weak player. To see this we assume, without loss of generality, that n is a weak player and n is odd.

Then

$$\begin{aligned} v(N-\{n\}) &= v(\{1,2,\ldots,n-1\})\geqq \\ &\geqq v(\{1,2\})+\ldots+v(\{n-2,n-1\})= \\ &=\sum_{i=1}^{n-1} q_i=1-q_n>1 \end{aligned}$$

which is impossible.

THEOREM 2. [97] A quota game is 1-stable if and only if there is no weak player.

Proof. It is easy to see by direct substitution that if there is no weak player, then the pay-off configuration $(q_1, \ldots, q_n; \{1\}, \ldots, \{n\})$ is 1-stable.

Conversely, suppose there is a weak player, which by relabeling we may assume to be n, and let $(\mathbf{x},\tau)$ be a 1-stable pay-off configuration. Clearly $n\geqq 3$. Label the coalitions $T_1, \ldots, T_t$ of τ so that $n\in T_t$. For any $T_i\in\tau$, the 1-stability requirement (i) implies $v(T_i)\leqq x(T_i)$. Now, if $|T_i|$ is even, then T_i can be partitioned into $\frac{|T_i|}{2}$ non-overlapping two element coalitions, each of which has characteristic function value $v(\{i,j\})=q_i+q_j$. Thus $x(T_i)\geqq v(T_i)\geqq q(T_i)$. If $|T_i|>1$ and odd, then for any $k\in T_i$, $|T_i-\{k\}|$ is even, and so by the same argument

$x(T_i-\{k\})\geqq q(T_i-\{k\})$. Summing over all $k\in T_i$ we get

$$\sum_{k\in T_i} x(T_i-\{k\})=(|T_i|-1)x(T_i)\geqq$$

$$\geqq \sum_{k\in T_i} q(T_i-\{k\})=(|T_i|-1)q(T_i),$$

hence $x(T_i)\geqq q(T_i)$. If $|T_i|=1$, let $T_i=\{i\}$, and then for any $k\in N-\{i\}$, $\{i,k\}$ is a 1-critical coalition and so

$$x_i+x_k\geqq v(\{i,k\})=q_i+q_k\,.$$

Summing over all $k\in N-\{i\}$, we have

$$(n-2)x_i+x(N)\geqq(n-2)q_i+q(N)\,.$$

But $x(N)=q(N)=1$, so with $n\geqq 3$, $x_i\geqq q_i$. Since these inequalities hold for all $T_i\in\tau$ and since $x(N)=q(N)$, the equalities

$$(8)\qquad \begin{array}{ll} x(T_i)=q(T_i)=v(T_i), & \text{if } |T_i| \text{ is even} \\ x(T_i)=q(T_i) & \text{if } |T_i| \text{ is odd} \end{array}$$

must hold.

Next we show that if n is weak and $n\in T_t$, then $|T_t|$ is even. Suppose, on the contrary that $|T_t|$ is odd. If $|T_t|>1$, then by the partitioning argument $v(T_t)\geqq v(T_t-\{n\})\geqq q(T_t-\{n\})$ since $|T_t-\{n\}|$ is even. But we know that $x(T_t)=q(T_t)$, and since n is weak, $q_n<0$, so $v(T_t)\geqq q(T_t)-q_n>x(T_t)$ which violates the 1-stability condition (i). If $|T_t|=1$, then $T_t=\{n\}$ and we have shown above that $x_n=q_n<0$, which is impossible. Thus $|T_t|$ is even.

By (8) it is clear that in $N-T_t$ there is at least one k such that $q_k\geqq x_k$. Consider the 1-critical coalition $T_t\cup\{k\}$. Since $|T_t|$ is even, so is $|(T_t\cup\{k\})-\{n\}|$, and so we may partition that coalition into non-overlapping two-element coalitions:

$$v(T_t\cup\{k\})\geqq v[(T_t\cup\{k\})-\{n\}]\geqq q(T_t)+q_k-q_n\,.$$

But $q_n<0$ and $q_k\geqq x_k$, so

$$v(T_t\cup\{k\})>x(T_t)+x_k=x(T_t\cup\{k\}),$$

which violates the assumption that $(\mathbf{x}, \tau)$ is 1-stable. Thus we must conclude that there is no weak player. ∎

COROLLARY. All quota games with an odd number of players are 1-stable.

Proof. Theorem 2 coupled with the observation that when n is odd, there is no weak player.

THEOREM 3. [97] Let $\Gamma = \{N, v\}$ be a k-stable quota game and let $(\mathbf{x}, \tau)$ be a k-stable pay-off configuration. If n is odd or if n is even and $k \geqq 2$, then $\mathbf{x} = \mathbf{q}$. If n is even and $k = 1$, then either $\mathbf{x} = \mathbf{q}$ or $|T|$ is even and $v(T) = q(T) = x(T)$ for every $T \in \tau$. There are quota games with n even and $k = 1$ in which $\mathbf{x} \neq \mathbf{q}$.

Proof. Suppose $(\mathbf{x}, \tau)$, where $\tau = (T_1, \ldots, T_t)$, is 1-stable and that for some r, $x_r \neq q_r$. From the proof of Theorem 2 we know that for each $T_i \in \tau$, $x(T_i) = q(T_i)$. It follows, therefore, that in some T_i, say T_t, there exist r and s such that $x_r > q_r$ and $x_s < q_s$. Now assume that for $i \neq t$, $|T_i|$ is odd, then $T_i \cup \{s\}$ has an even number of elements and is 1-critical, so

$$x(T_i \cup \{s\}) \geqq v(T_i \cup \{s\}) \geqq q(T_i \cup \{s\}) =$$
$$= q(T_i) + q_s > x(T_i \cup \{s\})$$

which is impossible. Thus $|T_i|$ is even. If n is even, then so is $|T_t|$. Suppose n, and therefore $|T_t|$ is odd. Since we know (from the proof of Theorem 2) that if $|T_t| = \{r\}$, then $q_r = x_r$, it follows that $|T_t| > 1$. Since $|T_t - \{r\}|$ is even, $v(T_t - \{r\}) \geqq q(T_t - \{r\}) > x(T_t - \{r\})$, which is impossible. Thus if $(\mathbf{x}, \tau)$ is 1-stable either $\mathbf{x} = \mathbf{q}$ or $|T|$ is even for $T \in \tau$. Since any k-stable pair is also 1-stable, the conclusion also holds for k-stable pay-off configurations. If $|T|$ is even we know from the proof of Theorem 2 that $v(T) = q(T) = x(T)$.

Next, let us assume that n is even and $k \geqq 2$, and suppose $(\mathbf{x}, \tau)$ is k-stable and $\mathbf{x} \neq \mathbf{q}$. Thus there exists $r \in T_i$, for some i, such that $x_r > q_r$, for any $j \neq i$, and for any $j \neq i$, there exists $s \in T_j$, such that

$x_s \leqq q_s$. Consider $(T_i - \{r\}) \cup \{s\}$ which is k-critical for $k \geqq 2$ and which has an even number of elements since T_i does. Thus, by the partitioning argument

$$v[(T_i - \{r\}) \cup \{s\}] \geqq q(T_i) - q_r + q_s > x[(T_i - \{r\}) \cup \{s\}],$$

which is impossible. Thus, $\mathbf{x} = \mathbf{q}$.

It remains to give an example of a quota game in which $(\mathbf{x}, \tau)$ is 1-stable and $\mathbf{x} \neq \mathbf{q}$. Our example is also symmetric and is given by the following data:

$$n = 6, \quad q_i = \frac{1}{6}, \quad v(2) = \frac{4}{12}, \quad v(3) = \frac{5}{12},$$

$$v(4) = \frac{8}{12}, \quad v(5) = v(6) = 1.$$

It is easy to show that the pay-off configuration

$$\left(\frac{1}{12}, \frac{3}{12}, \frac{2}{12}, \frac{2}{12}, \frac{2}{12}, \frac{2}{12}; \{1,2\}, \{3,4\}, \{5,6\}\right)$$

is 1-stable. ∎

Additional properties of quota games in terms of k-stability can be found in [97].

In the definition of Ψ-stability a pay-off configuration is considered to be stable if coalitions given by the function Ψ cannot object successfully against the particular distribution of wealth. However, stability may be achieved if each objection can be turned down by an "effective counterobjection" by some coalition threatening to decrease the pay-off of the objecting coalition. The various *bargaining sets* defined and studied by Aumann, Maschler and Peleg are mathematical models of a bargaining process in which stability is achieved by a "balance" of objections and counterobjections.

We will restrict the set of possible pay-off configurations in game $\Gamma=\{N, v\}$ by making the following "rationality" assumptions, on $(\mathbf{x}, \mathfrak{C})$

(9) $\quad x_i \geqq v(\{i\})$ for all $i \in N$.

This is called "*individual rationality*" and it seems to be natural to accept it.

(10) $\quad x(L) \geqq v(L)$ for all $L \subset M_k \in \mathfrak{C}, \quad k=1, \ldots, m$.

This requirement, called "*coalitional rationality*" states that in a stable pay-off configuration no sub-coalition can assure more for its members than the pay-off they get in $(\mathbf{x}, \mathfrak{C})$.

Let now $\mathfrak{C}=\{M_1, \ldots, M_m)$ be a coalition structure and C a nonempty coalition. The set

$$P(C, \mathfrak{C})=\{i \mid i \in M_k, M_k \cap C \neq \emptyset\}$$

is called "*the partners* of C" in $\mathfrak{C}$. Thus player i is a partner of C in $\mathfrak{C}$ if he belongs to the same coalition M_k as some member of K. For the members of C to get their share from the pay-off configuration $(\mathbf{x}, \mathfrak{C})$ they only need the consent of their partners.

Let D and E be two disjoint subcoalitions of $M_k \in \mathfrak{C}$. Then, an *objection* of D against E in $(\mathbf{x}, \mathfrak{C})$ is defined to be a pay-off configuration $(\mathbf{y}, \mathfrak{D})$ which satisfies

(a) $P(D, \mathfrak{D}) \cap E=\emptyset$,
(b) $y_i > x_i$ for all $i \in D$,
(c) $y_i \geqq x_i$ for all $i \in P(D, \mathfrak{D})$.

Let $(\mathbf{x}, \mathfrak{C})$ be a pay-off configuration and D, E and objection $(\mathbf{y}, \mathfrak{D})$ as defined previously. Then, a counterobjection of E against D is defined to be a pay-off configuration $(\mathbf{z}, \mathfrak{E})$ satisfying the relations

(a) $D \not\subset P(E, \mathfrak{E})$,
(b) $z_i \geqq x_i$ for all $i \in P(E, \mathfrak{E})$,
(c) $z_i \geqq y_i$ for all $i \in P(E, \mathfrak{E}) \cap P(D, \mathfrak{D})$.

We can explain briefly these definitions in the following way: Members of D, when objecting against E, claim that they can get more by changing to a new pay-off configuration and that their new partners will not be worse off. Coalition E can counter-object if they are able to find a third pay-off configuration in which they and their partners receive at least as much as their original share. If they need some of D's partners for this, they give these players at least as much as in the objection.

A pay-off configuration $(\mathbf{x}, \mathfrak{C})$ is said to be *stable* if any objection can be counter-objected. The *bargaining set* $\mathfrak{M}$ is the set of all stable pay-off configurations.

Define $\mathfrak{M}_1$ as the set of pay-off configurations $(\mathbf{x}, \mathfrak{C})$ such that, whenever a coalition D objects against coalition E, there is at least one member of E who can counter-object.

Similarly, let $\mathfrak{M}_2$ be the set of pay-off configurations $(\mathbf{x}, \mathfrak{C})$ such that, whenever a trivial coalition $\{i\}$ has an objection against E coalition E can counter-object against player i.

The relations $\mathfrak{M} \subset \mathfrak{M}_1$ and $\mathfrak{M} \subset \mathfrak{M}_2$ are obvious but the relationship between $\mathfrak{M}_1$ and $\mathfrak{M}_2$ is not clear.

If we require only individual rationality and drop (10), then we get bargaining sets $\overline{\mathfrak{M}}$, $\overline{\mathfrak{M}}_1$, $\overline{\mathfrak{M}}_2$ by the above definitions replacing the term pay-off configuration with individually rational pay-off configuration.

None of the above bargaining sets is empty. The trivial pay-off configuration $(\mathbf{x}, \mathfrak{C})$ where $\mathbf{x} = \mathbf{0}$ and $\mathfrak{C} = (\{1\}, \ldots, \{n\})$ can easily be shown to be stable. However, the question of the existence of a vector $\mathbf{x}$ such that $(\mathbf{x}, \mathfrak{C})$ is stable for a particular coalition structure $\mathfrak{C}$ cannot be answered trivially. To establish such an existence theorem for $\overline{\mathfrak{M}}_1$ we need to define a relation between the pair of players i and k. We say that i is *stronger* than k in the pay-off configuration $(\mathbf{x}, \mathfrak{C})$ if i has an objection against k but k does not have any objection against i. We denote this by $i \gg k$. If neither $i \gg k$ nor $k \gg i$ holds, then we say that i and k are equally strong and denote it by $i \sim k$.

We remind the reader here that the amount $e(C)=v(C)-x(C)$ is called the *excess* of coalition C.

THEOREM 4. [131], [138]. Let $\Gamma=\{N, v\}$ be an n-person cooperative game and $\mathfrak{C}=\{M_1, \ldots, M_m\}$ a coalition structure. Then there exists at least one vector $\mathbf{x}$ such that $(\mathbf{x}, \mathfrak{C}) \in \overline{\mathfrak{M}}_1$.

Proof. Denote $x(\mathfrak{C})$ the set of those $\mathbf{x}$ for which $(\mathbf{x}, \mathfrak{C})$ is individually rational. We divide the proof into three main steps.

1. Let $c_1(\mathbf{x}), \ldots, c_n(\mathbf{x})$ be nonnegative continuous real-valued functions defined for each $\mathbf{x} \in X(\mathfrak{C})$. Assume that for each $\mathbf{x} \in X(\mathfrak{C})$ and $M_j \in \mathfrak{C}$ there exists a player $i \in M_j$ such that $c_i(\mathbf{x}) \geqq x_i$. We will prove that there exists at least one $\boldsymbol{\xi}=(\xi_1, \ldots, \xi_n) \in X(\mathfrak{C})$ such that $c_i(\boldsymbol{\xi}) \geqq \xi_i$ for all $i=1, \ldots, n$.

Let $\mathbf{x} \in X(\mathfrak{C})$ and $i \in N$ be arbitrary and define

$$d_i=\begin{cases} x_i-c_i(\mathbf{x}) & \text{if} \quad x_i \geqq c_i(\mathbf{x}) \\ 0 & \text{if} \quad x_i<c_i(\mathbf{x}). \end{cases}$$

If $i \in M_j$, then define

$$y_i=x_i-d_i+\frac{1}{|M_j|} d(M_j).$$

Thereby we have defined a mapping $\mathbf{x} \to \mathbf{y}$ for each $\mathbf{x} \in X(\mathfrak{C})$.

Clearly, this mapping is continuous and, since x_i, d_i and $c_i(\mathbf{x})$ are nonnegative, $y_i \geqq 0$ for all $i=1, \ldots, n$. Moreover

$$y(M_j)=x(M_j)=v(M_j) \quad \text{for all} \quad j=1, \ldots, m,$$

so $\mathbf{y} \in X(\mathfrak{C})$ implying that $\mathbf{x} \to \mathbf{y}$ is a mapping of $X(\mathfrak{C})$ into itself.

Suppose now that $x_i>c_i(\mathbf{x})$. Then $d_i>0$. By our assumption there is a $k \in M_j$, such that $x_k \leqq c_k(\mathbf{x})$ implying $d_k=0$. Thus

$$y_k \geqq x_k+\frac{d_i}{|M_j|}>x_k$$

which means that our mapping has no fixed point. However $X(\mathfrak{C})$ is closed, bounded and convex therefore any continuous mapping defined on it must have at least one fixed point by Brouwer's theorem. Having reached a contradiction we can state that $x_i > c_i(\mathbf{x})$ cannot hold.

2. Let $(\mathbf{x}, \mathfrak{C})$ be an individually rational pay-off configuration. We will prove that the relation $\gg$ is not cyclic.

By the definition of an objection if i and k belong to different coalitions, then $i \sim k$. Suppose now that there is a coalition M_j containing players $1, 2, \ldots, t$ such that $1 \gg 2 \gg \ldots \gg t \gg 1$. This means that each player $i (1 \leqq i \leqq t)$ has an objection against $i+1$ $(t+1=1)$, while $i+1$ cannot counter-object. Let $P(i, \mathfrak{F}_i)$ be the set of i's partners in the "objecting" coalition structure $\mathfrak{F}_i$. Clearly, $P(i, \mathfrak{F}_i) \in \mathfrak{F}_i$. Let $P(i_0, \mathfrak{F}_{i_0})$ have the maximal excess for $i=1, \ldots, t$. We assert that i_0 can counter-object against i_0-1 (if $i_0=1$, then $i_0-1=t$) through the same coalition structure $\mathfrak{F}_{i_0}$. Since in his objection i_0-1 can offer at most $e[P(i_0-1), \mathfrak{F}_{i_0-1}]$ to his partners and since $e[P(i_0, \mathfrak{F}_{i_0})] \geqq e[P(i_0-1), \mathfrak{F}_{i_0-1}]$, therefore i_0 can offer to the common partners of i_0-1 and i_0 in his counterobjection at least as much as i_0-1 can. To prove the validity of i_0's counterobjection we only have to show that property (a) holds, i.e. $i_0-1 \notin P(i_0, \mathfrak{F}_{i_0})$. Assuming, on the contrary, $i_0-1 \in P(i_0, \mathfrak{F}_{i_0})$ we would have $P(i_0-1, \mathfrak{F}_{i_0}) = P(i_0, \mathfrak{F}_{i_0})$ and $e[P(i_0-1), \mathfrak{F}_{i_0}] = e[P(i_0, \mathfrak{F}_{i_0})]$. Repeating this argument we would get $i_0-2 \in P[(i_0-1), \mathfrak{F}_{i_0}] = P(i_0, \mathfrak{F}_{i_0})$ and finally $i_0+1 \in P(i_0, \mathfrak{F}_{i_0})$ which contradicts to condition (a) of an objection.

3. Now, using the results of the two previous steps, we will prove the theorem. Let $(\mathbf{x}, \mathfrak{C})$ be an individually rational pay-off configuration. Denote $(\mathbf{y}^{M_j}, \mathbf{x}^{N-M_j}, \mathfrak{C})$ the individually rational pay-off configuration which is obtained from $(\mathbf{x}, \mathfrak{C})$ by holding x_i fixed for $i \in N - M_j$ and replacing x_k by y_k for $k \in M_j$. Furthermore $y_k \geqq 0$, $\sum_{k \in M_j} y_k = v(M_j)$ where $M_j \in \mathfrak{C}$ and the y_k's are treated as variables.

Denote by $E_j^{(i)}(\mathbf{x})$ the set of those $\mathbf{y}^{M_j}$ of pay-off configuration $(\mathbf{y}^{M_j}, \mathbf{x}^{N-M_j}, \mathfrak{C})$ for which player $i \in M_j$ is not weaker than any other one in M_j.

The set $E_j^{(i)}(\mathbf{x})$ is obviously closed and it is nonempty. Clearly, and $y_i = 0$, and

$$\sum_{\substack{i \neq k \\ k \in M_j}} y_k = \sum_{k \in M_j} x_k = v(M_j)$$

belongs to $E_j^{(i)}(\mathbf{x})$, since i can always counter-object through the coalition structure where he himself is alone and the others get the same pay-offs as before.

Consider now the functions

$$c_i(\mathbf{x}) = x_i + \max_{\mathbf{y}^{M_j} \in E_j^{(i)}(\mathbf{x})} \min_{k \in M_j} (x_k - y_k), \quad i \in M_j.$$

These functions are obviously continuous, and we will show that $c_i(\mathbf{x}) \geqq 0$, $(i \in M_j)$. Let, namely, $\bar{y}_i = 0$ and $\bar{y}_k \geqq x_k$, $(k \in M_j,\ k \neq i)$.

Since player i can always counter-object, $y^{M_j} \in E_j^{(i)}(\mathbf{x})$. The equality

$$x_i + \sum_{k \neq i} x_k = \sum_{k \neq i} \bar{y}_k$$

implies $x_i = \sum_{k \neq i} \bar{y}_k - x_k$, therefore $\bar{y}_k - x_k \leqq x_i$ holds for any k, i.e. $x_k - \bar{y}_k \geqq -x_i$ for any $k \in M_j$ from which $c_i(\mathbf{x}) \geqq 0$ follows.

By the assertion of step 2, for any $\mathbf{x} \in X(\mathfrak{C})$ and $M_j \in \mathfrak{C}$ there is an i not weaker than any other $k \in M_j$. Hence $\mathbf{x}^{M_j} \in E_j^{(i)}(\mathbf{x})$ and thus $c_i(\mathbf{x}) \geqq x_i$. We have proved in step 1 that, in this case, there is a $\boldsymbol{\xi} \in X(\mathfrak{C})$ such that $c_i(M) \geqq \xi_i$ for any $i \in N$. Since $\sum_{k \in M_j} \xi_k = \sum_{k \in M_j} y_k$, and $\min_{k \in M_j} (\xi_k - y_k) \leqq 0$ implying $c_i(\boldsymbol{\xi}) \leqq \xi_i$. Hence $c_i(\boldsymbol{\xi}) = \xi_i$ for any $i \in N$. This, in turn, means that for any player i there is a $\mathbf{y}$ such that $\mathbf{y} \in E_j^{(i)}(\boldsymbol{\xi})$ and $y_k = \xi_k$. Thus $\boldsymbol{\xi}^{M_j} \in E_j^{(i)}(\boldsymbol{\xi})$ for all i and j which means that in the pay-off configuration $(\boldsymbol{\xi}, \mathfrak{C})$ no player is weaker than any other i.e. $(\boldsymbol{\xi}, \mathfrak{C}) \in \overline{\mathfrak{M}}_1$. ∎

Before giving an example we define a *simple game*. A (0,1) normalized game $\Gamma = \{N, v\}$ is said to be simple if for each coalition $M \subset N$ either $v(M)=0$ or $v(M)=1$ holds. Coalition M is called *winning* if $v(M)=1$ and *loosing* if $v(M)=0$.

Consider now a five-person simple game in which the winning coalitions are $\{1,2\}$, $\{1,3\}$, $\{1,4\}$, $\{1,5\}$ and $\{2,3,4,5\}$.

Suppose now that the coalition $\{2,3,4,5\}$ forms. The individually rational pay-off configurations $(\mathbf{x}, \mathfrak{C})$ with respect to the coalition structure $\mathfrak{C} = (\{1\}, \{2,3,4,5\})$ are nonnegative vectors $\mathbf{x}$ with sum 1 and $x_1 = 0$. Assume $x_2 > x_3$. Then player 3 can object against 2 with $(1 - x_3 - \varepsilon, 0, x_3 + \varepsilon, 0, 0)$ where $0 < \varepsilon < x_2 - x_3$. It is easy to verify that 2 has no counter-objection. By symmetry, we have that $\overline{\mathfrak{M}}_1$ consists of the single pay-off:

$$x_1 = 0, \quad x_2 = x_3 = x_4 = x_5 = \frac{1}{4}.$$

25. A bargaining model of cooperative games

In previous chapters our investigation of a cooperative game $\Gamma=\{N, v\}$ was based on two fundamental assumptions:

(i) each coalition S of players can assure itself a particular amount $v(S)$ of resource no matter what the remaining players do,

(ii) any coalition may divide what it receives among its players in a completely arbitrary manner i.e. there is no restriction on side payments between players.

The aim of this chapter is to present an elaborate model of the bargaining process as a result of which the amount $v(S)$ is distributed among members of S. We will formalize the bargaining procedure as a multi-stage process. In Chapter 24, we have dealt with a two-stage process where objections and counter-objections have been permitted to be raised only once. In this respect, the model to be discussed in this chapter can be considered more general. The model is due to Weber [200].

At the beginning of a stage, an imputation $\mathbf{x}$ is given to represent the proposal for final allocation which is presently under consideration. All coalitions which wish to amend this proposal by making an objection to $\mathbf{x}$ declare the action they wish to take. A permissible action for a coalition S is the suggestion of an allocation $\mathbf{y}^S$ among the players of S of a total amount not exceeding $v(S)$, where this allocation is strictly preferred by all members of S to their present shares in $\mathbf{x}$. We assume that external factors determine which of the (possibly more than one) objecting coalitions is actually "given the floor" to make its suggestion. Next, the players in the complementary coalition $N-S$ respond to the action of S by agreeing on an allocation $\mathbf{z}^{N-S}$ of the remaining resources $v(N)-y^S(S)$ among themselves. Thus a new proposed allocation $(\mathbf{y}^S, \mathbf{z}^{N-S})$ is constructed, and the next stage of the bargaining process commences with this new proposal replacing $\mathbf{x}$. When no coalition opposes a proposal at some stage, it is considered to be accepted by all players, and the final division of resources occurs accordingly.

It should be noticed the particular final agreed-upon outcome may depend on the initial imputation from which the first stage begins. Thus the collection of all possible final outcomes forms a set which is, in a sense, stable and which can be interpreted as a "standard of behaviour" to which the players will conform. (This interpretation is quite similar to the one give by von Neumann and Morgenstern for stable-sets. See Chapter 23.)

When the players of a game consider a proposed allocation of resources among themselves, it is possible that several coalitions will wish to raise objections to this proposal. If several coalitions do indeed wish to act, there must be some mechanism that decides which coalition is given the chance to state its objection. In the real world, this selection may be governed by a number of factors: random choice, the size of the coalition etc.

We shall assume the mechanism, representing all external factors, depends solely on the proposal being considered and on the collection of objecting coalitions, and therefore the mechanism is independent of time, history, and experience. The mechanism may be of a probabilistic nature and may include the possibility that no objecting coalition is given the floor.

Formally, we define a coalitional *hierarchy* H to be a function which assigns to every collection $\mathfrak{S}=\{S, T, \ldots\}$ of coalitions a "sub-probability distribution" over $\mathfrak{S}$. That is, to each S in $\mathfrak{S}$, $H(\mathfrak{S})$ assigns a probability $P_{H(\mathfrak{S})}(S)$, so that $\sum_{S\in\mathfrak{S}} P_{H(\mathfrak{S})}(S)\leqq 1$. The probability that the hierarchy selects no coalition from the objecting coalitions $\mathfrak{S}$ is $H_0(\mathfrak{S})=1-\sum_{S\in\mathfrak{S}} P_{H(\mathfrak{S})}(S)$.

A *hierarchical structure* for a game Γ is a mapping $\mathfrak{H}$ which associates a hierarchy $H_{\mathbf{x}}$ to each imputation $\mathbf{x}$ in the imputation space $X \subset \mathbb{R}^n$. Hence a hierarchical structure represents a choice mechanism which chooses between objecting coalitions at any proposal which arises in the course of the bargaining process, and abstracts all external factors which play a role in such a choice mechanism.

Several types of hierarchical structures have special intuitive appeal. One type is the *uniform* structure, which assigns

$$P_{H_{\mathbf{x}}(\mathfrak{S})}(S) = \frac{1}{|S|}$$

for all $\mathbf{x}$ in X and $H_{\mathbf{x}}$ in $\mathfrak{H}$. Another type is the *excess* structure, in which for each $\mathbf{x}$ in X, $H_{\mathbf{x}}(\mathfrak{S})$ assigns equal probability to all coalitions S in $\mathfrak{S}$ which maximize $v(S) - x(S)$ (and zero probability to all other coalitions). A third type is the *linear* structure, in which a weight w_i is assigned to each player i in the player set N, and each $H_{\mathbf{x}}(S)$ assigns equal probability to all coalitions in $\mathfrak{S}$ of equal maximal total weight. In the following we shall restrict ourselves to considering games with the uniform hierarchical structure.

In the model of bargaining previously discussed, when a player acts as a member of a coalition, he lacks determinate knowledge of what the ultimate result of this action will be because of the uncertainty with regard to other players' actions. The most that a player can do is to anticipate the probable result of his actions. Therefore, in trying to analyze the problem of what actions a player will take, it is necessary to have some knowledge of his preferences over probabilistic outcomes.

The anticipated result of a player's actions can be considered as a probability distribution over the space of possible outcomes. For the sake of simplicity we shall be concerned only with finite or discrete distributions. We assume that each player is concerned only with the amount he personally receives in any imputation. Therefore, the discrete probability distributions which represent probabilistic outcomes to a player may be described by sequences of the form $(x_1, p_1; x_2, p_2; \ldots)$, where the x_i are distinct real numbers, the p_i are positive and sum to 1. The meaning of such a sequence is that the probability of the player receiving x_i in an outcome is p_i. Thus each discrete distribution over the imputation space X induces a discrete distribution for each player.

Given distributions $A=(x_1, p_1; x_2, p_2; \ldots)$ and $B=(y_1, q_1; y_2, q_2; \ldots)$, for any $0 \leqq \lambda \leqq 1$ we define the distribution $\lambda A+(1-\lambda)B= =(z_1, r_1; z_2, r_2; \ldots)$ by $r_i=\lambda P_A(z_i)+(1-\lambda)P_B(z_i)$ where $\{z_1, z_2, \ldots\}=\{x_1, x_2 \ldots\}\cup\{y_1, y_2, \ldots\}$. A preference ordering for a player is a total ordering "$\succsim$" of the space of all discrete distributions for which the following three axioms hold. Let A, B and C be any discrete distributions.

I. If $A \sim C$, then for any $0 \leqq \lambda \leqq 1$ and B

$$(\lambda A+(1-\lambda)B) \sim (\lambda C+(1-\lambda)B).$$

II. If $A \succ C$, then for any $0<\lambda \leqq 1$ and B

$$(\lambda A+(1-\lambda)B) \succ (\lambda C+(1-\lambda)B).$$

III. If x and y are real numbers, then $(x, 1) \succ (y, 1)$.

The expected value ordering "$\succsim_E$" is of special importance. We define $(x_1, p_1; x_2, p_2; \ldots) \succ_E (y_1, q_1; y_2 q_2; \ldots)$ if and only if $\Sigma x_i p_i > \Sigma y_i q_i$.

Assume now that to each player i in a coalition S there is an associated preference ordering "$\succsim_i$". Let $\mathfrak{A}$ (resp. $\mathfrak{B}$) be a discrete distribution over a set of imputations, and let A_i (resp. B_i) be the distribution induced for i. Then S is said to prefer $\mathfrak{A}$ to $\mathfrak{B}$, written $\mathfrak{A} \succ_S \mathfrak{B}$, if $A_i \succ_i B_i$ for every player i in S. Relation "$\succsim_S$" satisfies axioms I–III but is generally not a total order.

The bargaining process outlined previously can be modelled in several ways. One of the difficulties encountered when setting up a formal bargaining model is the treatment of "stopping rules". There are several reasons for a bargaining procedure to terminate. Just to mention a few: a satisfactory agreement has been reached, an external time limit forces termination of the proceedings, the participants have reached a point of exhaustion etc. We shall define games with termination rules representative of the "time limit" and "exhaustion" stopping rules.

We first define the "time limit" bargaining game. The definition is of a recursive nature. Let $\Gamma = \{N, v\}$ be the cooperative game under consideration, and $\mathfrak{H}$ be the hierarchical structure associated with the imputation space X of this game. Let $\mathbf{x} \in X$ and $\mathbf{c} \in \mathbb{R}^n$ which represents conflict pay-offs to the players.

The bargaining game $B(N, v, \mathfrak{H}, \mathbf{c}, \mathbf{x}, 0)$ is the "null"-game in which each player i in N receives the pay-off x_i.

Let T be a positive integer. Then the *bargaining game* $B(N, v, \mathfrak{H}, \mathbf{c}, \mathbf{x}, T)$ is played in the following manner. Each player i in N declares for each coalition S containing i, a vector $\mathbf{y}^{i,S} \in \mathbb{R}^{|S|}$. All declarations by all players are made simultaneously. Let

$$\mathfrak{S} = \{S \mid \mathbf{y}^{i,S} = \mathbf{y}^{j,S} \equiv \mathbf{y}^S \text{ for all } i, j \in S,\ \mathbf{y}^S \text{ dominates } \mathbf{x} \text{ through } S\}$$

be the collection of all coalitions whose players unanimously declare an allocation which dominates the current proposal. The hierarchy $H_{\mathbf{x}}$ is used to select a coalition from $\mathfrak{S}$. If the hierarchy fails to select a coalition (as will always be the case if $\mathfrak{S}$ is empty), the game ends with final pay-off vector $\mathbf{x}$. On the other hand, if a coalition W in $\mathfrak{S}$ is selected by the hierarchy, then the response-bargaining game $\bar{B}(N, v, \mathfrak{H}, \mathbf{c}, W, \mathbf{y}^W, T)$ ensues.

Take $N, v, \mathfrak{H}, \mathbf{c}$ and T as previously defined. Let W be a coalition in N, and $\mathbf{y}^W \in R^W(v(W))$ be a vector. $R^W(v(W)) = \{\mathbf{y} \in \mathbb{R}^{|W|} \mid y_k \geqq v(\{k\})$ for all $k \in W$, $\sum_{k \in W} y_k \leqq v(W)\}$ (i.e. $R^W(v(W))$ is a projection of the imputation space X). The *response-bargaining game* is played as follows. Each player i in $N - W$ declares a vector $\mathbf{z}^{i, N-W} \in \mathbb{R}^{|N-W|}$ for which $(\mathbf{y}^W, \mathbf{z}^{i, N-W}) \in X$. The declarations by all players are made simultaneously. If for some pair of players i and j in $N - W$, $\mathbf{z}^{i, N-W} \neq \mathbf{z}^{j, N-W}$ then the game ends with final pay-offs y_k^W for all players k in W and pay-offs c_k for all k in $N - W$. On the other hand, if all $\mathbf{z}^{i,N-W} \equiv \mathbf{z}^{N-W}$, then the bargaining game $B(N, v, \mathfrak{H}, \mathbf{c}, (\mathbf{y}^W, \mathbf{z}^{N-W}), T-1)$ ensues.

We next define the bargaining game which terminates upon "exhaustion" of the players. The definition is again recursive. Let N, v, $\mathfrak{H}$ and $\mathbf{c}$ be as previously defined. Let $0<\delta\leq 1$ be a real number and $\mathbf{x} \in X$. The bargaining game $E(N, v, \mathfrak{H}, \mathbf{c}, \mathbf{x}, \delta)$ is played similarly to the game $B(N, v, \mathfrak{H}, \mathbf{c}, \mathbf{x}, 1)$, with the following differences. With probability δ a chance event occurs before the players make their declarations, and the game ends with final pay-off vector $\mathbf{x}$. Otherwise, if the players in coalition W unanimously declare $\mathbf{y}^W$, and W is selected by the hierarchy $H_{\mathbf{x}}$, then the response-bargaining game $\bar{E}(N, v, \mathfrak{H}, \mathbf{c}, W, \mathbf{y}^W, \delta)$ ensues. This response-bargaining game is similar to the game $\bar{B}(N, v, \mathfrak{H}, \mathbf{c}, W, \mathbf{y}^W, 1)$ except that if the players in $N-W$ unanimously declare $\mathbf{z}^{N-W}$, then the bargaining game $E(N, v, \mathfrak{H}, \mathbf{c}, (\mathbf{y}^W, \mathbf{z}^{N-W}), \delta)$ ensues. Since $\delta>0$ implies that the bargaining game has probability 1 of terminating after a finite number of moves, we arbitrarily assign pay-offs of 0 to all players in the case of infinite play. The stopping probability δ represents the chance that the players will be so exhausted after any stage of the game as to forego their possible objections to the current proposal.

In both types of bargaining games, the information structure is such that each player has full knowledge of the parameters of the game, and remembers the full history of the game as it progresses, including all players' declarations at each previous stage. In order to make things as simple as possible, of the two types of bargaining games discussed, we shall deal only with the type based on the "exhaustion" stopping rule. The "time-limit" game can be treated similarly though not every result carries over automatically to this case.

As a bargaining game is being played, a sequence of declarations (by the players) and chance selections (by the hierarchical structure) is generated. Any play of the game up to a particular moment in time (at which the play of a bargaining or response-bargaining subgame commences) may be described by the initial parameters of the game and such a sequence. We call such a description a *history* of the bargaining game.

A *bargaining situation* $L(N, v, \mathfrak{H}, \mathbf{c}_0, \delta_0)$ is the collection of all bargaining and response-bargaining games which are defined in terms of N, v, $\mathfrak{H}$, with conflict pay-off vector $\mathbf{c}$ satisfying $c_{0i} \leqq c_i < v(\{i\})$ for all i in N and with stopping probability $0 < \delta \leqq \delta_0$. Thus a bargaining situation consists of all games in a "neighbourhood" of the game with conflict pay-offs $c_i = v(\{i\})$ and with stopping probability 0.

We wish to consider systems of behaviour for a player which describe how he will act in any game in a given bargaining situation. Therefore, define a *global pure strategy* for a player i in a bargaining situation as a function which maps each game in the situation, and each associated history of that game, into an action by i of the type called for in the resulting subgame (depending on this subgame, such an action is a collection of declarations $\{\mathbf{v}^{i,S}\}_{i \in S}$, a declaration $\mathbf{y}^{i,N-W}$, or a "null action" if the subgame is a response-bargaining game in which i has no move). Note that this definition requires a full plan of action for every game in the given bargaining situation.

We shall actually work with the more general concept of a global behavioural strategy. Such a strategy for a player in a bargaining situation maps each bargaining game and associated history into a finite (probabilistic) sample space and a function from the sample space into the set of actions which the player may probably take. Each sample space and associated function correspond to a "random experiment" performed by the player to select his action at a particular stage of the game, and for later simplicity we assume that all such experiments are independently repeatable. We require the sample space to be finite for reasons of both theoretical and notational convenience.

A *strategy n-tuple* $\boldsymbol{\sigma} = (\sigma_1, \ldots, \sigma_n)$ for a bargaining situation $L(N, v, \mathfrak{H}, \mathbf{c}_0, \delta_0)$ is a collection of global behavioural strategies, with σ_i being the strategy of player i. Because each player's strategy at a stage of the game is a randomization over a finite number of actions, any strategy n-tuple $\boldsymbol{\sigma}$ associates to every game in a bargaining situation a discrete distribution over the outcome space of the game.

We shall pay special attention to certain special strategies, called reactive strategies. A strategy for a player in a bargaining situation $L(N, v, \mathfrak{H}, \mathbf{c}_0, \delta_0)$ is a *reactive strategy* if it specifies the same action for the player in all bargaining and response-bargaining games which differ only in their conflict pay-offs, stopping probabilities, and histories leading up to the games. Thus, when playing a reactive strategy in a given situation, a player "reacts" only to the vector $\mathbf{x}$ in a game $E(N, v, \mathfrak{H}, \mathbf{c}, \mathbf{x}, \delta)$, and only to the vector $\mathbf{y}^W$ in a game $\bar{E}(N, v, \mathfrak{H}, \mathbf{c}, W, \mathbf{y}^W, \delta)$ – with no regard for the circumstances which led to the play of the game, or for the values of $\mathbf{c}$ and δ.

From the collection of all strategy n-tuples in a game, it is desirable to be able to single out those which exhibit some form of stability. A general approach to this is to apply Nash's idea of an equilibrium n-tuple of strategy, an approach used throughout our discussions of noncooperative games. For defining equilibrium for bargaining games of the kind we have been dealing with we shall use an extension of Nash's concept.

Let $\boldsymbol{\sigma} = (\sigma_1, \ldots, \sigma_n)$ be an n-tuple of global behavioural strategies for the situation $L(N, v, \mathfrak{H}, \mathbf{c}_0, \delta_0)$. As discussed previously, to any particular game G in the bargaining situation L, $\boldsymbol{\sigma}$ associates a discrete probability distribution $\psi(\boldsymbol{\sigma}, G)$ over the imputation space of the game, and this in turn induces a collection of distributions $\{\psi_i(\boldsymbol{\sigma}, G)\}_{i \in N}$, one for each player over his outcome space. Let σ_i' be a strategy for player i, and let $\boldsymbol{\sigma}' = (\sigma_1, \ldots, \sigma_{i-1} \sigma_i', \sigma_{i+1}, \ldots, \sigma_n)$. σ_i' is a *better response* than σ_i in σ if for some game G in L, $\psi_i(\boldsymbol{\sigma}', G) \succ_i \psi_i(\boldsymbol{\sigma}, G)$, and for every G in L, $\psi_i(\boldsymbol{\sigma}', G) \succsim_i \psi_i(\boldsymbol{\sigma}, G)$. Thus a better response for a player with respect to an n-tuple of strategies for a bargaining situation is a strategy change which benefits him in some game of the situation, and which hurts him in no game of the situation. An n-tuple of strategies $\boldsymbol{\sigma}$ is an *individual equilibrium point* for a bargaining situation if no player has a better response than his strategy in $\boldsymbol{\sigma}$.

Let S be a fixed coalition, and $\boldsymbol{\sigma}$ an individual equilibrium point for the bargaining situation $L(N, v, \mathfrak{H}, \mathbf{c}_0, \delta_0)$. A strategy n-tuple $\boldsymbol{\tau}=(\tau_1, \ldots, \tau_n)$ is *related to* $\boldsymbol{\sigma}$ by S if the following conditions are satisfied. First, for every player i not in S, $\sigma_i=\tau_i$. Second, for every player i in S, τ_i is arbitrary with respect to all response-bargaining subgames arising in L, but in any bargaining subgame σ_i and τ_i differ only in the specification of declarations of the form $\mathbf{y}^{i,S}$. That is, the sample spaces and functions $\{\tau_i\}_{i\in S}$ associated with each particular bargaining subgame are such that the induced distributions and correlations between declarations made by the players, with respect to coalitions other than S, remain unchanged. An individual equilibrium point $\boldsymbol{\sigma}$ is a *coalitional equilibrium point* for the bargaining situation $L(N, v, \mathfrak{H}, \mathbf{c}_0, \delta_0)$ if there is no coalition S and n-tuple $\boldsymbol{\tau}$ related to $\boldsymbol{\sigma}$ by S for which $\psi_S(\boldsymbol{\tau}, G) \succ_S \psi_S(\boldsymbol{\sigma}, G)$ for some G in L and $\psi_S(\boldsymbol{\tau}, G) \succsim \psi_S(\boldsymbol{\sigma}, G)$ for all G in L.

If we restrict our considerations to reactive strategies, then we can define analogously *individual and coalitional reactive equilibrium points* resp.

Let now $\boldsymbol{\sigma}$ be a reactive coalitional equilibrium point and let G be a bargaining game (*not* a response-bargaining game) in the associated bargaining situation, and let $\mathbf{x}$ be the proposal associated with G. Further let $\mathfrak{S}$ be the collection of all coalitions which take effective unanimous action (raise a valid, enforceable objection) against $\mathbf{x}$. Then $\boldsymbol{\sigma}$ is *motivated* at $\mathbf{x}$ if for at least one coalition S in $\mathfrak{S}$, that coalition prefers its result from its objection at $\mathbf{x}$ to the particular outcome of $\mathbf{x}$. If $\boldsymbol{\sigma}$ is motivated at every $\mathbf{x} \in X$, then $\boldsymbol{\sigma}$ is a *motivated reactive coalitional equilibrium point.*

In the following we restrict our attention to n-tuples of reactive strategies which form motivated coalitional equilibrium points. The following theorem allows us to restrict our search for equilibria to only those situations in which all coalitions always cooperate fully in all response-bargaining games.

THEOREM 1. [200] If $\boldsymbol{\sigma}$ is a reactive equilibrium point, then in every response-bargaining subgame in which the players of a coalition S are to move, all of these players make the same declaration with probability 1.

Proof. We first show that for any player i and imputation $\mathbf{x}$ with $x_i \geqq c_i$ (c_i is the conflict pay-off to player i), player i prefers the result of following $\boldsymbol{\sigma}$, in the bargaining game beginning at $\mathbf{x}$, to the outcome of receiving c_i with certainty. This follows immediately from the observation that i can refuse to cooperate in objections to $\mathbf{x}$ and also refuse to cooperate in all responses arising from objections to $\mathbf{x}$, and in this way he can assure himself of at least c_i in all eventualities.

Assume that the players of S fail to cooperate in some response in $\boldsymbol{\sigma}$. Then in their noncooperative response, each player i receives c_i. However, by dividing the amount available in response equally, each player i in S receives at least $v(\{i\})$ immediately, and expects no less than c_i eventually (by the preceding paragraph). Thus the cooperative response is a better response for all players in S than their strategies in $\boldsymbol{\sigma}$, and $\boldsymbol{\sigma}$ cannot be in equilibrium. ∎

From the proof of the above theorem, we immediately have a guarantee of individual rationality at all imputations to which no objections are made. Thus, if the hierarchical structure gives positive probability to the recognition of some coalition whenever objections are raised, then every imputation $\mathbf{x}$, to which no objection is made, is individually rational (satisfies $x_i \geqq v(\{i\})$ for all players i).

Our next theorem shows that each coalition, if it has positive probability of raising an effective objection to a proposal $\mathbf{x}$, might as well object to $\mathbf{x}$ with probability 1.

THEOREM 2. [200] If $\boldsymbol{\sigma}$ is a reactive equilibrium point, $\mathbf{x}$ is an imputation, and S is a coalition of players who have in $\boldsymbol{\sigma}$ a positive probability of all making the same objection (as players of S) against $\mathbf{x}$, then there exists an equilibrium point in which the players of S

correlate their play so that they are unanimous in their objection with probability 1.

Proof. Assume the players of S cooperate with probability p in raising an objection to $\mathbf{x}$. Let q be the probability that they are selected by the hierarchy at $\mathbf{x}$ when they make a unanimous objection. Since $\boldsymbol{\sigma}$ is an equilibrium point, we must have the event

A: with probability $p(1-q)$, S objects and some other coalition is given the floor; with probability pq, S objects and is given the floor; with probability $(1-p)$, S does not raise an objection to $\mathbf{x}$,

preferred (although not necessarily strictly preferred) by each player of S to the event

B: with certainty, S does not raise an objection to $\mathbf{x}$.

But then, by axiom II, all players in S prefer a certain $(p=1)$ objection by S to the event A. Changing the strategies of the players of S in $\boldsymbol{\sigma}$ to conform in this manner with the statement of the theorem yields the required new equilibrium point. ∎

We are now prepared to formally define a theory of bargaining solutions. The definitions will be so given as to characterize motivated coalitional equilibrium in reactive strategies.

There are given an n-person cooperative game $\Gamma=\{N, v\}$, a hierarchical structure $\mathfrak{H}$ on the imputation space X, and a system of individual preferences $\{\succsim_i\}_{i\in N}$ from which coalitional preferences may be derived. For any coalition S, define

$$R^S=\{\mathbf{x}\in\mathbb{R}^n \mid x_i=0 \quad \text{for all} \quad i\notin S\}$$

and

$$R^S(\alpha)=\left\{\mathbf{x}\in\mathbb{R}^S \,\middle|\, \sum_{i\in S} x_i\leqq\alpha\right\}.$$

We denote by $F(A)$ and $D(A)$ the set of all finite and discrete probability distributions resp., on a set A. If ρ is any such distribution then, for any $a \in A$, $P_\rho(a)$ is the probability assigned to a by ρ, and $\bar{\rho}$ is defined as

$$\bar{\rho} = \{a \in A \mid P_\rho(a) > 0\}.$$

A *coalitional strategy* σ_S for a coalition S is a pair of functions (σ_S^1, σ_S^2), such that

$$\sigma_S^1 \colon X \to F(R^S)$$

$$\sigma_S^2 \colon R^{N-S}(v(N-S)) \to F(R^S).$$

For each $\mathbf{x} \in X$, $\sigma_S^1(\mathbf{x})$ is assumed to satisfy either

(a) $\mathbf{y} \in \overline{\sigma_S^1(\mathbf{x})}$ implies $\mathbf{y}$ dominates $\mathbf{x}$ with respect to S, or

(b) $\overline{\sigma_S^1(\mathbf{x})} = \{\mathbf{x}^S\}$.[1]

For any $\mathbf{x} \in R^{N-S}(v(N-S))$, $\sigma_S^2(\mathbf{x})$ is assumed to satisfy

$$\mathbf{y} \in \sigma_S^2(\mathbf{x}) \quad \text{implies} \quad (\mathbf{x}, \mathbf{y}) \in X.$$

The strategy σ_S^1 is the "objection strategy" of the coalition S, and σ_S^2 is the "response strategy". The conditions on $\sigma_S^1(\mathbf{x})$ are that either (a) S raises a dominating objection to $\mathbf{x}$, or (b) S does not object to $\mathbf{x}$. The condition on $\sigma_S^2(\mathbf{x})$ is simply that, after an objection and response, the resulting proposal must be an imputation.

Let $\sigma \in \{\sigma_S\}_{S \subset N}$ be a collection of coalitional strategies. For each $\mathbf{x} \in X$, define

$$\eta(\mathbf{x}) = \{S \mid \bar{\sigma}_S^1(\mathbf{x}) \neq \{x^S\}\}.$$

Thus $\eta(\mathbf{x})$ is the collection of coalitions which, in σ, raise objections to $\mathbf{x}$. σ induces a *transition map*

$$\theta_\sigma \colon X \to F(X)$$

defined by

$$\overline{\theta_\sigma(\mathbf{x})} = \bigcup_{S \in \eta(\mathbf{x})} \{\mathbf{w} \mid \mathbf{w} = (\mathbf{y}, \mathbf{z}), \quad \text{where} \quad \mathbf{y} \in \overline{\sigma_S^1(\mathbf{x})} \quad \text{and}$$

[1] $\mathbf{x}^S$ is the vector of those components x_i of $\mathbf{x}$ for which $i \in S$.

$\mathbf{z}\in\overline{\sigma^2_{N-S}(\mathbf{y})}\}\cup\{\mathbf{w}\,|\,\mathbf{w}=\mathbf{x}$, and either $\eta(\mathbf{x})=\emptyset$ or $(H_{\mathbf{x}})_0(\eta(\mathbf{x}))>0\}$, where for each $\mathbf{w}\in\overline{\theta_\sigma(\mathbf{x})}$, either $\mathbf{w}\neq\mathbf{x}$, and

$$P_{\theta_\sigma(\mathbf{x})}(\mathbf{w})=\sum_{\substack{(\mathbf{y},\mathbf{z})=\mathbf{w}\\ \mathbf{y}\in\overline{\sigma^1_S(\mathbf{x})}\\ \mathbf{z}\in\overline{\sigma^2_{N-S}(\mathbf{y})}}} P_{\sigma^1_S(\mathbf{x})}(\mathbf{y})\cdot P_{\sigma^2_{N-S}(\mathbf{y})}(\mathbf{z})\cdot P_{H_{\mathbf{x}}(\eta(\mathbf{x}))}(S)$$

or

$$\mathbf{w}=\mathbf{x},\quad\text{and}\quad P_{\theta_\sigma(\mathbf{x})}(\mathbf{x})=1-\sum_{\substack{\mathbf{w}\in\overline{\theta_\sigma(\mathbf{x})}\\ \mathbf{w}\neq\mathbf{x}}} P_{\theta_\sigma(\mathbf{x})}(\mathbf{w})\,.$$

The set $\overline{\theta_\sigma(\mathbf{x})}$ is the collection of all imputations which might arise in the stage of the bargaining game immediately following the stage in which $\mathbf{x}$ is proposed. The last set in the definition serves only to include cases in which the imputation $\mathbf{x}$ results from itself (i.e., the game ends).

For any $0<\delta\leqq 1$, σ also induces a *valuation map*

$$\varphi_{\sigma,\delta}\colon\ X\to D(X)$$

defined by

$$\overline{\varphi_{\sigma,\delta}(\mathbf{x})}=\bigcup_{k=0}^{\infty}\overline{(\theta_\sigma)^k(\mathbf{x})}\,,$$

where for each $\mathbf{y}\in\overline{\varphi_{\sigma,\delta}(\mathbf{x})}$,

$$P_{\varphi_{\sigma,\delta}(\mathbf{x})}(\mathbf{y})=\sum_{k=0}^{\infty}P_{(\theta_\sigma)^k(\mathbf{x})}(\mathbf{y})\cdot\delta(1-\delta)^k\,.$$

In this definition, $(\theta_\sigma)^k(\mathbf{x})$ is the iterated distribution over X which arises k stages after the proposal $\mathbf{x}$ is made. Thus $\varphi_{\sigma,\delta}$ assigns to each imputation the distribution which arises after the play of a bargaining game, with stopping probability δ, in which all players follow the strategies in σ.

Similarly, for each $0<\delta\leqq 1$ and each $\mathbf{x}\in X$, σ induces a *response-valuation map*

$$\varphi_{\sigma,\delta,\mathbf{x}}\colon \eta(\mathbf{x})\to D(X)$$

where $\varphi_{\sigma,\delta,\mathbf{x}}(S)$ is the distribution which arises after the play of a bargaining game, beginning at $\mathbf{x}$ with coalition S having just been selected by the hierarchy $H_{\mathbf{x}}$ to raise an objection.

We say that σ is a *bargaining solution* to the cooperative game $\Gamma=\{N, v\}$, with respect to the given hierarchical structure and system of preferences, if the following conditions are satisfied.

1. There exists $0<\delta_0\leqq 1$ such that for every $0<\delta<\delta_0$, $i\in N$, $S\ni i$, $\mathbf{y}\in R^{N-S}(v(N-S))$, and $\mathbf{z}\in\sigma_S^2(y)$,

$$\varphi_{\sigma,\delta}(\mathbf{y},\mathbf{z})\gtrsim_i(v(\{i\}),1)\,.$$

(Recall that $(v(\{i\}), 1)$ signifies the event "receiving $v(\{i\})$" with probability 1.)

This condition is derived from Theorem 1, and guarantees that σ is in equilibrium with respect to response strategies.

2. For any $i\in N$, let τ be a collection of coalitional strategies for which $\tau_S^1=\sigma_S^1$ for all $S\not\ni i$, and $\tau_S^2=\sigma_S^2$ for all S. Further assume that for each $\mathbf{x}\in X$ and S containing but not equal to i, $P_{\tau_S^1(\mathbf{x})}(\mathbf{y})\leqq$ $\leqq P_{\sigma_S^1(\mathbf{x})}(\mathbf{y})$ for all $\mathbf{y}\neq\mathbf{x}^S$. Then there is no $0<\delta_0\leqq 1$ for which, for all $\mathbf{x}\in X$ and $0<\delta<\delta_0$,

$$\varphi_{\tau,\delta}(\mathbf{x})\gtrsim_i\ \varphi_{\sigma,\delta}(\mathbf{x})\,,$$

and for which, for some $\mathbf{x}\in X$ and for every $0<\delta_0'<\delta_0$, there is a $0<\delta<\delta_0'$ such that

$$\varphi_{\tau,\delta}(\mathbf{x})\succ_i\varphi_{\sigma,\delta}(\mathbf{x})\,.$$

This condition is merely a restatement of the requirement that σ be in equilibrium, with respect to objection strategies, for each individual $i\in N$ and all stopping probabilities "sufficiently close" to zero.

3. For any coalition S, let τ be a collection of coalitional strategies for which $\tau_W = \sigma_W$ for all $W \neq S$. Then there is no $0 < \delta_0 \leqq 1$ for which, for all $\mathbf{x} \in X$ and $0 < \delta < \delta_0$,

$$\varphi_{\tau,\delta}(\mathbf{x}) \succsim_S \varphi_{\sigma,\delta}(\mathbf{x}),$$

and for which, for some $\mathbf{x} \in X$ and for every $0 < \delta_0' < \delta_0$, there is a $0 < \delta < \delta_0'$ such that

$$\varphi_{\tau,\delta}(\mathbf{x}) \succ_S \varphi_{\sigma,\delta}(\mathbf{x}).$$

This requires coalitional equilibrium in all bargaining subgames with a sufficiently small stopping probability. Further, we require the analogous condition in all such response-bargaining subgames.

4. There is a $0 < \delta_0 \leqq 1$ for which, for every $\mathbf{x} \in X$ with $\eta(\mathbf{x}) \neq \emptyset$ and every $0 < \delta < \delta_0$, there exists a coalition $S \in \eta(\mathbf{x})$ with

$$\Phi_{\sigma,\delta,\mathbf{x}}(S) \succ (\mathbf{x}^S, 1).$$

This condition restates the requirement that some coalition which objects to $\mathbf{x}$ be motivated in its objection.

5. There is a positive integer b such that, for all $\mathbf{x} \in X$, every sequence of imputations $\{\mathbf{y}_k\}_{k=0}^{b}$ with $\mathbf{y}_0 = \mathbf{x}$, which satisfies

$$y_k \neq y_{k-1} \quad \text{for all} \quad k = 1, \ldots, b-1$$

and

$$\mathbf{y}_k \in \overline{\theta_\sigma(y_{k-1})} \quad \text{for all} \quad k = 1, \ldots, b,$$

also satisfies

$$\mathbf{y}_b = \mathbf{y}_{b-1}.$$

This simply requires that, in σ, every bargaining subgame must end, after at most b stages, with a proposal to which no coalition objects.

Consider any bargaining solution σ to a game. Associated with a bargaining-solution σ is the set of all imputations to which no coalition objects. Formally, the *stationary set* T_σ of σ is defined by

$$T_\sigma = \{\mathbf{x} \in X \mid \overline{\theta_\sigma(\mathbf{x})} = \{\mathbf{x}\}\}.$$

Each element of T_σ is a *stationary imputation* of σ. Condition 5 implies the existence of at least one stationary imputation. This is because, regardless of the initial proposal, after at most b stages of the bargaining game all objections must end. The set

$$T = \bigcap_{\sigma \in \Sigma} T_\sigma$$

can be defined as the stationary set of the bargaining game, where Σ is the set of all bargaining solutions. Thus T is the set of imputations which are "stable" in a certain sense, i.e., to which no valid, enforcable objections can be raised provided any coalition sticks to some bargaining solution σ of the game.

Conditions under which Σ and T are nonempty are not available yet. Investigation of many concrete games shows (see [200]) that stationary sets have a close relationship to von Neumann–Morgenstern stable sets (defined in Chapter 23).

26. The solution concept of Nash for n-person cooperative games[1]

Let $\Gamma=\{\Sigma_1, \ldots, \Sigma_n, K_1, \ldots, K_n\}$ be an n-person game given in normal form. Let $\Sigma=\Sigma_1\times\ldots\times\Sigma_n$ and

$$(1) \qquad \Phi=\{\mathbf{f}\mid\mathbf{f}=(K_1(\mathbf{x}), \ldots, K_n(\mathbf{x})),\ \mathbf{x}\in\Sigma\}.$$

Φ is the set of all possible pay-offs. Let L be the convex hull of Φ and we allow any element of L as a pay-off. Let, e.g.,

$$(K_1(\mathbf{x}^{(i)}), \ldots, K_n(\mathbf{x}^{(n)}))\in\Phi, \quad (i=1, \ldots, m)$$

and $0\leqq\lambda_i\leqq 1$, $(i=1, \ldots, m)$, $\left(\sum_{i=1}^{m}\lambda_i=1\right)$ real numbers. Then the vector

$$\boldsymbol{\varphi}=\left(\sum_{i=1}^{m}\lambda_i K_1(\mathbf{x}^{(i)}), \ldots, \sum_{i=1}^{m}\lambda_i K_n(\mathbf{x}^{(i)}\right)\in L$$

can be realized by playing the n-tuple of strategies $\mathbf{x}^{(i)}$ with probability λ_i, $(i=1, \ldots, m)$ and thus $\boldsymbol{\varphi}$ is the expected pay-off to the players.

We assume furthermore that a pay-off vector $\mathbf{f}^*\in\mathbb{R}^n$, called the "status quo" point is given. If the players cannot come to an agreement, then they receive the components of $\mathbf{f}^*$ as pay-offs. We call the pair $(L, \mathbf{f}^*)$ the *extended form* of the cooperative game Γ. The players would like to increase their components in $\mathbf{f}^*$ and to achieve an $\mathbf{f}\in L$ for which $\mathbf{f}\geqq\mathbf{f}^*$. Of course, the different position and

[1] Nash developed his concept for two-person games. The generalization discussed in this chapter is due to F. Szidarovszky [182].

strength of the players should be reflected in the pay-off vector $\mathbf{f}$ they agree to realize.

We approach the problem of assigning a unique vector $\mathbf{f}$ to a pair $(L, \mathbf{f}^*)$ in an axiomatic manner. Define a vector valued function $\boldsymbol{\psi}$ whose domain is the set of pairs $(L, \mathbf{f}^*)$, where $L \subset \mathbb{R}^n$ is a closed, bounded, convex set, $\mathbf{f}^* \in \mathbb{R}^n$ such that there exists an $\mathbf{f} \in L$ for which $\mathbf{f} \geqq \mathbf{f}^*$ and its range is a subset of $\mathbb{R}^n$.

We assume that $\boldsymbol{\psi}$ also satisfies the following axioms:

1. $\boldsymbol{\psi}(L, \mathbf{f}^*) \in L$, (feasibility),
2. $\boldsymbol{\psi}(L, \mathbf{f}^*) \geqq \mathbf{f}^*$, (rationality),
3. $\mathbf{f} \in L$, $\mathbf{f} \geqq \boldsymbol{\psi}(L, \mathbf{f}^*)$ imply $\mathbf{f} = \boldsymbol{\psi}(L, \mathbf{f}^*)$, (Pareto-optimality),
4. If $L_1 \subseteqq L$ and $\boldsymbol{\psi}(L, \mathbf{f}^*) \in L_1$, then $\boldsymbol{\psi}(L, \mathbf{f}^*) = \boldsymbol{\psi}(L_1, \mathbf{f}^*)$, (independence from unfavourable alternatives),
5. Let $\alpha_k > 0$, β_k, $(k = 1, \ldots, n)$ be arbitrary constants and

$$\mathbf{f}^{*\prime} = (\alpha_1 f_1^* + \beta_1, \ldots, \alpha_n f_n^* + \beta_n),$$

$$L' = \{(\alpha_1 l_1 + \beta_1, \ldots, \alpha_n l_n + \beta_n) | (l_1, \ldots, l_n) \in L\}.$$

Then $\boldsymbol{\psi}(L', \mathbf{f}^{*\prime}) = (\alpha_1 \psi_1 + \beta_1, \ldots, \alpha_n \psi_n + \beta_n)$, (independence from monotone increasing linear transformations),

6. If there exist indices i, j such that $\mathbf{f} = (f_1, \ldots, f_n) \in L$ if and only if $\boldsymbol{\varphi} = (\varphi_1, \ldots, \varphi_n) \in L$, $(\varphi_k = f_k, k \neq i, k \neq j, \varphi_i = f_j, \varphi_j = f_i)$ $f_i^* = f_j^*$ for $\mathbf{f}^* = (f_1^*, \ldots, f_n^*)$, imply $\psi_i = \psi_j$ for the vector $\boldsymbol{\psi}(L, \mathbf{f}^*) = (\psi_1, \ldots, \psi_n)$ (symmetry).

THEOREM 1. [182] There is a unique function ψ satisfying axioms 1–6.

Proof. We divide the rather lengthy proof into five steps.

(a) Let L, $\mathbf{f}^*$ satisfy axiom 1. Define $r \geqq 0$ as the largest number for which there exists a $\boldsymbol{\varphi} = (\varphi_1, \ldots, \varphi_n) \in L$ such that $\boldsymbol{\varphi} \geqq \mathbf{f}^* = (f_1^*, \ldots, f_n^*)$, $\varphi_{i_k} > f_{i_k}^*$, $(k = 1, \ldots, r)$ for suitable indices $i_1, \ldots, i_r$. We now prove that there is no $\bar{\boldsymbol{\varphi}} = (\bar{\varphi}_1, \ldots, \bar{\varphi}_n) \in L$ to satisfy $\bar{\boldsymbol{\varphi}} \geqq \mathbf{f}^*$ and $\bar{\varphi}_i > f_i^*$ for some $i \neq i_k$, $(k = 1, \ldots, r)$. If there existed a $\bar{\boldsymbol{\varphi}} \in L$ with the above property, then by the convexity of L we would have

$$\tilde{\boldsymbol{\varphi}}=(\tilde{\varphi}_1, \ldots, \tilde{\varphi}_n)=\frac{1}{2}(\bar{\boldsymbol{\varphi}}+\boldsymbol{\varphi})\in L$$

and

$$\tilde{\varphi}_{i_k}=\frac{1}{2}\bar{\varphi}_{i_k}+\frac{1}{2}\varphi_{i_k}>\frac{1}{2}f^*_{i_k}+\frac{1}{2}f^*_{i_k}=f^*_{i_k}, \quad (k=1, \ldots, r)$$

$$\tilde{\varphi}_i=\frac{1}{2}\bar{\varphi}_i+\frac{1}{2}\varphi_i>\frac{1}{2}f^*_i+\frac{1}{2}f^*_i=f^*_i$$

contradicting the definition of r.

(b) Let $L, f^*, r, i_1, \ldots, i_r$ be as defined in step (a) of the proof and consider the nonlinear programming problem:

$$(2) \qquad g(\tilde{\boldsymbol{\varphi}})=g(u_1, \ldots, u_r)=\prod_{k=1}^{r}(u_k-f^*_{i_k})\to\max$$

$$\tilde{\boldsymbol{\varphi}}\in L$$

$$\tilde{\boldsymbol{\varphi}}\geqq\mathbf{f}^*$$

$$\tilde{\varphi}_{i_k}=u_k, \quad (k=1, \ldots, r)$$

$$\tilde{\varphi}_j=f^*_j, \quad (j\neq i_k;\ k=1, \ldots, r)$$

$$\tilde{\boldsymbol{\varphi}}=(\varphi_1, \ldots, \varphi_n)\,.$$

The feasible set of (2) is nonempty, closed and bounded thus there exists an optimal solution. We will prove that the optimal solution is unique. Assume, on the contrary, that $(u_1, \ldots, u_r)$ and $(u'_1, \ldots, u'_r)$ are two distinct optimal solutions. The optimal objective functions value of (2) is positive since $\boldsymbol{\varphi}$ as defined in step (a) is a feasible solution. Thus $u_k>f^*_{i_k}$, $u'_k>f^*_{i_k}$, $(k=1, \ldots, r)$. Let now $a_k=u_k-f^*_{i_k}$, $b_k=u'_k-f^*_{i_k}$ and define the feasible solution $u''_k=\frac{1}{2}(u_k+u'_k)$, $(k=1, \ldots, r)$. We now obtain

$$(3) \qquad g(u_1'', \dots, u_r'') = \prod_{k=1}^{r} \frac{a_k + b_k}{2} \geqq \prod_{k=1}^{r} \sqrt{a_k b_k} =$$

$$= \sqrt{\prod_{k=1}^{r} a_k \prod_{k=1}^{r} b_k} = g(u_1 \dots, u_r).$$

By the optimality of $(u_1, \dots, u_r)$ we must have equality in (3) implying $a_k = b_k$, i.e., $u_k = u_k'$ for $k = 1, \dots, r$.

(c) Let $(u_1, \dots, u_r)$ be the optimal solution of (2) and $\tilde{\boldsymbol{\varphi}} = (\tilde{\varphi}_1, \dots, \tilde{\varphi}_n)$, $\tilde{\varphi}_{i_k} = u_k$, $(i = 1, \dots, r)$, $\tilde{\varphi}_j = f_j^* (j \neq i_k, k = 1, \dots, r)$. Define the function $\tilde{\boldsymbol{\varphi}}$ as $\tilde{\boldsymbol{\varphi}}(L, \mathbf{f}^*) = \tilde{\boldsymbol{\varphi}}$. We will prove that $\tilde{\boldsymbol{\varphi}}$ satisfies axioms 1–6. By construction axioms 1–3 are satisfied. The validity of axiom 4 follows from the fact that if $\tilde{\boldsymbol{\varphi}}(L, \mathbf{f}^*)$ is optimal on L, then it is also optimal on its subset $L_1 \subset L$. To prove that 5 is satisfied let us take constants $\alpha_k > 0$, β_k, $(k = 1, \dots, n)$ and let L, $\mathbf{f}^*$, r, $i_1, \dots, i_r$ be as defined in step (a). If L', $\mathbf{f}^{*\prime}$ are as defined in axiom 5, then $r' = r$ and the indices $i_1, \dots, i_r$ have the same properties as for L, $\mathbf{f}^*$. Since $g(u_1', \dots, u_r') = \alpha_{i_1} \dots \alpha_{i_r} g(u_1, \dots, u_r)$ the validity of axiom 5 readily follows. Assume now that indices i, j satisfy the conditions of axiom 6. By the assumption of axiom 6, $\varphi_k^* = f_k^*$, $(k \neq i,\ k \neq j)$, $\varphi_i^* = f_j^*$, $\varphi_j^* = f_i^*$ imply $\boldsymbol{\varphi}^* = \mathbf{f}^*$ if $\boldsymbol{\varphi}^* = (\varphi_1^*, \dots, \varphi_n^*)$. Let $\tilde{\boldsymbol{\varphi}} = (\tilde{\boldsymbol{\varphi}}_1, \dots, \tilde{\boldsymbol{\varphi}}_n)$ be the optimal solution of (3) and $\tilde{\boldsymbol{\psi}} = (\tilde{\psi}_1, \dots, \tilde{\psi}_n)$, $(\tilde{\psi}_k = \tilde{\varphi}_k,\ k \neq i,\ k \neq j,\ \tilde{\psi}_i = \tilde{\varphi}_j,\ \tilde{\psi}_j = \tilde{\varphi}_i)$. Then $\tilde{\boldsymbol{\varphi}} \in L$ if and only if $\tilde{\boldsymbol{\psi}} \in L$ and $i \in \{i_1, \dots, i_r\}$ if and only if $j \in \{i_1, \dots, i_r\}$. If $i, j \in \{i_1, \dots, i_r\}$, then $\tilde{\varphi}_i = \tilde{\varphi}_j = f_i^*$ by the constraints of (3). If $i, j \in \{i_1, \dots, i_r\}$, then $\tilde{\varphi}_i = \tilde{\varphi}_j$ (i.e. $\tilde{\boldsymbol{\varphi}} = \tilde{\boldsymbol{\psi}}$) follows from $\boldsymbol{\varphi}^* = \mathbf{f}^*$ and from the uniqueness of the optimal solution to (3).

(d) Let once again L, $\mathbf{f}^*$, r, $i_1, \dots, i_r$ be as defined in step (a). We will prove that the optimal solution $\tilde{\boldsymbol{\varphi}} = (\tilde{\varphi}_1, \dots, \tilde{\varphi}_n)$ of (3) also maximizes the function

$$h(\boldsymbol{\varphi}) = \sum_{k=1}^{r} \left(\prod_{\substack{j=1 \\ j \neq k}}^{r} (\tilde{\varphi}_{i_j} - f_{i_j}^*) \right) \varphi_{i_k}$$

on the set L. Suppose on the contrary that there is a $\boldsymbol{\varphi} = (\varphi_1, \dots, \varphi_n) \in L$ for which $h(\boldsymbol{\varphi}) > h(\tilde{\boldsymbol{\varphi}})$ holds.

Define

$$\bar{\boldsymbol{\varphi}}=\tilde{\boldsymbol{\varphi}}+\varepsilon(\boldsymbol{\varphi}-\tilde{\boldsymbol{\varphi}}), \quad (0<\varepsilon<1).$$

It can easily be seen that

$$g(\bar{\boldsymbol{\varphi}})=\prod_{j=1}^{r}[\tilde{\varphi}_{i_j}-f^*_{i_j}+\varepsilon(\varphi_{i_j}-\tilde{\varphi}_{i_j})]=g(\tilde{\boldsymbol{\varphi}})+$$

$$+\varepsilon h(\boldsymbol{\varphi}-\tilde{\boldsymbol{\varphi}})+a_2\varepsilon^2+\ldots+a_r\varepsilon^r,$$

where $a_2, \ldots, a_r$ are constants (independent from ε). By the linearity of h we have $h(\boldsymbol{\varphi})-h(\tilde{\boldsymbol{\varphi}})=h(\boldsymbol{\varphi}-\tilde{\boldsymbol{\varphi}})>0$. Thus for small enough ε we obtain $g(\bar{\boldsymbol{\varphi}})>g(\tilde{\boldsymbol{\varphi}})$ contradicting the optimality of $\tilde{\boldsymbol{\varphi}}$.

(e) Finally, we prove that any function $\boldsymbol{\psi}$ satisfying axioms 1–6 is necessarily $\tilde{\boldsymbol{\varphi}}$, i.e., for a pair $(L, \mathbf{f}^*)$ the unique vector satisfying axioms 1–6 is the solution of the programming problem (3). Using the notation of the preceding steps define

$$H=\{\boldsymbol{\varphi}\,|\,h(\boldsymbol{\varphi})\leqq h(\tilde{\boldsymbol{\varphi}}),\quad \boldsymbol{\varphi}=(\varphi_1, \ldots, \varphi_n)\geqq \mathbf{f}^*,$$

$$\varphi_i=f^*_i, \qquad (i\neq i_k, k=1, \ldots, r)\},$$

and

$$L_1=L\cap\{\boldsymbol{\varphi}\,|\,\boldsymbol{\varphi}\geqq\mathbf{f}^*\}.$$

Obviously $L_1\subseteqq H$. Consider now the linear transformation

$$f'_{i_k}=\frac{\varphi_{i_k}-f^*_{i_k}}{\tilde{\varphi}_{i_k}-f^*_{i_k}}, \quad (k=1, \ldots, r),$$

$$f'_j=\varphi_j, \quad (j\neq i_k,\ k=1, \ldots, r).$$

Then, by step (c) of our proof $\mathbf{f}^*$ and $\tilde{\boldsymbol{\varphi}}$ are transformed as

$$f'_{i_k}=0, \quad (k=1, \ldots, r), \quad f^{*\prime}_j=f^*_j, \qquad (j\neq i_k, k=1, \ldots, r),$$

$$\tilde{\varphi}'_{i_k}=1, \quad (k=1, \ldots, r), \quad \tilde{\varphi}'_j=\tilde{\varphi}_j, \qquad (j\neq i_k, k=1, \ldots, r).$$

Thus H is transformed to

$$H' = \left\{ \boldsymbol{\varphi}' \;\middle|\; \sum_{k=1}^{i} \varphi'_{i_k} \leqq r, \quad \varphi'_{i_k} \geqq 0, \quad \varphi'_i = f_i^*, \qquad (i \neq i_k),\ k = 1, \ldots, r \right\},$$

which is symmetric in the indices i_k. Thus by axiom 6 for $\boldsymbol{\psi}' = (\psi'_1, \ldots, \psi'_n) = \boldsymbol{\psi}(H', \mathbf{f}^*)$ we have $\psi'_{i_k} = \psi'_{i_l} = 1$, $(k, l = 1, \ldots, r)$. Transforming back we immediately get $\boldsymbol{\psi}(H, \mathbf{f}^*) = \tilde{\boldsymbol{\varphi}}$ by axiom 5. Since $\tilde{\boldsymbol{\varphi}} \in L_1$ and $L_1 \subseteqq H$ we get $\boldsymbol{\psi}(L_1, \mathbf{f}^*) = \tilde{\boldsymbol{\varphi}}$ by axiom 4. But, by axiom 2, $\boldsymbol{\psi}(L, \mathbf{f}^*) \in L_1$, therefore $\boldsymbol{\psi}(L, \mathbf{f}^*) = \boldsymbol{\psi}(L_1, \mathbf{f}^*) = \tilde{\boldsymbol{\varphi}}$ by axiom 4, which was to be proved. ∎

We have seen that finding an "acceptable" (in the sense of Nash's axioms) pay-off vector amounts to solving a mathematical programming problem. For two person games this can be done easily but it gets harder as the number of players grows.

It is also questionable how the status quo point should be defined. Different authors come up with various proposals for a "rational" status quo point. Shapley, e.g., suggests the security level determined by a maximin strategy of each player to be the status quo point (i.e., the characteristic function value of the single-player coalitions). Equilibrium points can also be thought of as status quo points.

27. Examples of cooperative games

27.1. A LINEAR PRODUCTION GAME [132]

There are n producers (players) having the amounts $\mathbf{b}^i = (b_1^i, \ldots, b_m^i)$ of m different resources at their disposal. We assume $\mathbf{b}^i \geqq \mathbf{0}$, $(i = 1, \ldots, n)$. Using these resources p different goods can be produced. The amount used from the kth resource for producing one unit of the jth good is a_{kj}, $(j = 1, \ldots, p;\ k = 1, \ldots, m)$ while a unit of jth good yields profit c_j, $(j = 1, \ldots, p)$.

Let S be a coalition of producers, i.e., a subset of $N = \{1, \ldots, n\}$. If the members of S unit their resources they have the amount

$$\mathbf{b}(S) = \sum_{i \in S} \mathbf{b}^i$$

at their disposal. The production programs available for coalition S are described by the system of linear inequalities

$$(1) \qquad \begin{aligned} \mathbf{x} &\geqq \mathbf{0} \\ \mathbf{A}\mathbf{x} &\leqq \mathbf{b}(S) \end{aligned}$$

where $\mathbf{x} = (x_1, \ldots, x_p)$ denotes the amounts of the different goods to be produced, and $\mathbf{A} = [a_{kj}]$, $(k = 1, \ldots, m;\ j = 1, \ldots, p)$. If coalition S wants to maximize the overall profit, then the linear function $\mathbf{c}\mathbf{x}$, $(\mathbf{c} = (c_1, \ldots, c_p))$ has to be maximized subject to (1). Let us denote by $v(S)$ the optimal objective function's value of the linear program

$$(2) \qquad \begin{aligned} \mathbf{c}\mathbf{x} &\to \max \\ \mathbf{x} &\geqq \mathbf{0} \\ \mathbf{A}\mathbf{x} &\leqq \mathbf{b}(S) \end{aligned}$$

This problem has always a feasible solution and we assume that $v(S) < \infty$. We define $v(\emptyset) = 0$. Thus $\Gamma = \{N, v\}$ can be considered as a cooperative game given in characteristic function form. The game Γ is super-additive. Let, namely, S and T be two disjoint coalitions. Then

$$\mathbf{b}(S \cup T) = \mathbf{b}(S) + \mathbf{b}(T)$$

and if $\mathbf{x}_1^0$, $\mathbf{x}_2^0$ are optimal production programs for coalitions S and T resp., then

$$\mathbf{x}_1^0 \geqq \mathbf{0}, \quad \mathbf{x}_2^0 \geqq \mathbf{0}$$

$$\mathbf{A}(\mathbf{x}_1^0 + \mathbf{x}_2^0) \leqq \mathbf{b}(S) + \mathbf{b}(T),$$

and thus

$$v(S \cup T) \geqq v(S) + v(T) = \mathbf{c}(\mathbf{x}_1^0 + \mathbf{x}_2^0).$$

We will now show that Γ is balanced.

Let $\mathfrak{P}$ be a balanced set with weights γ_S, $(S \in \mathfrak{P})$. Then

$$\sum_{S \in \mathfrak{P}} \gamma_S \mathbf{b}(S) = \sum_{S \in \mathfrak{P}} \sum_{i \in S} \gamma_S \mathbf{b}^i = \sum_{i \in N} \left(\sum_{\substack{S \in \mathfrak{P} \\ i \in S}} \gamma_S \right) \mathbf{b}^i.$$

Since $\mathfrak{P}$ is balanced we have

$$\sum_{\substack{S \in \mathfrak{P} \\ i \in S}} \gamma_S = 1.$$

Hence

$$\sum_{i \in N} \left(\sum_{\substack{S \in \mathfrak{P} \\ i \in S}} \gamma_S \right) \mathbf{b}^i = \sum_{i \in N} \mathbf{b}^i = \mathbf{b}(N). \tag{3}$$

If $\mathbf{x}(S)$ is an optimal solution of (2), then $v(S) = \mathbf{c}\mathbf{x}(S)$ and

$$\sum_{S \in \mathfrak{P}} \gamma_S v(S) = \sum_{S \in \mathfrak{P}} \gamma_S \mathbf{c}\mathbf{x}(S).$$

Introducing the notation

$$\hat{\mathbf{x}} = \sum_{S \in \mathfrak{P}} \gamma_S \mathbf{x}(S)$$

we get

$$\sum_{S \in \mathfrak{P}} \gamma_S v(S) = \mathbf{c}\hat{\mathbf{x}} . \tag{4}$$

Since

$$\mathbf{A}\left(\sum_{S \in \mathfrak{P}} \gamma_S \mathbf{x}(S)\right) \leqq \sum_{S \in \mathfrak{P}} \gamma_S \mathbf{b}(S)$$

we obtain, using (3),

$$\mathbf{A}\hat{\mathbf{x}} \leqq \mathbf{b}(N) ,$$

$$\hat{\mathbf{x}} \geqq \mathbf{0} .$$

Thus $\hat{\mathbf{x}}$ satisfies (1) for the grand coalition $S = N$. By the definition of $v(N)$ we have

$$v(N) \geqq \mathbf{c}\hat{\mathbf{x}} ,$$

which together with (4) implies

$$\sum_{S \in \mathfrak{P}} \gamma_S v(S) \leqq v(N)$$

which means that game $\Gamma = \{N, v\}$ is balanced. By Theorem 5 of Chapter 23 we deduce that Γ has a nonempty core. In fact, we can easily compute a point in the core of Γ.

Consider the dual of (2)

$$\begin{aligned} &\mathbf{y} \geqq \mathbf{0} \\ &\mathbf{y}\mathbf{A} \geqq \mathbf{c} \\ &\mathbf{y}\mathbf{b}(S) \rightarrow \min . \end{aligned} \tag{5}$$

It is well known that the optimal objective function's value of (5) is $v(S)$. Let $\mathbf{y}^*$ be an optimal solution of (5) in case of $S = N$.

Obviously

$$(6) \qquad v(N) = \mathbf{y}^* \mathbf{b}(N)$$

and for any coalition S

$$(7) \qquad v(S) \leqq \mathbf{y}^* \mathbf{b}(S).$$

Let $u_i = \mathbf{b}^i \mathbf{y}^*$, $(i = 1, \ldots, n)$. Then, for any coalition S

$$(8) \qquad \sum_{i \in S} u_i = \sum_{i \in S} \mathbf{b}^i \mathbf{y}^* = \sum_{k=1}^{m} \sum_{i \in S} b^i_k y^*_k = \mathbf{b}(S)\mathbf{y}^*.$$

Because of (6),

$$\sum_{i \in N} u_i = v(N)$$

and (7) and (8) imply

$$\sum_{i \in S} u_i \geqq v(S)$$

for any coalition S. Thus $\mathbf{u} = (u_1, \ldots, u_n)$ is an imputation in the core of Γ.

We have seen that a point in the core can be determined by solving the linear program (5) for $S = N$. The reverse is not true, i.e., there might exist points in the core not obtainable as solutions of (5) as shown in [132].

27.2. A MARKET GAME [163]

Markets give rise to cooperative games in a fairly natural way. Of course, markets should be thought of as very abstract entities when analyzed by game theoretic means. Specifically a *market* M is defined to be a quadruple $M = \{T, G, A, U\}$, where

T is a finite set (the set of traders),

G is the nonnegative orthant of a finite dimensional Euclidean space (the space of goods),

$A = \{\mathbf{a}^i \in G \mid i \in T\}$ is an indexed finite subset of G (initial stocks),

$U=\{u^i \mid i \in T\}$ is a finite collection of concave, continuous functions defined on G (utility functions).

Let S be a subset of T and

$$X^S=\left\{\mathbf{x}^i \in G \mid i \in T, \sum_{i\in S} \mathbf{x}^i = \sum_{i\in S} \mathbf{a}^i\right\}.$$

We call X^S a feasible S-allocation.

Now, the market $M=\{T, G, A, U\}$ induces a game $\Gamma_M=\{N, v\}$ in the following way:

$$\text{(9)} \qquad N=T$$

$$v(S)=\max_{\mathbf{x}^i \in X^S} \sum_{i\in S} u^i(\mathbf{x}^i)$$

for any subset (coalition) of N.

The game Γ_M is called a *market game.* Note that Γ_M is not necessarily super-additive. Generally, any game $\Gamma=\{N, v\}$ obtainable from a market M is called a market game. We shall show that, under the assumptions specified above, Γ_M has a nonempty core. To this end we have to prove two theorems.

THEOREM 1. [163] If $\Gamma_M=\{N, v\}$ is a market game, c is an additive set function and λ is a nonnegative scalar, then $\Gamma_{M'}= =\{N, \lambda v + c\}$ is also a market game.

Proof. Using the definition of a market game, direct calculation shows that the market generating the game Γ_M gives rise to the game $\Gamma_M{}'$ if the utility functions in M are replaced by $\lambda u^i(\mathbf{x})+c(\{i\})$. ∎

THEOREM 2. [163] If $\Gamma_{M'}=\{N, v'\}$ and $\Gamma_{M''}=\{M, v''\}$ are market games, then $\Gamma_{M'''}=\{N, v'+v''\}$ is also a market game.

Proof. Let $M'=\{N, G', A', U'\}$ and $M''=\{N, G'', A'', U''\}$ be the markets which generate $\Gamma_{M'}$ and $\Gamma_{M''}$ resp. Let $G'''=G' \times G''$, $A'''=A' \times A''$ and define U''' as the set of utility functions

$$u^i(\mathbf{x}', \mathbf{x}'')=u'^i(\mathbf{x}')+u''^i(\mathbf{x}'')$$

where $\mathbf{x}' \in G', \mathbf{x}'' \in G'', u'^i \in U', u''^i \in U''$. The elements of U''' are continuous, concave functions. Thus $M''' = \{N, G''', A''', U'''\}$ is a market which gives rise to the market game $\Gamma_{M'''}$. ∎

Now we establish the existence of the core for market games.

THEOREM 3. [163] Any market game has a nonempty core.

Proof. Consider the market game $\Gamma_M = \{N, v\}$ generated by the market $M = \{N, G, A, U\}$. Let $B = \{\mathbf{b}^i \mid i \in N\}$ be the set of feasible N-allocations which give $v(N)$ in (9), i.e., which maximize $\sum_{i \in N} u^i(\mathbf{x}^i)$ subject to $\sum_{i \in N} \mathbf{x}^i = \sum_{i \in N} \mathbf{a}^i$, $\mathbf{x}^i \in G$, $(i = 1, \ldots, n)$. Thus there exists a vector of Lagrange multipliers (prices) $\mathbf{p}$, such that for each $i \in N$ the function

$$\text{(10)} \qquad u^i(\mathbf{x}^i) - \mathbf{p}(\mathbf{x}^i - \mathbf{a}^i)$$

attains its maximum at $\mathbf{x}^i = \mathbf{b}^i$ subject to $\mathbf{x}^i \in G$. Now consider the imputation $\mathbf{z} = (z_1, \ldots, z_n)$, where

$$z_i = u^i(\mathbf{b}^i) - \mathbf{p}(\mathbf{b}^i - \mathbf{a}^i), \quad (i = 1, \ldots, n).$$

We shall prove that $\mathbf{z}$ is in the core of Γ_M. Let $S \subset N$ be a coalition and let Y^S be the set of feasible S-allocations which define $v(S)$ in (9), i.e., $v(S) = \sum_{i \in S} u^i(\mathbf{y}^i)$, $\mathbf{y}^i \in G$, $(i \in S)$. Since $\mathbf{b}^i$ maximizes (10), therefore

$$z_i \geqq u^i(\mathbf{y}^i) - \mathbf{p}(\mathbf{y}^i - \mathbf{a}^i).$$

Summing up for all $i \in S$, we have

$$\sum_{i \in S} z_i \geqq \sum_{i \in S} u^i(\mathbf{y}^i) - \mathbf{p}\mathbf{0} = v(S)$$

which together with $\sum_{i \in N} z_i = v(N)$ implies that $\mathbf{z}$ is in the core of Γ_M.

∎

27.3. THE COOPERATIVE OLIGOPOLY GAME [179]

In this chapter we treat the oligopoly game defined in Chapter 6 as a cooperative game, i.e., the manufacturers are allowed to form coalitions and maximize their total profit-functions. We deal only with the special case[1] when $M=1$, i.e., with the single-product case. Since in most solution concepts the characteristic function plays a crucial role, we first determine $v(S)$ for any coalition S of players. We assume that the oligopoly game satisfies the conditions of Theorem 8 of Chapter 6, and that the price function P is differentiable on the interval $[0, \xi]$. Let $S=\{i_1, \ldots, i_r\}$ be a subset of the player set $N=\{1, \ldots, n\}$. By the definition of the characteristic function

$$(11)\qquad v(S)=\max_{\substack{x_i\\ i\in S}} \min_{\substack{x_j\\ j\notin S}} \sum_{i\in S} \varphi_i(\mathbf{x}).$$

To compute $v(S)$ we have to determine

$$\min_{\substack{x_j\\ j\notin S}} \sum_{i\in S} \varphi_i(\mathbf{x})=\psi_S(x_{i_1}, \ldots, x_{i_r}).$$

We will distinguish between two cases

(a) If $\sum_{j\notin S} L_j \geqq \xi$, then obviously

$$(12)\qquad \psi_S(x_{i_1}, \ldots, x_{i_r})=-\sum_{i\in S} C_i(x_i).$$

In this case ψ_S attains its maximum at $x_{i_1}=\ldots=x_{i_r}=0$ since the cost functions C_j are monotonically increasing. Thus $v(S)= =-\sum_{i\in S} C_i(0)$.

(b) If $\sum_{j\notin S} L_j<\xi$, then

[1] Throughout this chapter we use the notations of Chapter 6.

$$(13)\qquad \psi_S(x_{i_1},\ldots,x_{i_r})=\left(\sum_{i\in S}x_i\right)P\left(\sum_{j\notin S}L_j+\sum_{i\in S}x_i\right)-\sum_{i\in I}C_i(x_i).$$

Denote $L_S=\sum_{i\in S}L_i$, $\tilde{L}_S=\sum_{j\notin S}L_j$. Then by (13)

$$(14)\qquad \psi_S(x_{i_1},\ldots,x_{i_r})=t_SP(\tilde{L}_S+t_S)-\sum_{i\in S}C_i(x_i),$$

where $t_S=\sum_{i\in S}x_i$. Now solve the concave programming problem

$$(15)\qquad \begin{aligned}&\sum_{i\in S}C_i(x_i)\to\min\\&0\leqq x_i\leqq L_i,\quad (i\in S)\\&\sum_{i\in S}x_i=t_S,\quad (t_S\in[0,L_S]).\end{aligned}$$

Let $Q_S(t_S)$ denote the optimal objective function's value of (15) and consider the programming problem

$$(16)\qquad \begin{aligned}&g(t_S)=t_SP(\tilde{L}_S+t_S)-Q_S(t_S)\to\max\\&0\leqq t_S\leqq L_S.\end{aligned}$$

It can be proved that g is concave on the interval $[0, U_S]$, where $U_S=\min\{\xi, L_S\}$. Thus an optimal solution t_S^* can be obtained by the rule

$$(17)\qquad t_S^*=\begin{cases}0, & \text{if}\quad g'(+0)\leqq 0,\\ U_S, & \text{if}\quad g'(U_S-0)\geqq 0,\\ 0<v<U_S, & \text{if}\quad g'(t-0)\geqq 0\geqq g'(t+0).\end{cases}$$

The derivatives on the left (right) exist by the continuity and concavity of g. The optimal value v can be determined by routine numerical methods.

Thus, we have shown that

$$v(S) = \begin{cases} -\sum_{i \in S} C_i(0), & \text{if } \tilde{L}_S \geq \xi, \\ g(t_S^*) & \text{otherwise}, \end{cases}$$

for any coalition S.

With the characteristic function at hand, the nucleolus and the Shapley-value can be computed by the methods described in Chapter 23. It would be interesting to know what the core of the game (if it is nonempty) and the von Neumann–Morgenstern stable sets look like (provided there exist any). No answers to these questions are known yet.

The cooperative Nash-solution (see Chapter 26) can also be obtained by solving a mathematical programming problem. To find out how this programming problem can be set up is left to the reader as an exercise.

27.4. A GAME THEORETIC APPROACH FOR COST ALLOCATION: A CASE [176]

It happens quite often that a group of people or institutions (we will refer to them as players) undertake a joint venture in the hope that the costs (benefits) so incurred will be less than when working independently or in smaller groups. How should they share the common costs "fairly"? This has been an intriguing and unresolved problem of economic theory for a long time. Among other possible approaches game theory also has something to say about cost allocation. We do not pretend to analyze cost allocation in all its complexity since the problem itself would deserve a whole book. Within the limits of an example taken from water resource development we will illustrate an application of cooperative game theory in assigning "fair" costs and benefits to the participants in a cooperative venture.

The case we are going to present is that of the Kanagawa prefecture in Japan. In Japan, the water resources in a given area are or can be exploited by two kinds of entities: agricultural associations and city water services. While agricultural associations view the existing water supply ample as to their irrigation needs, current or future demand for water in cities is not met by existing resources.

Let $N=\{1, 2, \ldots, n\}$ be the complete listing of the n entities, $N=A\cup B$, $A\cap B=\emptyset$, where A is the subset of agricultural associations and B is the subset of all city services. Each city $i\in B$ needs an additional annual quantity of water. Denote δ_i the additional annual quantity of water required by i ($\delta_i=0$ if $i\in A$). Technologically, a city $i\in B$ might acquire the additional water δ_i it needs by

(1) constructing a dam with or without the cooperation of other cities,

(2) arranging with the agricultural associations for the direct diversion of water from them to the city,

(3) combining the two former possibilities.

The main motive of the entities to form a coalition to carry out the necessary construction projects is saving money since the costs in a joint project will be less than the sum of the costs each member of the coalition has to bear when going alone. Once this is recognized, the members of the coalition need only be assured that the resulting cost savings are distributed among the participants in some "fair" way. This is the point where cooperative game theory comes into the picture. We will consider the Shapley-value and the nucleolus as a "fair" division scheme and compute it for a concrete problem. But first of all we have to determine the characteristic function of the game.

Define $v(S)$ to be the minimum cost of meeting all the additional water needs of the members of S, assuming no cooperation from players outside of S.

If $S\subseteqq A$, then $v(S)=0$ since the agricultural associations have no additional water needs. If $S\subseteqq B$, then $v(S)$ must be the cost of constructing a dam just capable of meeting the additional water needs of cities in S. If $S\cap A\neq\emptyset$ and $S\cap B\neq\emptyset$, then $v(S)$ the minimum cost of meeting the additional water needs of the cities in S, may be the sum of three separate costs:

(i) the cost of constructing a dam capable of meeting part of the additional demand for water of the members in $S \cap B$,

(ii) the cost of diverting water from the agricultural associations to the cities to meet the remaining demand for additional water,

(iii) the cost of compensating the agricultural associations for any decrease in agricultural production resulting from diversion.

To determine $v(S)$ for each possible coalition S is not always easy. In many problems where the form of the cost function in (i)–(iii) is known or can be approximated the determination of $v(S)$ becomes a mathematical programming problem. We get a computationally tractable problem if we assume proportional costs.

Define the following variables and parameters:

- y_j annual amount of water supplied by dam to player j,
- d^S dam construction cost for a unit of water in a year by coalition S,
- x_{ij} annual amount of water diverted from agricultural association i to city j,
- e_{ij}^S unit cost of diversion from agricultural association i to city j provided entities in S cooperate,
- w_i annual quantity of excess water for agricultural association $i \in A$,
- c_i value of agricultural production lost per unit water lost over and above excess.

Also define the (loss) function

$$F_i\left(\sum_{j \in S \cap B} x_{ij}\right) = \begin{cases} c_i\left(\sum_{j \in S \cap B} x_{ij} - w_i\right) & \text{if } \sum_{j \in S \cap B} x_{ij} > w_i \\ 0 & \text{otherwise.} \end{cases}$$

Thus $v(S)$ will be the optimal objective function value of the following programming problem:

$$d^S \sum_{j\in S\cap B} y_j + \sum_{i\in S\cap A} \sum_{j\in S\cap B} e_{ij}^S x_{ij} + \\ + \sum_{i\in S\cap A} F_i\left(\sum_{j\in S\cap B} x_{ij}\right) \to \min$$

$$y_j + \sum_{i\in S\cap A} x_{ij} \geqq \delta_j \quad \text{for all} \quad j\in S\cap B$$

$$y_j \geqq 0 \quad \text{for all} \quad j\in S\cap B$$

$$x_{ij} \geqq 0 \quad \text{for all} \quad i\in S\cap A \text{ and } j\in S\cap B.$$

Once $v(S)$ has been determined for each coalition $S\subseteqq N$, the problem has been characterized as an n-person cooperative game $G=\{N, v\}$ in characteristic function form.[1]

In the case of Kanagawa prefecture we assume that a dam will be constructed on the River Sakawa and diversion of water from agricultural use to city service is carried out in the basins of Rivers Sakawa and Sagami. We have 5 players in the game

1. The agricultural association of River Sakawa
2. The agricultural association of River Sagami
3. Kanagawa city water service authority
4. Yokohama city water service authority
5. Kawasaki city water service authority

The parameters of the cost function are shown in Tables 1 and 2. The game G is given in Table 3.

[1] Since we treated games in characteristic function form in a very general setting, the "cost" (instead of "gain") – character of v must not cause any difficulty.

Table 1

Parameters w_i and δ_j		
	Estimated Quantity of Excess Water w_i[m^3/year]	Quantity of Additional Water which must be developed δ_j[m^3/year]
Player 1	1.67×10^8	0
Player 2	1.28×10^8	0
Player 3	0	1.48×10^8
Player 4	0	2.28×10^8
Player 5	0	1.94×10^8

Table 2

Cost Coefficients			
Coalitions	Unit Cost of Dam Construction	Unit Cost of Construction for Diversion	
S_B	d^S[yen/m^3/year]	e_{ij}^S[yen/m^3/year]	
3	330.9	$e_{13}^{3}=297.3$	$e_{23}^{3}=328.3$
4	327.9	$e_{14}^{4}=299.7$	$e_{24}^{4}=170.6$
5	386.5	$e_{15}^{5}=353.1$	$e_{25}^{5}=200.9$
3, 4	294.3	$e_{13}^{34}=225.7$	$e_{14}^{34}=257.5$
		$e_{23}^{34}=328.3$	$e_{24}^{34}=170.6$
3, 5	324.1	$e_{13}^{35}=227.4$	$e_{15}^{35}=304.4$
		$e_{23}^{35}=328.3$	$e_{25}^{35}=200.9$
4, 5	286.5	$e_{14}^{45}=202.8$	$e_{15}^{45}=232.7$
		$e_{24}^{45}=170.6$	$e_{25}^{45}=200.9$
3, 4, 5	272.8	$e_{13}^{345}\equiv 215.8$	$e_{14}^{345}=194.8$
		$e_{15}^{345}=224.7$	$e_{23}^{345}=328.3$
		$e_{24}^{345}=170.6$	$e_{25}^{345}=200.9$

Profit Coefficient of the Agricultural Products $c_1=c_2=14.15$ [yen/m^3/year]

Table 3

Characteristic Function and Allocation of the Developed Water Resource

Coalition S	$v(S)$ [$\times 10^8$ yen]	Developed Quantity of Water x_{ij}, y_j [$\times 10^8$ m^3/year]								
		x_{13}	x_{23}	y_3	x_{14}	x_{24}	y_4	x_{15}	x_{25}	y_5
1	0									
2	0									
3	489.7			1.48						
4	747.6						2.28			
5	749.8									1.94
12	0									
13	440.0	1.48		0						
14	700.5				1.67		0.61			
15	694.0							1.67		0.27
23	486.4		1.28	0.20						
24	546.3					1.28	1.00			
25	512.2								1.28	0.66
34	1106.5			1.48			2.28			
35	1108.3			1.48						1.94
45	1209.0						2.28			1.94
123	440.0	1.48	0	0						
124	518.1				1.67	0.61	0			
125	490.2							0.66	1.28	0'
134	998.0	1.48		0	0.19		2.09			
135	961.5	1.48		0				0.19		1.75
145	1069.2				1.67		0.61	0		1.94
234	948.2		0	1.48		1.28	1.00			
235	950.7		0	1.48					1.28	0.66
245	1069.2					1.28	1.00		0	1.94
345	1554.3			1.48			2.28			1.94
1234	865.4	0.67	0	0.81	1.00	1.28	0			
1235	803.8	1.48	0	0				0.19	1.28	0.47
1245	940.9				1.00	1.28	0	0.67	0	1.27
1345	1424.6	0		1.48	1.67		0.61	0		1.94
2345	1424.3		0	1.48		1.28	1.00		0	1.94
12345	1307.9	0.67	0	0.81	1.00	1.28	0	0	0	1.94
				1085						

The optimal allocation of the water resource developed by the grand coalition N shows that Kanagawa city water service authority should acquire the water from the diversion and the dam, while Yokohama city water service authority should acquire their quantity only from the diversion and Kawasaki city water service authority only from the dam.

The Shapley-value and the nucleolus was calculated for the game G and is shown on Table 4.

Table 4

	The Assigned Cost				
	Players				
	1	2	3	4	5
Shapley-value	−71,7	−100,9	395,1	554,3	531,1
Nucleolus	−70,9	−71,2	412,4	549,6	488

The absolute value of the assigned cost to player i ($i \in A$) means the dividend paid in addition to the compensation for the decrease of the agricultural products. The city water service authority can reduce its cost of the development by a cooperative venture as much as $C_i - X_i^*$, where X_i^* is the assigned cost to player i ($i \in B$) and C_i is the cost if going alone.

The structure of the assigned costs is pretty similar for both solution concepts except for the slight difference in the amounts assigned to players 2 and 5.

27.5. COMMITTEE DECISION MAKING AS A GAME [166]

In this example we will study the *committee decision making* problem (briefly CDM) in a game theoretical framework. By a committee we mean a finite group of people (players, committee members) $N = \{1, 2, \ldots, n\}$ who have to pick one option from a given finite set of alternatives (outcomes) $X = \{a_0, a_1, \ldots, a_m\}$

where $m \geqq 1$ and a_0 is a distinguished element called the *status quo outcome.*

We assume that our players are situated in one room and they arrive at a decision after careful deliberations. This presupposes that N is relatively small in contrast to the theory of elections where the players (voters) are numerous and spread out extensively.

We will now define the rule (not necessarily the straight majority rule) by which the committee will arrive at a decision. The rules are designed so that the decision of the committee consists of a unique outcome.

As usual we call nonempty subsets of N coalitions. Let 2^N denote the set of nonempty coalitions and 2^X the set of all subsets of X. The rules by which the committee members arrive at a decision are given by the *characteristic function*[1] $v: 2^N \to 2^X$ and v is required to satisfy the following conditions:

$$(18) \quad R_1 \supset R_2 \Rightarrow v(R_1) \supset v(R_2) \quad \text{for any} \quad R_1, R_2 \in 2^N,$$

$$(19) \quad v(N) = X,$$

$$(20) \quad R_1 \cap R_2 = \emptyset, \quad v(R_1) \neq \emptyset, \quad v(R_2) \neq \emptyset \Rightarrow \quad \text{for any} \quad R_1, R_2 \in 2^N$$
$$\Rightarrow v(R_1) = v(R_2) \quad \text{and} \quad |v(R_1)| = 1.$$

$v(R)$ denotes the subset of outcomes that coalition R can achieve if all members of R come to an agreement. Condition (18) requires that the set of achievable outcomes be larger for a larger coalition while condition (19) stipulates that if the whole committee is unanimous, then the choice of any outcome is open to it. Condition (20) ensures that the committee decision consists of at most one outcome.

For each $i \in N$, player i has a preference relation $>_i$ on X that is a *weak order.* (It is asymmetric, transitive and negatively transitive.)

[1] Unlike in previous chapters of the book v is not a real-valued function here since side-payments are not allowed in this model.

We assume that the members make decisions under conditions of complete information, i.e., anyone is completely aware of his own and everyone else's preference relation.

Now we define the *n-person committee game* as $\Gamma=(N, X, v, \{>_i\}_{i\in N}\}$. A committee game is said to be *simple* if for each $R \in 2^N$

$$v(R)=\emptyset$$

$$\text{or} \quad v(R)=\{a_0\}$$

$$\text{or} \quad v(R)=X\,.$$

If $v(R)=\emptyset$, , then coalition R can realize no outcome at all whereas if $v(R)=\{a_0\}$, then R can only achieve the status quo outcome. In both cases R is called a *losing coalition* while if $v(R)=X$, then R is said to be *winning.* If $v(R)=\{a_0\}$, then R is also called a *blocking* coalition. Denote $\mathfrak{W}$ the set of all winning coalitions, $\mathfrak{L}$ the set of all loosing coalitions and $\mathfrak{B}$, the set of all blocking coalitions. Player i is said to be a *dictator* if $\{i\} \in \mathfrak{W}$. If $\bigcap_{R\in\mathfrak{W}} R \neq \emptyset$ then any member of this coalition is called a *veto* player.

The characteristic function v and the preference relations induce a natural dominance relation on the set of outcomes. Let a, $b \in X$ and R a coalition. We say *a dominates b* through R denoted by $a \succ_R b$ if

(21) $\quad a >_i b \quad$ for any $\quad i \in R \quad$ and

(22) $\quad a \in v(R)$

i.e. a is preferable to b for each member of R and coalition R can realize it. Obviously $\succ_R$ is asymmetric and transitive for any R. Just like for games with side-payment given in characteristic function form we can define the binary dominance relation $\succ$ on X:

(23) $\quad a \succ b$

if there exists a coalition R such that $a \succ_R b$.

Though $\succ$ may not be transitive but we have

THEOREM 4. [166] The binary relation $\succ$ is asymmetric for all committee games.

Proof. Assume it is not. Then there exist $a, b \in X$ to satisfy $a \succ b$ and $b \succ a$, i.e., there are coalitions R_1, R_2 such that $a \succ_{R_1} b$ and $b \succ_{R_2} a$. Since the preference relations $>_i$ are asymmetric we have $R_1 \cap R_2 = \emptyset$. Also we have $a \in v(R_1)$, $b \in v(R_2)$ contradicting to (20). ∎

In the following we will study some solution concepts for committee games. To this end we introduce some notation. For any $a \in X$, let $D(a) = \{x \in X \mid a \succ x\}$ and $U(a) = X - D(a)$. Also for any nonempty $B \in 2^X$ let $D(B) = \bigcup_{x \in B} D(x)$ and $U(B) = X - D(B)$.

For a committee game $\Gamma = (N, X, v, \{>_i\}_{i \in N})$ an outcome $\alpha \in X$ is said to be the *dominant* (Condorcet) *solution* if $\{\alpha\} = U(\alpha)$, i.e., it dominates every other outcome in X.

From the asymmetric property of the relation $\prec$ we can deduce that if the dominant solution exists, then it is unique.

We can also define the *core* of a committee game $\mathfrak{C} = U(X)$, i.e., the set of all undominated outcomes. Also the asymmetry of $\prec$ implies that if the dominant solution α exists, then $\mathfrak{C} = \{\alpha\}$.

Unfortunately neither the dominant solution nor the core exist always for any committee game. Therefore we introduce two other solution concepts well-suited for committee games which are more likely to be nonempty than the dominant solution and the core.

Let Γ be a committee game. A pair (i, x) is said to be a *proposal* if $i \in N$ and $x \in X$. A proposal represents a motion x introduced by player i. Denote $P = N \times X$ the (finite) set of all proposals. Define

$$\hat{C}^{(i)} = \{(i, x) \in P \mid x \text{ is not dominated through any coalition } N - \{i\}\}.$$

$\hat{C}^{(i)}$ represents the set of proposals made by i that are undominated assuming player i does not cooperate in any effort to dominate his proposal. We also define $C^{(i)}$ the maximal (best) proposals in the set $\hat{C}^{(i)}$ for player i

(24) $C^{(i)}=\{(i,x)\in\hat{C}^{(i)} \mid y\leqq_i x \text{ for any } (i,y)\in\hat{C}^{(i)}\}.$

The *one-core* $\mathfrak{C}_1$ of the committee game Γ is then defined by

(25) $\mathfrak{C}_1=\bigcup_{i\in N} C^{(i)}$

which consists of all undominated maximal proposals assuming that the player who makes the proposal does not cooperate in any effort to beat his own proposal. The connection between the one-core and the dominant solution is established by the following theorem.

THEOREM 5. [166]. If the dominant solution α exists for a committee game Γ, then the one-core is given by

$$\mathfrak{C}_1=\{(1,\alpha),\ldots,(n,\alpha)\}$$

i.e. any player proposes the dominant solution if it is his turn to make a proposal.

Proof. Since the dominant solution is not dominated by any outcome we have $(i,\alpha)\in\hat{C}^{(i)}$ for each $i\in N$. We claim that $C^{(i)}\supset\{i,\alpha\}$ for any $i\in N$. Assume not. Let $(i,x)\in\hat{C}^{(i)}$ such that $x\neq\alpha$ and $x>_i\alpha$. Since α is the dominant solution $\alpha\succ_R x$ for some coalition R. If $i\in R$, then $\alpha>_i x$ and this is a contradiction. If $i\notin R$, then $(i,x)\notin\hat{C}^{(i)}$ also a contradiction. Using exactly the same argument as before, it can be shown that $(i,x)\in C^{(i)}, x\neq\alpha$ leads to a contradiction. Hence $C^{(i)}=\{(i,\alpha)\}$ implying the validity of the assertion of the theorem. ∎

There is also a strong connection between the one-core and the core.

THEOREM 6. [166]. Let Γ be a committee game for which the core $\mathfrak{C}$ is not empty. Then $C^{(i)}\neq\emptyset$ for any $i\in N$ and $(i,x)\in C^{(i)}$ implies $y\leqq_i x$ for any $y\in\mathfrak{C}$.

Proof. $\mathfrak{C}\neq\emptyset$ implies $\hat{C}^{(i)}\neq\emptyset$ for any $i\in N$. Since $P=N\times X$ is a finite set $C^{(i)}\neq\emptyset$. Thus $\mathfrak{C}_1\neq\emptyset$ and the maximum property of $C^{(i)}$ implies the second assertion of the theorem. ∎

This theorem states that each player by choosing his proposal from the one-core does at least as good as any outcome in the core.

Unfortunately neither the dominant solution nor the one-core exist for all committee game. This is because of the strict stability requirements inherent in the above solution concepts. By loosening these requirements we define another solution concept, the *bargaining set* which uses some of the ideas presented in Chapter 24 for characteristic function games but is quite different from it in that it is relevant to the context of a committee game.

Assume a player $i\in N$ introduces a motion $x\in X$ in the form of a proposal (i, x). The proposal is debated by the members. At the end of the debate there are three possible courses of action:

(1) Player i withdraws his motion. Another motion is introduced and the process continues.

(2) The proposal (i, x) is accepted and becomes the decision of the committee.

(3) Another member $j\in N$ introduces another motion $y\in X$ and y will be the current proposal if a vote will decide so.

We define an *objection* against proposal (i, x) as a triple (j, S, y) such that

(26) $\quad j\in S\in 2^{N}, \quad y\in X$

(27) $\quad i\notin S$

(28) $\quad y\succ_S x\,.$

While debating the merits of proposal (i, x), player j is considering to introduce a motion y against (i, x). He can rightly do so since $y\succ_S x$ which means that he can expect the members of S to vote for y and win. In this effort to beat (i, x) he cannot expect i to join him (see (27)) and of course $i\neq j$.

A *counterobjection* against the objection (j, S, y) to proposal (i, x) is defined to be a triple (k, T, z) such that

(29) $k \in T \in 2^N, \quad z \in X,$

(30) either $k = i$ or $x >_k y,$

(31) $z \succ_T y$

(32) $x >_j z$

(33) $x \leqq_k z.$

The counterobjection is made either by player i or by a player who stands to lose if the objection is carried out (see (30)). The counterobjection is a threat by player k against j. We may visualize him to tell player j: If you carry out your objection and win, then in the next round I will introduce motion z against y and will win (see (31)). This will put you in a worse position than you were before (see (32)) and I will do no worse than what I started with (see (33)).

A proposal (i, x) is said to be $\mathfrak{M}$-*stable* if every objection has a counterobjection. An $\mathfrak{M}$-stable proposal (i, a) is said to be *maximal* if $x \leqq_i a$ for any $x \in X$ such that (i, x) is $\mathfrak{M}$-stable. The *bargaining set* $\mathfrak{M}$ is the set of all maximal $\mathfrak{M}$-stable proposals.

We will now study the question of existence of the bargaining set. It is a natural requirement that if the dominant solution exists, then the bargaining set should contain it.

THEOREM 7. [166] If the dominant solution α exists for a committee game Γ, then the bargaining set $\mathfrak{M}$ is given by $\mathfrak{M} = \{(1, \alpha), \ldots, (n, \alpha)\}$.

Proof. It is clear that (i, α) is $\mathfrak{M}$-stable for any $i \in N$. Thus we have only to show that there is no other $\mathfrak{M}$-stable solution. Consider a proposal (i, x) such that $x \neq \alpha$. Since α is the dominant solution $\alpha \succ_R x$ for some coalition R. If $i \notin R$, then (j, R, α), $(j \in R)$ is an objection to (i, x) for which there is no counterobjection. Hence

$(i, x) \notin \mathfrak{M}$. If $i \in R$, then if (i, x) is $\mathfrak{M}$-stable it is not maximal since (i, α) is also $\mathfrak{M}$-stable and $\alpha >_i x$. ∎

The following two theorems relate the bargaining set to the core and one-core resp.

THEOREM 8. [166] If the core $\mathfrak{C}$ of a committee game Γ is not empty, then

(1) for each $i \in N$, there exists $x \in X$ such that $(i, x) \in \mathfrak{M}$

(2) $(i, x) \in \mathfrak{M} \Leftrightarrow y \leqq_i x$ for each $y \in \mathfrak{C}$.

Proof. Clearly, (i, x) is $\mathfrak{M}$-stable if $i \in N$ and $x \in \mathfrak{C}$. Since X is finite there exist maximal $\mathfrak{M}$-stable proposals and assertion (1) holds. The third assertion follows from the fact that $\mathfrak{M}$ consists of only maximal $\mathfrak{M}$-stable proposals. ∎

THEOREM 9. [166] If the one-core $\mathfrak{C}_1$ is nonempty for a committee game Γ, then we have

(1) $\mathfrak{M} \neq \emptyset$,

(2) $(i, x) \in \mathfrak{M} \Rightarrow y \leqq_i x$ for each $y \in X$ such that $(i, y) \in \mathfrak{C}_i$.

Proof. The same as for Theorem 8. ∎

Although the bargaining set is broader than the core and the one-core, it is not always nonempty. There can be given a five-person simple game with an empty bargaining set [166].

To illustrate the above solution concepts let us consider a game Γ where $N = \{1, 2, 3\}$, $X = \{\mathbf{x} = \{x_1, x_2, x_3\} \in \mathbb{R}^3 \mid x_1 + x_2 + x_3 = 9$, $x_i \geqq 0$ and integer for $i = 1, 2, 3\}$. The function v is given by

$$v(R) = \begin{cases} X & \text{if} \quad |R| \geqq 2 \\ \emptyset & \text{if} \quad |R| = 1 \end{cases}$$

and $\mathbf{x} >_i \mathbf{y}$, $\mathbf{x} \geqq_i \mathbf{y}$ whenever $x_i > y_i$, $x_i \geqq y_i$ for all $i \in N$.

For this game simple calculation reveals that

(1) the core $\mathfrak{C} = \emptyset$,

(2) the one-core $\mathfrak{C}_1 = \{(i, x) \mid x_i = 1,\ x \in X,\ i = 1, 2, 3\}$,

i.e., any proposal where a player proposes himself 1 and arbitrary numbers for the others is an element of the one-core.

This means that in this game it is a major drawback to propose first. But in our model we did not include the question of agenda and its effect on the outcomes.

Much more realistic solution is provided by the bargaining set. In fact the bargaining set $\mathfrak{M}$ is given by

$$(34) \qquad \mathfrak{M} = \{(1, (3, 3, 3)), (2, (3, 3, 3)), (3, (3, 3, 3))\},$$

i.e., no matter which player is to propose he will propose the "fair" distribution (3, 3, 3).

To see that $\mathfrak{M}$ is really of the form given by (34) consider the proposal (1, (3, 3, 3)). An objection by player 2 is (2, (2, 3), (0, 4, 5)). A counterobjection by player 1 is (1, (1, 3), (3, 0, 6)). It is easy to show this kind of stability for any objection and to see that (1, (3, 3, 3)) is also maximal.

28. Game theoretical treatment of multicriteria decision making problems

We are facing a multicriteria decision making problem (briefly MDM) whenever we are to select "the best one" out of a set of alternatives H each element of which is evaluated by a number of incommensurable criteria. Very often we cannot characterize some of the criteria numerically and even stochastic elements can enter the problem making the choice more difficult and complex.

MDM plays an important role in several fields, especially in operations research. Attention has been focused on it because of the abundance of unsolved mathematical and conceptual problems. In order to be able to realize the applicational possibilities basic concepts of MDM must first be laid on a solid foundation. In this respect the situation seems to be rather "chaotic". Besides Pareto-optimality there is not essentially any generally accepted theoretically sound and operationally useful solution concept for MDM.

However, requiring Pareto-optimality allows too much freedom for the decision maker: the selection among Pareto-optimal solutions still remains to be done. Many proposals [151], [211] (even computer programs) have been made to set up "reasonable" selection procedures but they are mostly too flexible, the underlying heuristics is too hard for a user to accept without reservation and without (usually justified) suspicion. Primarily it is the lack of theoretical foundation which is most disturbing.

One of the possible ways to devise a theoretically sound solution concept can be outlined in the following fashion: Set up requirements which seem intuitively rational and which must be met by a "solution". The system of requirements should preferably be strict enough to single out a unique solution which is then considered to be a "best compromise".

We have seen in previous chapters of the book that this approach is successfully applied in the theory of cooperative games. Good examples are the Shapley-value for games given in characteristic function form (see Chapter 23) and the Nash–Szidarovszky solution for games without side-payments (see Chapter 26).

In the following we will formulate MDM as a game such that the transition from MDM to the game can be well interpreted and intuitively acceptable. Then we apply (relatively) sound solution concepts of cooperative game theory to the models thus obtained and will reinterpret the solutions in terms of the MDM.

Of course, we cannot deal with the MDM without imposing mathematically tractable structure on the set of alternatives H. We will consider only the deterministic case and assume H to be finite

$$H=\{A_1, A_2, \ldots, A_r\}$$

and suppose that each alternative is characterized numerically by m criteria, i.e., the set of the possible outcomes H' is given by a finite set of m-vectors

$$H'=\{\mathbf{a}_1, \mathbf{a}_2, \ldots, \mathbf{a}_r\}.$$

We also assume that $a_{ij}>a_{ik}$ implies that A_j is preferred to A_k by criterion i and $a_{ij}=a_{ik}$ if none of them is preferred to the other. (Here a_{ij} and a_{ik} denote the ith component of $\mathbf{a}_j$, $\mathbf{a}_k$ resp.) Although there is a one-to-one correspondence between H and H', theoretically a distinction must be made between the alternatives and their consequences.

Finiteness and numerical characterization are really strong restrictions but even in this relatively simple case almost all of the conceptual difficulties of an MDM are present and a wide range of practical applications can be covered with this model.

As a first step we extend the set of feasible outcomes H'. We allow mixed outcomes, i.e., the set of feasible outcomes will be the polytope P spanned by vectors $\mathbf{a}_1, \mathbf{a}_2, \ldots, \mathbf{a}_r$.

$$P=\left\{\mathbf{x}\mid\mathbf{x}=\sum_{i=1}^{r}\lambda_i\mathbf{a}_i,\ \lambda_i\geqq 0,\ (i=1,\dots,r),\ \sum_{i=1}^{r}\lambda_i=1\right\}.$$

This can be interpreted in the following way: if the ("pure") outcomes are chosen with probabilities $\lambda_1, \lambda_2, \dots, \lambda_r$, then the expected vector of "consequences" is

$$\mathbf{x}=\sum_{i=1}^{n}\lambda_i\mathbf{a}_i .$$

When determining the "best" vector of consequences the standard of "goodness" is set up in the form of rational axioms a "best" vector is supposed to satisfy. To this end we imbed our problem into a game theoretical framework.

Assign to each criterion a "player" whose aim is to choose an alternative giving as large a numerical value as it is possible as to the criterion represented by that particular player. To put it more precisely: the strategy set of each player is the finite set $H=\{A_1, A_2, \dots, A_r\}$ and the pay-off function f_i of player i is defined as

$$f_i(A_{j_1}, A_{j_2}, \dots, A_{j_r})=\begin{cases}a_{ij} & \text{if } j_1=j_2=\dots=j_r=j\\ -\alpha_i & \text{otherwise}\end{cases},\qquad (i=1,\dots,m)$$

where α_i is a suitable (generally large) positive number, $(i=1, \dots, m)$.

In other words this means that if all players choose the same alternative, say A_j, then they get the corresponding entries of $\mathbf{a}_j$ as pay-offs. But if at least one of them "deviates", then the pay-off for everybody gets very "bad", there is a penalty α_i for lack of consensus imposed on every player i, $(i=1, \dots, m)$.

For the game $G=\{H, H, \dots, H; f_1, f_2, \dots, f_r\}$ thus defined we look for a cooperative "solution". Solution concepts allowing exchange value in form of side-payments have to be left out of consideration since we have assumed that the criteria are incommensurable.

However, by allowing probabilistic mixtures of alternatives a peculiar kind of "side-payment" has been made possible. If the players come to an agreement to choose $\mathbf{x} \in P$ and $\mathbf{x} = \sum_{i=1}^{r} \lambda_i \mathbf{a}_i$, then they have implicitly agreed to choose alternatives $A_1, A_2, \ldots, A_r$ with probabilities $\lambda_1, \lambda_2, \ldots, \lambda_r$ resp. Thus their side-payment to each other is their commitment to possibly "play" unfavourable strategies (alternatives) with positive probability.

Now we apply Nash's solution concept for game G as discussed in Chapter 26. To this end we have to assume that a status quo point $\mathbf{f}^* \in \mathbb{R}^m$ is given which represents the pay-off, the players get if they cannot come to an agreement, i.e., if no decision has been made in the MDM. Given the status quo point Nash's axioms assure the existence of a function ψ assigning to the pair $(P, \mathbf{f}^*)$ a unique point $\psi(P, \mathbf{f}^*)$ which is considered to be the "solution" of the cooperative game G, i.e., "the best compromise" outcome of the MDM. The axioms (feasibility, rationality, Pareto-optimality, independence of unfavourable alternatives, independence of monotone increasing linear transformations, symmetry) are very hard not to accept even in the context of MDM.

In Chapter 26 we also gave a method for finding the best compromise vector. It is the unique solution of the programming problem

$$(1) \qquad \prod_{k=1}^{m} (x_k - f_k^*) \to \max$$

$$\mathbf{x} = (x_1, \ldots, x_m) \in P$$

$$\mathbf{x} \geqq \mathbf{f}^* .$$

provided there is an $\mathbf{f} \in P$ such that $\mathbf{f} > \mathbf{f}^*$.

Since the components of the status quo point $\mathbf{f}^*$ have been interpreted as the consequences (penalties) of disagreement it is only natural to choose $\mathbf{f}^* = -(\alpha_1, \ldots, \alpha_m)$.

Let $\boldsymbol{\alpha}=(\alpha_1, \ldots, \alpha_m)$ satisfy $\mathbf{x} > -\boldsymbol{\alpha}$ for any $\mathbf{x} \in P$, i.e., any (mixed) decision is better than the disagreement. We can now rewrite (1) as

$$(2) \qquad \prod_{k=1}^{m} (x_k+\alpha_k) \to \max$$

$$\mathbf{x}=\mathbf{A}\boldsymbol{\lambda}$$

$$\boldsymbol{\lambda} \geqq \mathbf{0}$$

$$\mathbf{1}\boldsymbol{\lambda}=1\,,$$

where $\mathbf{A}=(\mathbf{a}_1, \ldots, \mathbf{a}_r)$. If the rows of $\mathbf{A}$ are $\mathbf{b}_1, \ldots, \mathbf{b}_m$, then (2) can be written as

$$(3) \qquad \prod_{k=1}^{m} (\mathbf{b}_k\boldsymbol{\lambda}+\alpha_k) \to \max$$

$$\boldsymbol{\lambda} \geqq \mathbf{0}$$

$$\mathbf{1}\boldsymbol{\lambda}=1\,.$$

By axiom 5 of Chapter 26 we may assume without loss of generality that $\mathbf{A} > \mathbf{0}$. The objective functions of (2) and (3) are both positive on the feasible sets resp. Both are products of concave functions therefore they are explicitly quasiconcave (see [107] Theorem 47, p. 61). Thus any local maximumpoint is also global. Since $\mathbf{A} > \mathbf{0}$ the gradient vectors of the objective functions of both (2) and (3) are positive which together with explicit quasiconcavity implies that any critical point (generalized stationary point) is a global maximumpoint. This enables us to apply efficient local methods of mathematical programming (e.g. gradient methods) for solving (2) and (3).

Taking the logarithm of (2) (this is a monotonous transformation which does not affect the maximum points) we obtain the following separable concave programming problem:

$$(4) \qquad \sum_{k=1}^{m} \log(x_k+\alpha_k) \to \max$$
$$\mathbf{x}=\mathbf{A}\boldsymbol{\lambda}$$
$$\boldsymbol{\lambda} \geqq \mathbf{0}$$
$$\mathbf{1}\boldsymbol{\lambda}=1\,.$$

Problem (4) can be linearized by standard techniques of separable programming (see [37]).

It is worth noting that the controversial issue of assigning "weights" to the criteria has not emerged explicitly in our game theoretical treatment of MDM. It seems to us, however, that without incorporating parameters providing explicit or implicit information about the relative "importance" of criteria, the decision maker cannot come to a meaningful solution of the MDM problem. Unlike "traditional" methods using weights of criteria our approach heavily relies on the reasonable choice of the penalty vector $\boldsymbol{\alpha}$. The relative importance of the criteria comes into effect through the choice of the vector $\boldsymbol{\alpha}$.

We now mention a few possibilities for choosing $\boldsymbol{\alpha}$ which seem to be "rational". Of course, in actual decision situations it must be thoroughly thought over which one (or possibly something else) should be applied.

1. Let us suppose that the decision maker has to improve a "situation" characterized by a positive vector $\mathbf{a}_0$ and to this end he has r alternatives to choose from. These are also given by vectors $\mathbf{a}_1, \ldots, \mathbf{a}_r$ of the same dimension as $\mathbf{a}_0$. We assume further that the situation can really be improved, i.e., there is a convex linear combination $\hat{\mathbf{a}}$ of $\mathbf{a}_1, \ldots, \mathbf{a}_r$ to satisfy $\mathbf{a}_0 < \hat{\mathbf{a}}$. Then we set $\boldsymbol{\alpha} = \mathbf{a}_0$ which can be interpreted in a straightforward way: if no decision has been made because of disagreement among players (criteria), then the situation will not be any better, it still remains to be characterized by $\mathbf{a}_0$.

2. If the decision maker cannot choose from among the alternatives, then a random mechanism will do so according to a

probability distribution p which is known or can be estimated. If $\mathbf{Ap}>\mathbf{0}$ and there is an $\hat{\mathbf{a}}\in P$ to satisfy $\mathbf{Ap}<\hat{\mathbf{a}}$, then the choice $\boldsymbol{\alpha}=\mathbf{Ap}$ is possible.

3. Let $\mathbf{A}>\mathbf{0}$ and $\alpha_i=\min a_{ij}$, $(i=1,\ldots,m)$. The vector $\boldsymbol{\alpha}=(\alpha_1,\ldots,\alpha_m)$ thus defined will be considered to be the penalty vector. The rational behind this choice is the following. If we do not know anything about the consequences of the failure to reach a consensus, then each "player" (criterion) must consider even the worst case. Thus going as far as possible from an "ideally bad point" which may never realize but its components express real dangers might be desirable for the decision maker.

4. Finally we consider the case when lack of consensus is absolutely out of question, some choice among alternatives should be made and disagreement is to be treated only formally as a mathematical device. In other terms this means that we are looking for a solution (if there exists any) which can be obtained if the "penalty" tends to infinity. In particular, we assume that

$$\boldsymbol{\alpha}=\alpha\mathbf{r}$$

where $\mathbf{r}$ is a positive vector representing the relative share of the players from the penalty and α measures its magnitude. We will investigate what happens if α tends to infinity.

Denote $\mathbf{x}(\alpha)$ the (unique) optimal solution of (2) if $\boldsymbol{\alpha}=\alpha\mathbf{r}$ and take a sequence of real numbers $\alpha_1,\alpha_2,\ldots$ tending to infinity. Then the elements of the sequence $\{\mathbf{x}(\alpha_k)\}$ are uniquely determined and the sequence has at least one cluster point since P is closed and bounded. However, it is far from being trivial that it has only one cluster point. This conceptual difficulty is resolved by the following theorem.

Define the programming problem:

$$(5)\qquad F(\alpha):\prod_{k=1}^{m}(x_k+\alpha_k)\rightarrow\max$$

$$\mathbf{x}=\mathbf{A}\boldsymbol{\lambda}$$

$$\boldsymbol{\lambda} \geqq \mathbf{0}$$

$$\mathbf{1}\boldsymbol{\lambda} = 1 .$$

THEOREM 1 [55] If $\lim_{k \to \alpha} \alpha_k = \infty$, then the sequence $\{\mathbf{x}(\alpha_k)\}$ has exactly one cluster point.

Proof. The objective function of $F(\alpha)$ is a polynomial of order m of the positive parameter α. Let this polynomial be

$$s(\mathbf{x}, \alpha) = h_m(\mathbf{x})\alpha^m + h_{m-1}(\mathbf{x})\alpha^{m-1} + \dots$$

$$\dots + h_1(\mathbf{x})\alpha + h_0(\mathbf{x}) .$$

We know that for any fixed positive α $s(\mathbf{x}, \alpha)$ is quasi-concave on the positive orthant $\mathbb{R}^+$. Let $K \subset \mathbb{R}^+$ be an arbitrary convex set. Let j be the largest index for which $h_j(\mathbf{x})$ is not constant on K. We claim that $h_j(\mathbf{x})$ is quasi-concave on K. Suppose on the contrary that there exist $\mathbf{x}_1, \mathbf{x}_2 \in K$ and λ $(0 < \lambda < 1)$ such that

$$h_j(\lambda\mathbf{x}_1 + (1-\lambda)\mathbf{x}_2) < \min \{h_j(\mathbf{x}_1),\ h_j(\mathbf{x}_2)\} .$$

This means that for sufficiently large α

$$s(\lambda\mathbf{x}_1 + (1-\lambda)\mathbf{x}_2, \alpha) < \min \{s(\mathbf{x}_1, \alpha),\ s(\mathbf{x}_2, \alpha)\}$$

contradicting to the fact that $s(\mathbf{x}, \alpha)$ is quasiconcave on $\mathbb{R}^+$.

Let $P^m \equiv P$ and P^k be the set of optimal solutions to the following programming problem:

$$\text{(6)} \qquad h_k(\mathbf{x}) \to \max$$

$$\mathbf{x} \in P^{k+1}$$

for $k = m-1, \dots, 1, 0$. Obviously

$$P^m \supseteqq P^{m-1} \supseteqq \dots \supseteqq P^1 \supseteqq P^0$$

and problem (6) is solvable for any k since P is closed and bounded and h_k is continuous $(k = 0, \dots, m)$ being a polynomial. The set P^m is convex and $h_j(\mathbf{x})$, $(j \geqq k+1)$ is constant on P^{k+1}, therefore $h_k(\mathbf{x})$ is quasiconcave on P^{k+1}. This implies that P^k is convex for any k, as well. P^0 consists of a single point since the last problem is

$$(7)\qquad h_0(\mathbf{x}) \equiv x_1 x_2 \ldots x_m \to \max$$
$$\mathbf{x} \in P^1$$

which is equivalent to

$$\log x_1 + \log x_2 + \ldots + \log x_m \to \max$$
$$\mathbf{x} \in P^1$$

which has a strictly concave objective function.

We assert that the only element of P_0, say $\mathbf{x}_0$, is the unique cluster point of the sequence $\{\mathbf{x}(\alpha_k)\}$. Assume on the contrary that there is a cluster point $\mathbf{x}_1$ of $\{\mathbf{x}(\alpha_k)\}$ for which $\mathbf{x}_1 \neq \mathbf{x}_0$. It suffices to show that $s(\mathbf{x}_0, \alpha) = s(\mathbf{x}_1, \alpha)$ which is impossible since P^0 has only one element.

Suppose that $s(\mathbf{x}_0, \alpha) \not\equiv s(\mathbf{x}_1, \alpha)$ and j is the largest index for which $h_j(\mathbf{x}_0) > h_j(\mathbf{x}_1)$. Then there exists an α_0 such that for any $\alpha \geqq \alpha_0$ we have $s(\mathbf{x}_0, \alpha) > s(\mathbf{x}_1, \alpha)$. Since $\mathbf{x}_1$ is a cluster point, therefore in any ε-neighbourhood $K_1(\mathbf{x}_1, \varepsilon)$ there are infinitely many points $\mathbf{x}(\alpha_k)$. The radius ε can be chosen so small that $h_j(\mathbf{x}_0) > h_j(\mathbf{x}(\alpha_k))$ and $s(\mathbf{x}_0, \alpha_0) > s(\mathbf{x}(\alpha_k), \alpha_0)$ hold for any $\mathbf{x}(\alpha_k) \in K(\mathbf{x}_1, \varepsilon)$. This implies that $s(\mathbf{x}_0, \alpha) > s(\mathbf{x}(\alpha_k, \alpha_0))$ holds for any $\alpha \geqq \alpha_0$ and $\mathbf{x}(\alpha_k) \in K(\mathbf{x}_1, \varepsilon)$. For sufficiently large k $\alpha_k \geqq \alpha_0$ and hence $s(\mathbf{x}_0, \alpha_k) > s(\mathbf{x}(\alpha_k), \alpha_k)$ which contradicts to the assumption that $\mathbf{x}(\alpha_k)$ is an optimal solution of $F(\alpha_k)$. ∎

From the proof it turns out that to determine the unique cluster point we have to solve at most m programming problems having quasiconcave objective functions. Since

$$h_{m-1}(\mathbf{x}) = \sum_{i=1}^{m} \left(\prod_{j \neq i} r_j \right) x_i$$

is linear therefore the first problem to be solved

$$h_{m-1}(\mathbf{x}) \to \max$$
$$\mathbf{x} \in P$$

is a linear programming problem which generally (except in the case of dual-degeneration) has a unique solution. This solution is a vertex of P, i.e., it is an original ("pure") alternative.

The proportion vector $\mathbf{r}$ has a crucial role in this model.

We may choose $\mathbf{r}$ to represent the magnitude of the numerical values characterizing the criteria. The simplest idea is to set

$$r_i = \frac{1}{r} \sum_{j=1}^{r} a_{ij} \qquad (i = 1, \ldots, m)$$

i.e. the penalties tend to infinity proportionally to the average values.

It is worth noting that uniqueness of the optimal solution to (2) does not imply the uniqueness of $\boldsymbol{\lambda}$. It can happen that the unique consequence vector can be achieved by several mixtures of pure alternatives. This may give rise to a certain selection: alternatives which do not have positive weights in any mixture resulting in the unique "best" consequence vector may be left out of further considerations.

As an example let us consider the following four alternatives to choose from:

$$\begin{matrix} A_1 & A_2 & A_3 & A_4 \\ \begin{bmatrix} 2 \\ 3 \end{bmatrix} & \begin{bmatrix} 1 \\ 6 \end{bmatrix} & \begin{bmatrix} 3 \\ 1 \end{bmatrix} & \begin{bmatrix} 3 \\ 2 \end{bmatrix} \end{matrix}. \tag{8}$$

The strategy set of each player is $S = \{A_1, A_2, A_3, A_4\}$. The pay-off functions are:

$$f_1(A_1, A_1) = 2$$

$$f_1(A_2, A_2) = 1$$

$$f_1(A_3, A_3) = 4$$

$$f_1(A_4, A_4) = 3$$

$$f_1(A_i, A_j) = -\alpha_1, \quad (i \neq j),$$

$$f_2(A_1, A_1)=3$$

$$f_2(A_2, A_2)=6$$

$$f_2(A_3, A_3)=1$$

$$f_2(A_4, A_4)=2$$

$$f_2(A_i, A_j)=-\alpha_2, \quad (i \neq j).$$

Taking (1, 1) as status quo point (the worst possible value for both criteria is 1) the optimal solution to problem (2) turns out to be $\left(\frac{5}{2}, \frac{7}{2}\right)$ which means that alternatives A_2, A_3 should be chosen with probabilities $\frac{1}{2}, \frac{1}{2}$ resp.

The average of vectors (8) is $\left(\frac{5}{2}, 3\right)$ and taking the status quo point $-\alpha\left(\frac{5}{2}, 3\right)$ where α is a positive parameter we get (1, 6) as the best alternative whenever $\alpha \geqq \frac{26}{7}$.

In the following we formulate another model where the game we associate to the MDM is given in characteristic function form and we would like to determine "weights" for the criteria representing their relative importance. The MDM is defined exactly as before and the players also represent criteria. We also assume that the m by r matrix of outcomes $\mathbf{A}$ is positive.

Let M denote the set of all criteria (players) $M=\{1, 2, \ldots, m\}$ and consider a subset S of M. If $S=\emptyset$, then we define the characteristic function v to be $v(\emptyset)=0$ and if $S=M$, then $v(M)=1$. $v(M)=1$ means that the set of all criteria perfectly describes the decision making situation under consideration. If S is a proper subset of M, then according to the definition of the characteristic function $v(S)$ is supposed to measure the ability of criteria contained in S to describe the whole situation, i.e., $v(S)$ should measure the extent to which the criteria in S (forming a "coalition") represent the information contained in the rows controlled by the coalition $M-S$.

Since our aim is to determine weights it seems logical to assume that coalition S can generate any convex linear combination of the corresponding rows of **A** i.e. it can determine new "secondary" criteria by properly mixing the "pure" criteria. Of course, the complementary coalition $M-S$ may also do so. If S can generate any row of **A** under the control of coalition $M-S$ by suitable mixture of the rows belonging to S, then it seems rational to define $v(S)=1$. Similarly, the "more" S is able to generate any row under the control of $M-S$, the larger $v(S)$ should be. Of course, the term "more" must be defined exactly. We will measure the similarity of two criteria (pure or mixed) **a** and **b** by the cosine of the angle they make

$$\cos\varphi = \frac{\mathbf{ab}}{\|\mathbf{a}\|\,\|\mathbf{b}\|} \qquad (\mathbf{a}\neq\mathbf{0},\mathbf{b}\neq\mathbf{0}).$$

It is clear that if $\cos\varphi=1$ then **a** is a positive scalar multiple of **b**, i.e., both criteria represented by them orient in the same direction. The greater $\cos\varphi$ is the more we can consider **a** and **b** orienting in a similar direction.

Partition now matrix **A** as to coalitions S and $M-S$

S	$\mathbf{A}_1$	weight vector **p**
$M-S$	$\mathbf{A}_2$	weight vector **q**

If coalitions S and $M-S$ weigh the "pure criteria" by **p** and **q** resp. then the cosine of the angle made by vectors $\mathbf{pA}_1$ and $\mathbf{qA}_2$ thus obtained can be computed as

$$\cos\varphi = \frac{\mathbf{pA}_1\mathbf{A}_2^T\mathbf{q}}{\|\mathbf{pA}_1\|\,\|\mathbf{A}_2^T\mathbf{q}\|},$$

($\|\mathbf{pA}_1\| \neq 0, \|\mathbf{qA}_2\| \neq \mathbf{0}$ since $\mathbf{A}_1 > \mathbf{0}$, $\mathbf{A}_2 > \mathbf{0}$). According to our reasoning we define

$$v(S) = \max_{\mathbf{p}} \min_{\mathbf{q}} \frac{\mathbf{pA}_1 \mathbf{A}_2^T \mathbf{q}}{\|\mathbf{pA}_1\| \, \|\mathbf{A}_2^T \mathbf{q}\|}$$

for any coalition S ($S \subset M$, $S \neq \emptyset$). Here the vectors $\mathbf{p}$ and $\mathbf{q}$ traverse spaces of probability vectors of proper dimension.

For arbitrary fixed $\mathbf{p}$

$$\frac{\mathbf{pA}_1 \mathbf{A}_2^T \mathbf{q}}{\|\mathbf{pA}_1\| \, \|\mathbf{A}_2^T \mathbf{q}\|}$$

is a quasiconcave functions of $\mathbf{q}$ since the numerator is a positive linear function and the denominator is a positive convex function. It is known that a quasiconvex functions attains its maximum at some vertex of the feasible region. Since the vertices of the set of probability vectors are the unit vectors we have

$$\min_{\mathbf{q}} \frac{\mathbf{pA}_1 \mathbf{A}_2^T \mathbf{q}}{\|\mathbf{pA}_1\| \, \|\mathbf{A}_2^T \mathbf{q}\|} = \min_{j} \frac{\mathbf{pA}_1 \mathbf{A}_2^T \mathbf{e}_j}{\|\mathbf{pA}_1\| \, \|\mathbf{A}_2^T \mathbf{e}_j\|}.$$

Thus computing $v(S)$ amounts to solving the mathematical programming problem

$$\min_{j} \frac{\mathbf{pA}_1 \mathbf{A}_2^T \mathbf{e}_j}{\|\mathbf{pA}_1\| \, \|\mathbf{A}_2^T \mathbf{e}_j\|} \to \max$$

$$\mathbf{p} \geqq \mathbf{0}$$

$$\mathbf{1p} = 1$$

which is equivalent to solving

$$\alpha \to \max$$

$$\frac{\mathbf{pA}_1 \mathbf{A}_2^T \mathbf{e}_j}{\|\mathbf{pA}_1\| \, \|\mathbf{A}_2^T \mathbf{e}_j\|} \geqq \alpha \quad \text{for all} \quad j$$

$$\mathbf{p} \geqq \mathbf{0}$$

$$\mathbf{1p} = 1 .$$

By the quasiconcavity of the constraint functions the feasible set of the above problem is convex. Maximizing a linear function on a convex feasible set constitutes a "well-behaving" problem where numerous efficient local methods (e.g. gradient-like methods) are available to solve it.

As an illustration we consider the MDM where three alternatives are evaluated according to three criteria:

		Alternatives		
		A_1	A_2	A_3
	C_1	1	3	2
Criteria	C_2	2	1	4
	C_3	4	1	1

By elementary calculation (we give no details) we get the characteristic function

Empty coalition	$v(\emptyset)=0$
One-member coalitions	$v(\{1\})=0.567$
	$v(\{2\})=0.668$
	$v(\{3\})=0.567$
Two-member coalitions	$v(\{1,2\})=0.675$
	$v(\{1,3\})=0.812$
	$v(\{2,3\})=0.758$
Grand coalition	$v(\{1,2,3\})=1.$

There are several possibilities to assign "weights" or "values" to the players (criteria) representing certain ideas about "fairness", "strength" and "position" in the game. In our book we only dealt with the Shapley value and the nucleolus both realizing these ideas in a special way.

In our example we get the following results

	Criteria		
	C_1	C_2	C_3
Shapley-value	0.313	0.336	0.351
Nucleolus	0.3	0.4	0.3

It is not unexpected that we got slightly different numbers for the Shapley-value and the nucleolus since they define "fair distribution" in a different way.

Weighing the alternatives by the Shapley-value and the nucleolus we get

	Alternatives		
	A_1	A_2	A_3
Shapley-value	2.389	1.626	2.321
Nucleolus	2.300	1.600	2.500

When using the Shapley-value we get A_1 whereas when using the nucleolus we get A_3 as best. A_2 is worst in both cases.

Although explicitly no weights of the criteria or no ideally bad outcome was assumed to have been given when the decision took place we must not think that the numerical values of the criteria alone determine either a "best alternative" or a "fair weighing". Several assumptions about the behaviour, the preferences of the decision maker as well as a priori assumptions about a "fair and reasonable" selection procedure are all present implicitly in the above models. Just to mention a few out of these we refer to the rather arbitrary manner the characteristic function has been computed or the definition of the Shapley value which implies a certain kind of equality among players (criteria) by taking any order of players when forming the grand coalition equiprobable. The "fairness" realized by the nucleolus is also rather arbitrary, several other schemes of fair distribution can be and have been proposed.

In conclusion we can state that game theory is no panacea for all the troubles of MDM. Just as in other fields all it can do is provide more insight into the nature of the decision problem. In addition, in certain special cases it can also offer operational solutions if the necessary data are available.

29. Games with incomplete information

A game is called a *game with incomplete information* (I-game) if the players do not know exactly the game itself let it be given either in extensive or in normal form. They may not know completely the strategy sets and/or the pay-off functions, the rules of the game etc. I-games are not to be mistaken with games of imperfect information where the players may not remember their own and/or the other players' (including "chance") previous moves.

I-games can be analyzed with various models depending upon the additional assumptions. These models intend to reduce I-games to games with complete information (C-game). Depending on the assumptions made on the "behaviour" of the players and on the information structure conceptually entirely different models can be devised to convert I-games into C-games. Of these we will only deal with the basic features of the Harsanyi and Selten models [69].

Assume that n players participate in an I-game G. G is given in normal form:

$$G=\{\Sigma_1, \Sigma_2, \ldots, \Sigma_n; K_1, K_2, \ldots, K_n\}$$

where the strategy sets and the pay-off functions are denoted by Σ_i and K_i, $(i=1, 2, \ldots, n)$ resp. The functions K_i are defined on the set $\Sigma = \bigtimes_{i=1}^{n} \Sigma_i$. The incompleteness of the information is assumed to appear in two ways:

1. The players do not know precisely the sets Σ_i.
2. The functions K_i are not completely known to them.

Other kinds of information incompleteness can generally be reduced to one of the above two cases. Let us suppose, e.g., that player i does not know whether player j is informed about the occurrence of an event E. From a game theoretical point of view we are only interested in player i's ability to decide whether player j is in a position to use a strategy $\sigma_j^0 \in \Sigma_j$ which implies a specific course of action if E does occur and some other course of action if it does not. Therefore we may look at the situation as if player i did not completely know player j's strategy set Σ_j. Thus we have arrived at case 1.

We can even reduce case 1 to case 2 by appropriately redefining the pay-off functions. This is so because $\sigma_i^0 \notin \Sigma_i$ can be represented by defining $K_i(\sigma_1, \sigma_2, \ldots, \sigma_{i-1}, \sigma_i^0, \sigma_{i+1}, \ldots, \sigma_n) = -\infty$ for all $\sigma_j \in \Sigma_j, (j \neq i)$. Therefore, when defining strategy sets $\Sigma_i, (i=1, \ldots, n)$ we assume that $\Sigma_i, (i=1, \ldots, n)$ is the largest set of strategies σ_i that can conceivably be included in player i's strategy set in any player's opinion.

In the following we assume that game G is an I-game in the sense that the actual value of a finite number of parameters in the pay-off functions is not known to some players. The mathematical form of the pay-off functions is assumed to be known, only some parameters are considered to be unknown up to a probability distribution. Let us take, e.g., an oligopoly game. We take it for granted that the players want to maximize their profit but certain parameters of the cost and price functions (e.g., wages paid by the other players) are unknown.

For player i let $\mathbf{c}_i$ denote the parameter vector about which only limited information is available and let C_i be the set containing all the feasible $\mathbf{c}_i$'s. Thus an I-game is defined as the collection of the strategy sets, pay-off functions and parameter sets $C_i, (i=1, \ldots, n)$:

$$G = \{\Sigma_1, \Sigma_2, \ldots, \Sigma_n; K_1, K_2, \ldots, K_n; C_1, C_2, \ldots, C_n\}$$

The functions K_i map $\Sigma \times C$ onto the real line $\mathbb{R}$, where $C = \bigtimes_{i=1}^{n} C_i$.

29.1. THE HARSANYI-MODEL

In this model the I-game G is studied from the point of view of a particular player, say player j. Any player i is assumed to know his own "information vector" $\mathbf{c}_i$ but lacks full information about the other vectors $\mathbf{c}_k$, $(k \neq i)$. Player j from whose point of view the game is being analyzed is "Bayesian" and he considers the other players to be "Bayesian", too. This means that he assigns a subjective probability distribution

$$P_i(\mathbf{c}^i) = R_i(\mathbf{c}^i | \mathbf{c}_i)$$

where

$$\mathbf{c}^i = (\mathbf{c}_1, \mathbf{c}_2, \ldots, \mathbf{c}_{i-1}, \mathbf{c}_{i+1}, \ldots, \mathbf{c}_n)$$

$$\mathbf{c}_k \in C_k, \quad (k \neq i)$$

to any player (including himself). These distributions are in fact conditional ones since any player knows his own information vector $\mathbf{c}_i$. Distributions P_i, $(i = 1, \ldots, n)$ are estimations of player j as to the probability distributions entertained by player i, $(i = 1, \ldots, n)$. Certain parameters of this distribution which are unknown to player j are contained in vector $\mathbf{c}_i$.

An I-game is said to be given in standard form (as viewed by player j) by the ordered set

$$G^B = \{\Sigma_1, \ldots, \Sigma_n; K_1, \ldots, K_n; C_1, \ldots, C_n; R_1, \ldots, R_n\}$$

where $R_i = R_i(\mathbf{c}^i | \mathbf{c}_i)$ is a probability distribution over the set

$$C^i = C_1 \times \ldots \times C_{i-1} \times C_{i+1} \times \ldots \times C_n, \quad (i = 1, \ldots, n).$$

In order to avoid mathematical difficulties (which are irrelevant as to the working of the model anyway) the set C_i, $(i = 1, \ldots, n)$ is assumed to be finite. Thus probability distributions can always be defined on it.

The basic assumption of the Harsanyi-model is the existence of a probability distribution R^* on $C = \bigtimes_{i=1}^{n} C_i$ marginal distributions of which are the R_i's, i.e.,

$$(1) \qquad R_i(\mathbf{c}^i | \mathbf{c}_i) = R^*(\mathbf{c} | \mathbf{c}_i), \quad (i = 1, \ldots, n), (\mathbf{c} \in C).$$

In other words we assume that player j's subjective judgement on the distributions R_i is consistent. The existence of the basic probability distribution R^* can be interpreted in the following way. We may think of a lottery first assigning the actual vectors $\mathbf{c}_i$ according to the probability distribution R^*. Then the C-game with the original strategy sets and pay-off functions determined by the actual values of the parameters will be played. The C-game

$$G_*^B = \{\Sigma_1, \ldots, \Sigma_n; K_1, \ldots, K_n; C_1, \ldots, C_n; R^*\}$$

obtained in this manner is called the "Bayes equivalent" belonging to G^B. The strategy of player j in the original I-game is supposed to be determined by the strategy employed by him in G_*^B. Thus game G_*^B begins with a "chance-move" and then an ordinary C-game without any chance move is played.

Game G_*^B can be thought of in another way, too. Player i can be of k_i different "type" ($k_i = |C_i|$; $(i = 1, \ldots, n)$). Anyone knows his own "type" but lacks full information about the "type" of others. Now the lottery selects the actual "type" of the players and the players thus selected play an ordinary C-game. The "types" differ only in the parameter vector $\mathbf{c}_i$.

When reducing our I-game to a C-game crucial role is played by the distribution R^* and the hypothesis of the R_i's being consistent with R^* (see (1).) This assumption implies that in player j's opinion every player has consistent information on the nature of the mechanism producing the particular realizations of the vector $\mathbf{c}_i$, $(i = 1, \ldots, n)$.

It can be proved [69] that under some nondegeneracy (or rather "nondecomposability") conditions equation system (1) has at most one solution. So, if the R_i's are consistent, then R^* is uniquely determined.

29.2. THE SELTEN-MODEL

Let us consider a K-person game $\left(K = \sum_{i=1}^{n} k_i\right)$, where any player i of the original I-game is replaced by a "class of players" differing only in the actual value of the parameter vector $\mathbf{c}_i$. These "potential" players choose strategies and then a lottery selects the actual players who will play the C-game thus obtained. The lottery is performed according to a basic probability distribution R^*. The actual players participating in the C-game are committed to playing the strategies they had chosen before the lottery took place. They get pay-offs determined by their pay-off functions while the $K-n$ potential players not participating in the game get 0.

If there is no consistent basic distribution R^* to the conditional probabilities $R_1, \ldots, R_n$ the Selten model still works. In this case the actual players must be chosen by n different lotteries $R_1, \ldots, R_n$.

29.3. DYNAMIC PROCESSES AND GAMES WITH LIMITED INFORMATION ABOUT THE PAY-OFF FUNCTION

We have seen in previous sections that under suitable assumptions I-games can be converted to C-games. If the players are allowed to play the same I-game several times and they are able to observe the rate of change in the pay-off they get, it seems intuitively reasonable for each player to adjust his strategy according to the change in his pay-off during the previous runs of the game. In most cases the rate of change in the pay-off can be observed without knowing the precise form of the pay-off function. In the following we are going to study a dynamic process realizing the idea of continuous adjustment of strategy vectors for certain I-games.

Let Γ be a finite game with players $1, 2, \ldots, N$. The players have $m+1$, $n+1$, $\ldots$, $t+1$, $s+1$ pure strategies resp. and the pay-off function is determined by the constants $a_{ij,\ldots,kl}, b_{ij,\ldots,kl}, \ldots, c_{ij,\ldots,kl}, d_{ij,\ldots,kl}$. The (mixed) strategy spaces of the players are the corresponding simplices of probability vectors X'_{m+1}, Y'_{n+1}, ..., U'_{t+1}, V'_{s+1}.

We shall treat the game Γ in a somewhat different but equivalent form. We express, say, the first component of the strategy vector of each player as a function of the remaining components. Thus Γ will be given in the following normal form:

$$\Gamma = \{X_m, Y_n, \ldots, U_t, V_s; f(\mathbf{x}, \mathbf{y}, \ldots, \mathbf{u}, \mathbf{v}),$$
$$g(\mathbf{x}, \mathbf{y}, \ldots, \mathbf{u}, \mathbf{v}), \ldots, p(\mathbf{x}, \mathbf{y}, \ldots, \mathbf{u}, \mathbf{v}),$$
$$r(\mathbf{x}, \mathbf{y}, \ldots, \mathbf{u}, \mathbf{v})\}$$

where X_m, Y_n ... U_t, V_s are sets of nonnegative vectors whose sum is less than or equal to 1 and $f, g, \ldots, p, r$ are multilinear functions.

We shall consider a "modified" game Γ_α where the strategy spaces of the players will be extended to the whole Euclidean space of proper dimension, i.e., R^m, R^n, ..., R^t, R^s resp. but a strategy choice outside X_m, Y_n, ..., U_t, V_s will be penalized via a properly chosen "penalty function" added to the original pay-off function. Furthermore we shall apply a small perturbation which will prove to be useful for the treatment of certain singular cases. In choosing the particular form of the penalty function we have great freedom. Since we want to use certain differentiability properties of this function we could not choose as simple a function as it is possible. Now let us define Γ_α in precise terms. As pointed out earlier the strategy spaces are R^m, R^n, ..., R^t, R^s resp. The pay-off functions look like this

$$(2) \qquad f_\alpha(\mathbf{x}, \mathbf{y}, \ldots, \mathbf{u}, \mathbf{v}) = f(\mathbf{x}, \mathbf{y}, \ldots, \mathbf{u}, \mathbf{v}) -$$
$$-\alpha\left[\sum_{i=1}^{m} [\min\{0, x_i - \varepsilon\}]^4 - \left[\min\left\{0, 1 - \sum_{i=1}^{m} x_i - \varepsilon\right\}\right]^4\right] - \delta \sum_{i=1}^{m} x_i^2$$

. . . .

$$r_\alpha(\mathbf{x}, \mathbf{y}, \ldots, \mathbf{u}, \mathbf{v}) = r(\mathbf{x}, \mathbf{y}, \ldots, \mathbf{u}, \mathbf{v}) -$$

$$-\alpha\left[\sum_{l=1}^{s}[\min\{0, v_l-\varepsilon\}]^4-\left[\min\left\{0, 1-\sum_{l=1}^{s}v_l-\varepsilon\right\}\right]^4\right]-\delta\sum_{l=1}^{s}v_l^2$$

where α is a "sufficiently" large, δ is a "sufficiently" small positive number and

$$0<\varepsilon\leqq\frac{1}{M+1}$$

where $M=\max\{m, n, \ldots, t, s\}$.

Let us now consider the following dynamic system where the variable is denoted by t, $(t\geqq 0)$ and every $x, y, \ldots, u, v$ is a function of this single variable,

$$\text{(3)}\qquad \frac{dx_i}{dt}=h_i\frac{\partial f_\alpha}{\partial x_i},\ (i=1, \ldots, m;\ 0<h_i\leqq 1),\ \sum_{i=1}^{m}h_i=1$$

. . . .

$$\frac{dv_l}{dt}=q_l\frac{\partial r_\alpha}{\partial v_l},\ (l=1, \ldots, s;\ 0<q_l\leqq 1),\ \sum_{l=1}^{s}q_l=1.$$

System (3) means that every player changes his strategy proportionally to the change in pay-off with respect to the particular component of the strategy vector. Our primary goal in the following is to study system (3).

THEOREM 1 [54]

For every ε there exists an $\alpha=\alpha(\varepsilon)$ such that if the initial position of (3) satisfies[1]

[1] Since $0<\varepsilon\leqq\frac{1}{M+1}$ there is an initial position to satisfy (4). E.g., $x_{0i}=\frac{1}{m+1}$, $(i=1, \ldots, m)\ldots v_{0_l}=\frac{1}{s+1}$, $l=1, \ldots, s$ is an admissible initial point.

$$
\begin{aligned}
(4)\qquad & \mathbf{x}^0=\mathbf{x}(0)\geqq\varepsilon \quad \sum_{i=1}^{m} x_i^0 \leqq 1-\varepsilon \\
& \vdots \\
& \mathbf{v}^0=\mathbf{v}(0)\geqq\varepsilon \quad \sum_{l=1}^{s} v_l^0 \leqq 1-\varepsilon ,
\end{aligned}
$$

then the solution of (3) satisfies for every $t\geqq 0$

$$
\begin{aligned}
(5)\qquad & \mathbf{x}_\alpha(t)\in X_m \\
& \vdots \\
& \mathbf{v}_\alpha(t)\in V_s .
\end{aligned}
$$

Proof. It is sufficient to prove $\mathbf{x}_\alpha(t)\in X_m$, the rest of the proof goes along the same lines. The proof will be indirect. Let us denote by a the maximum absolute value of the constants of the multilinear forms $f, g, \ldots, p, r$. Let t_0 be the least number where one of the conditions

$$
\begin{aligned}
& x_{\alpha i}(t)\geqq 0, \quad (i=1,\ldots,m) \\
& \sum_{i=1}^{m} x_{\alpha i}(t)\leqq 1 \\
& \vdots \\
& v_{\alpha l}(t)\geqq 0, \quad (l=1,\ldots,s) \\
& \sum_{l=1}^{s} v_{\alpha l}(t)\leqq 1
\end{aligned}
$$

is "just about" to be violated. This means that at least one of the following relations holds

$$
(6)\qquad x_{\alpha i}(t_0)=0, \quad \left.\frac{\mathrm{d}x_{\alpha i}(t)}{\mathrm{d}t}\right|_{t=t_0}<0 \quad \text{for some } i;
$$

$$
(7)\qquad 1-\sum_{i=1}^{m} x_{\alpha i}(t_0)=0, \quad \left.\frac{\mathrm{d}\sum_{i=1}^{m} x_{\alpha i}(t)}{\mathrm{d}t}\right|_{t=t_0}>0;
$$

$$\vdots$$

(8) $\quad v_{\alpha l}(t_0)=0, \quad \left.\frac{dv_{\alpha l}(t)}{dt}\right|_{t=t_0}<0$ for some l;

(9) $\quad 1-\sum_{i=1}^{m} v_{\alpha i}(t_0)=0, \quad \left.\frac{d\sum_{l=1}^{s} v_{\alpha l}(t)}{dt}\right|_{t=t_0}>0;$

and for $t \leqq t_0$ (5) is valid. Since all the functions involved are continuous and $\mathbf{x}(0), \ldots, \mathbf{v}(0)$ satisfy (5), t_0 exists. Suppose (6) holds for some i. Then two cases can be distinguished

(a) $\quad \sum_{i=1}^{m} x_{\alpha i}(t_0)-1 \leqq -\varepsilon\,.$

Let us take the derivative of $x_{\alpha i}(t)$ at t_0. (In the following all the derivatives are taken at t_0 unless otherwise indicated.)

$$\frac{dx_{\alpha i}(t)}{dt}=h_i\Bigg[\frac{\partial f}{\partial x_{\alpha i}}-4\alpha[\min\{0, x_{\alpha i}-\varepsilon\}]^3+$$

$$+4\alpha\left[\min\left\{0, 1-\sum_{i=1}^{m} x_{\alpha i}-\varepsilon\right\}\right]^3-2\delta x_{\alpha i}\Bigg] \geqq$$

$$\geqq h_i[-aM^N+4\alpha\varepsilon^3] \geqq 0$$

if $\alpha \geqq \dfrac{aM^N}{4\varepsilon^3}=\alpha_1$ which contradicts (6).

(b) $\quad 1-\sum_{i=1}^{m} x_{\alpha i}(t_0)-\varepsilon<0$

Let the minimum of the function $\varphi(t)=1-\sum_{i=1}^{m} x_{\alpha i}(t)-\varepsilon$ on the interval $[0, t_0]$ be ε_0 and $\varphi(\bar{t})=\varepsilon_0$, $(0<\bar{t} \leqq t_0)$. Clearly $-\varepsilon \leqq \varepsilon_0<0$. For at least one $x_{\alpha i}(\bar{t}) \geqq \varepsilon$ which follows from $1-\sum_{i=1}^{m} x_{\alpha i}(\bar{t})<\varepsilon$, and the definition of ε. Let i_0 be an index for which $x_{\alpha i_0}(\bar{t}) \geqq \varepsilon$. We shall distinguish two subcases:

(b1) $-\varepsilon \sqrt[3]{\dfrac{M}{M+h_{i_0}}} \leqq \varepsilon_0 < 0.$

Then

$$\left.\frac{dx_{\alpha i}(t)}{dt}\right|_{t=t_0} = h_i\left[\frac{\partial f}{\partial x_{\alpha i}} - 4\alpha[\min\{0, x_{\alpha i}-\varepsilon\}]^3 + \right.$$

$$\left. + 4\alpha\left[\min\left\{0, 1 - \sum_{i=1}^{m} x_{\alpha i} - \varepsilon\right\}\right]^3 - 2\delta x_{\alpha i}\right] \geqq$$

$$\geqq h_i\left[-aM^N + 4\alpha\varepsilon^3 - 4\alpha\frac{\varepsilon^3 M}{M+h_{i_0}}\right] \geqq 0$$

if

$$\alpha \geqq \frac{aM^N(M+h_{i_0})}{4\varepsilon^3 h_{i_0}} \geqq \frac{aM^N\beta}{4\varepsilon^3} = \alpha_2,$$

where β is the minimum of the fractions $\dfrac{M+h_i}{h_i}, \ldots, \dfrac{M+q_l}{q_l}$ $(i=1, \ldots, m), \ldots (l=1, \ldots, s)$.

This also contradicts (6)

(b2) $-\varepsilon \sqrt[3]{\dfrac{M}{M+h_{i_0}}} > \varepsilon_0 \geqq -\varepsilon$

Then

$$\left.\frac{d\varphi(t)}{dt}\right|_{t=\bar{t}} = -\sum_{i=1}^{m} h_i\left[\frac{\partial f}{\partial x_{\alpha i}} - 4\alpha[\min\{0, x_{\alpha i}-\varepsilon\}]^3 + \right.$$

$$\left. + 4\alpha\left[\min\left\{0, 1 - \sum_{i=1}^{m} x_{\alpha i} - \varepsilon\right\}\right]^3 - 2\delta x_{\alpha i}\right] \geqq$$

$$\geqq \sum_{i\neq i_0} h_i\left[-aM^N - 4\alpha\varepsilon^3 + 4\alpha\frac{\varepsilon^3 M}{M+h_{i_0}}\right] -$$

$$-h_{i_0}aM^N + 4\alpha h_{i_0}\frac{\varepsilon^3 M}{M+h_{i_0}} = -aM^N -$$

$$-4\alpha\varepsilon^3 + 4\alpha\frac{\varepsilon^3 M}{M+h_{i_0}} + 4\alpha h_{i_0}\varepsilon^3 > 0$$

if

$$\alpha > \frac{aM^N h_{i_0}(M+h_{i_0}-1)}{4\varepsilon^3(M+h_{i_0})} \geqq \frac{aM^N\gamma}{4\varepsilon^3} = \alpha_3$$

where γ is the minimum of the fractions

$$\frac{h_i(M+h_i-1)}{M+h_i}, \ldots, \frac{q_l(M+q_l-1)}{M+q_l},$$
$$(i=1, \ldots, m) \ldots (l=1, \ldots, s).$$

But this contradicts the assumption that $\bar{t}$ is a minimumpoint of $\varphi(t)$.

Now it only remains to show that (7) is impossible for large enough α. Suppose (7) holds. Then

$$(10) \quad \left.\frac{d\varphi(t)}{dt}\right|_{t=t_0} = -\sum_{i=1}^{m} h_i \left[\frac{\partial f}{\partial x_{\alpha i}} - 4\alpha[\min\{0, x_{\alpha i}-\varepsilon\}]^3 + \right.$$
$$\left. + 4\alpha\left[\min\left\{0, 1-\sum_{i=1}^{m} x_{\alpha i}-\varepsilon\right\}\right]^3 - 2\delta x_{\alpha i}\right] \geqq$$
$$\geqq \sum_{i \neq i_1} h_i[-aM^N - 4\alpha\varepsilon^3 + 4\alpha\varepsilon^3] - h_{i_1}aM^N +$$
$$+ 4\alpha h_{i_1}\varepsilon^3 = -aM^N + 4\alpha h_{i_1}\varepsilon^3 \geqq 0$$

if

$$\alpha \geqq \frac{aM^N}{4\varepsilon^3 h_{i_1}} \geqq \frac{aM^N}{4\varepsilon^3} = \alpha_4 .$$

Here i_1 is an index for which $x_{\alpha i_1}(t_0) \geqq 0$. But (10) contradicts to (7). The proof is similar for the other relations up to (8) and (9). ∎

We have proved Theorem 1, and besides we have an estimation of α for any given ε

$$(11) \quad \alpha > \max\{\alpha_1, \alpha_2, \alpha_3, \alpha_4\} .$$

For further purposes we write α in the form $\alpha = \frac{K}{\varepsilon^3}$ where K is a suitable large number independent of ε. (This estimation for α is rather rough. Our primary goal here was to keep the formulas as simple as possible. We could have chosen, e.g., different values for ε and δ for each player.)

Now we define a "Δ approximate" equilibrium point of the original game Γ in the following way: the N-tuple $(\hat{\mathbf{x}}, \hat{\mathbf{y}}, \ldots, \hat{\mathbf{v}})$ is called a "Δ approximate" equilibrium point of Γ if

$$
\begin{aligned}
(12)\quad & f(\hat{\mathbf{x}}, \hat{\mathbf{y}}, \ldots, \hat{\mathbf{u}}, \hat{\mathbf{v}}) \geqq f(\mathbf{x}, \hat{\mathbf{y}}, \ldots, \hat{\mathbf{u}}, \hat{\mathbf{v}}) - \Delta, \quad \mathbf{x} \in X_m \\
& g(\hat{\mathbf{x}}, \hat{\mathbf{y}}, \ldots, \hat{\mathbf{u}}, \hat{\mathbf{v}}) \geqq g(\hat{\mathbf{x}}, \mathbf{y}, \ldots, \hat{\mathbf{u}}, \hat{\mathbf{v}}) - \Delta, \quad \mathbf{y} \in Y_n, (\Delta > 0) \\
& \vdots \\
& r(\hat{\mathbf{x}}, \hat{\mathbf{y}}, \ldots, \hat{\mathbf{u}}, \hat{\mathbf{v}}) \geqq r(\hat{\mathbf{x}}, \hat{\mathbf{y}}, \ldots, \hat{\mathbf{u}}, \mathbf{v}) - \Delta, \quad \mathbf{v} \in V_s .
\end{aligned}
$$

THEOREM 2. [54] If differential equation system (3) is asymptotically stable, i.e., $\lim_{t \to \infty} \mathbf{x}_\alpha(t), \ldots, \lim_{t \to \infty} \mathbf{v}_\alpha(t)$ exist, then for any $\Delta \geqq (M+1)K\varepsilon + \delta M$ the vectors $\bar{\mathbf{x}}_\alpha = \lim_{t \to \infty} \mathbf{x}_\alpha(t), \ldots, \bar{\mathbf{v}}_\alpha = \lim_{t \to \infty} \mathbf{v}_\alpha(t)$ give a "Δ approximate" equilibrium point of game Γ.

Proof. Assume system (3) is asymptotically stable. This means that

$$
\begin{aligned}
(13)\quad & \left.\frac{\partial f_\alpha}{\partial x_i}\right|_{x_i = \bar{x}_{\alpha i}} = 0, \quad (i = 1, \ldots, m) \\
& \vdots \\
& \left.\frac{\partial r_\alpha}{\partial v_l}\right|_{v_l = \bar{v}_{\alpha l}} = 0, \quad (l = 1, \ldots, s).
\end{aligned}
$$

But since $f_\alpha, \ldots, r_\alpha$ are all strictly concave functions in the variables $\mathbf{x}, \ldots, \mathbf{v}$ resp. while the others are fixed, it follows readily that $(\bar{\mathbf{x}}_\alpha, \ldots, \bar{\mathbf{v}}_\alpha)$ is an equilibrium point of Γ_α. Thus by the definition of an equilibrium point we get for any $\mathbf{x} \in X_m$

$$
\begin{aligned}
f(\mathbf{x}, \bar{\mathbf{y}}_\alpha, \ldots, \bar{\mathbf{v}}_\alpha) \leqq f(\bar{\mathbf{x}}_\alpha, \bar{\mathbf{y}}_\alpha, \ldots, \bar{\mathbf{v}}_\alpha) + \\
+\alpha \sum_{i=1}^{m} [\min\{0, x_i - \varepsilon\}]^4 - \\
-\alpha \left[\min\left\{0, 1 - \sum_{i=1}^{m} x_i - \varepsilon\right\}\right]^4 + \delta \sum_{i=1}^{m} x_1^2 - \\
-\alpha \sum_{i=1}^{m} [\min\{0, \bar{x}_{\alpha i} - \varepsilon\}]^4 + \\
+\alpha \left[\min\left\{0, 1 - \sum_{i=1}^{m} \bar{x}_{\alpha i} - \varepsilon\right\}\right]^4 - \delta \sum_{i=1}^{m} x_{\alpha i}^2 \leqq \\
\leqq f(\bar{\mathbf{x}}_\alpha, \bar{\mathbf{y}}_\alpha, \ldots, \bar{\mathbf{v}}_\alpha) + (M+1)\alpha\varepsilon^4 + \delta M \leqq \\
\leqq f(\bar{\mathbf{x}}_\alpha, \bar{\mathbf{y}}_\alpha, \ldots, \bar{\mathbf{v}}_\alpha) + \Delta\,.
\end{aligned}
$$

Similar reasoning applies to the functions $g, \ldots, p, r$. ∎

COROLLARY. If system (3) is asymptotically stable for any ε, δ satisfying $0 < \varepsilon \leqq \varepsilon_0$, $0 < \delta \leqq \delta_0$ for given ε_0, δ_0, then any limit point of the sequence of N-tuples ("stable" solutions)

$$\{\bar{\mathbf{x}}_{\alpha(\varepsilon),\delta}, \bar{\mathbf{y}}_{\alpha(\varepsilon),\delta}, \ldots, \bar{\mathbf{v}}_{\alpha(\varepsilon),\delta}\}$$

is an equilibrium point of Γ if $\varepsilon \to 0$, $\delta \to 0$.

This means that proving the existence of an equilibrium point for a finite game amounts to proving the asymptotic stability of system (3) for sufficiently small values of ε and δ.

In the following we are going to use stability theorems due to Rosen [148] and Uzawa [188]. We recall them in the form of theorems

THEOREM 3. [148], Let $\dot{x}_i = f_i(\mathbf{x})$, $(i = 1, \ldots, m)$ be a dynamic system ($\dot{x}_i$ is the derivative with respect to the variable t) and $\mathbf{G}$ the Jacobian of f. If $\mathbf{G} + \mathbf{G}^T$ is negative definite for any $\mathbf{x}$, then the system is asymptotically stable.

Now we define quasi-stable processes. The process $\dot{x}_i = f_i(\mathbf{x})$, $(i = 1, \ldots, m)$ is said to be *quasi-stable* in the set $\Omega \in R^m$ if

(i) for any initial position $\mathbf{x}^0 \in \Omega$ the process has a solution $\mathbf{x}(t, \mathbf{x}^0)$ which is uniquely determined by $\mathbf{x}^0$ and is continuous with respect to $\mathbf{x}^0$, $(t \geqq 0)$,

(ii) the solution $\mathbf{x}(t, \mathbf{x}^0)$ is bounded, i.e., there exists a number K such that $-K \leqq \mathbf{x}(t, \mathbf{x}^0) \leqq K$ for all $t \geqq 0$,

(iii) every limit point of $\mathbf{x}(t, \mathbf{x}^0)$ as t tends to infinity is an equilibrium point, i.e., $f_i(\bar{\mathbf{x}})=0$, where $\bar{\mathbf{x}}$ is a limit point.

THEOREM 4. [188]. If the process satisfies (i) and for any initial position $\mathbf{x}^0$ in Ω the solution $\mathbf{x}(t, \mathbf{x}^0)$ is contained in a compact subset of Ω and the Jacobian of f is symmetric, then the process is quasi-stable.

In the following we will always assume that Ω is the set of all positive vectors.

We will apply these theorems for two special cases.

(i) Suppose all the pay-off functions $f, g, \ldots, p, r$ are bilinear in the strategies $\mathbf{x}, \mathbf{y}, \ldots, \mathbf{u}, \mathbf{v}$. For the sake of convenience we alter our notation. Let $\varphi_1, \ldots, \varphi_N$ denote the pay-off functions and $\mathbf{x}_1, \ldots, \mathbf{x}_N$ the strategy vectors. Then

$$\varphi_i(\mathbf{x}) = \sum_{j=1}^{N} [\mathbf{c}_{ij} + \mathbf{x}_i \mathbf{C}_{ij}]\mathbf{x}_j, \quad (j=1, \ldots, N)$$

where $\mathbf{c}_{ij}$'s are constant vectors, $\mathbf{C}_{ij}$'s are constant matrices.

Now the Jacobian of $\mathbf{G}$ system (3) is as follows

$$\mathbf{G} = \mathbf{DC}$$

where $\mathbf{D}$ is a diagonal matrix with the constants $h_1, \ldots, h_m, \ldots, q_1, \ldots, q_s$ as diagonal elements,

$$\mathbf{C} = \begin{bmatrix} \bar{\mathbf{C}}_{11} & \mathbf{C}_{12} & \cdots & \mathbf{C}_{1N} \\ \mathbf{C}_{21} & \bar{\mathbf{C}}_{22} & \cdots & \mathbf{C}_{2N} \\ \cdots & \cdots & \cdots & \cdots \\ \mathbf{C}_{N1} & \mathbf{C}_{N2} & \cdots & \bar{\mathbf{C}}_{NN} \end{bmatrix}$$

where

$$\bar{\mathbf{C}}_{ii}=\mathbf{C}_{ii}+\mathbf{S}_{ii}, \quad (i=1, \ldots, N)$$

$$\mathbf{S}_{ii}=-12\alpha\left[\min\left\{0, 1-\sum_j x_{ij}\right\}\right]^2 [1]-12\alpha\mathbf{R}_i .$$

$\mathbf{R}_i$ is a diagonal matrix with diagonal elements $[\min\{0, x_{ij}-\varepsilon\}]^2-2\delta$ (j goes through the indices of the vector $\mathbf{x}_i$), $(i=1, \ldots, N)$. It can be verified very easily that $\mathbf{S}_{ii}$ is negative definite if $\delta>0$. Thus as a direct corollary of Theorem 3 we get:

THEOREM 5. [54] A sufficient condition for system (3) to be asymptotically stable with constants $h_1, \ldots, q_s$ chosen 1 is that the game Γ be symmetric in the sense that $\mathbf{C}_{ij}=-\mathbf{C}_{ji}$ for all $i\neq j$ and all $\mathbf{C}_{ii}$ be negative definite. (If $h_1, \ldots, q_m$ are not all 1 then instead of $C_{ij}=C_{ji}$ the equation $G_{ij}=-G_{ji}$ should hold for all $i\neq j$.)

COROLLARY. If Γ is a two-person zero-sum game, then system (3) is asymptotically stable with all the constants chosen 1.

(ii) Assume Γ is a two-person general-sum game each player having only two pure strategies.

THEOREM 6. [54] The constants $h_1, \ldots, q_s$ can be chosen so that system (3) be quasi-stable.

Proof. If Γ is a two-person general-sum game where each player has only two pure strategies, then Γ can be given in normal form in the following way:

$$\Gamma=\{X, Y, f(x, y), g(x, y)\},$$

where X and Y is the closed interval [0, 1] and

$$f(x, y)=f_0+a_1x+b_1y+c_1xy,$$

$$g(x, y)=g_0+a_2x+b_2y+c_2xy.$$

The Jacobian of system (3) is as follows

$$\mathbf{G}=\begin{bmatrix} r & hc_1 \\ qc_2 & s \end{bmatrix},$$

where

$$r = -12\alpha h[\min\{0, 1-x-\varepsilon\}]^2 -$$
$$-12\alpha h[\min\{0, x-\varepsilon\}]^2 - 2\delta$$
$$s = -12\alpha q[\min\{0, 1-y-\varepsilon\}]^2 -$$
$$-12\alpha q[\min\{0, y-\varepsilon\}]^2 - 2\delta$$

(h and q are the constants of system (3)).

If one of the numbers c_1, c_2, is 0, then we face a trivial case since one of the equations of (3) will depend only on x or y and this particular equation means simply the maximization of the functions $f(x, y)$ or $g(x, y)$ with respect to x and y.

Thus we may assume that $c_1 \neq 0$, $c_2 \neq 0$. If sign $c_1 \neq$ sign c_2, then by choosing $h=1$, $q = \frac{c_1}{c_2}$ we get a skew-symmetric matrix **G** and Theorem 5 applies.

If sign c_1 = sign c_2, then letting $h=1$, $q = \frac{c_1}{c_2}$ **G** is symmetric and by Theorem 4 system (3) is quasi-stable. Notice that all the conditions of Theorem 5 are satisfied. It is well known from the theory of differential equations that if the right-hand side of (3) is continuous, then every solution is a continuous function of the initial position. Furthermore the way we chose α ensures that every solution of (3) is positive and contained in a compact subset of the space of positive vectors. ∎

COROLLARY. Combining Theorem 6 and the Corollary of Theorem 2 we get a proof of the existence of an equilibrium point for 2 by 2 bimatrix games.

If we want to approximate (3) by a difference equation system, then each player selects an ε, a δ, constants $h_1, \ldots, q_s$ and a suitable initial solution. Everybody observes his own pay-off and changes his strategy a "little bit". Then the first approximation of the derivatives $\frac{\partial f}{\partial x_i}, \ldots, \frac{\partial r}{\partial v_l}$ is available and the numerical solution of

(3) can proceed. The interesting feature of this process is that if the conditions specified in Theorems 5 and 6 are met, then the strategy vectors approach an "approximate" equilibrium point.

Theorems 2 and 4 have another implication for finite two-person general-sum games. Although we cannot expect our dynamic process to be asymptotically stable for any bimatrix game (in [161], Shapley gives an example where the fictitious play and consequently its continuous variant, too, fails to produce the unique equilibrium point) there is an at least $mn+m+n-1$ dimensional subset of the $2mn$ dimensional space of all m by n bimatrix games where the dynamic process described produces an equilibrium point. In other words, in addition to the special case when the game is constant-sum there are bimatrix games that can be solved by "fictitious play type" methods. The requirement expressed in matrix terms is the following.

THEOREM 7. [54] Let $(\mathbf{A}, \mathbf{B})$ be a bimatrix game. If there is a row i and a column j such that, for the matrices $\mathbf{A}'$ and $\mathbf{B}'$

$$\mathbf{SA}'=\mathbf{B}'\mathbf{R}$$

holds for some diagonal matrices $\mathbf{S}$ and $\mathbf{R}$ whose diagonal elements are all positive and

$$\mathbf{A}'=a_{ij}[\mathbf{1}]-\mathbf{1}\hat{\mathbf{a}}_i^T-\bar{\mathbf{a}}_j\mathbf{1}^T+\tilde{\mathbf{A}},$$

$$\mathbf{B}'=b_{ij}[\mathbf{1}]-\mathbf{1}\hat{\mathbf{b}}_i^T-\bar{\mathbf{b}}_j\mathbf{1}^T+\tilde{\mathbf{B}},$$

where $\hat{\mathbf{a}}_i$, $\hat{\mathbf{b}}_i$ and $\bar{\mathbf{a}}_j$, $\bar{\mathbf{b}}_j$ are row (column) vectors of $\mathbf{A}$ and $\mathbf{B}$ resp. with the component a_{ij} omitted and $\tilde{\mathbf{A}}$, $\tilde{\mathbf{B}}$ are obtained from $\mathbf{A}$, $\mathbf{B}$ by omitting row i and column j, then the dynamic process (3) results in an equilibrium point of the game $(\mathbf{A}, \mathbf{B})$.

Proof. The proof is straightforward and goes along the lines of Theorems 2 and 4. ∎

Epilogue

With its clear notions and problem setting theory of games can rightly be considered a unified mathematical discipline. Viewed in itself it can hardly be criticized on its merits. Yet, things are quite different if we look upon the background it originated from and examine its practical applicability, too. This is indispensable since there is a growing demand for mathematical tools capable of serving various sciences in a similar way they did mainly for physics in the past. Recently numerous new fields of mathematics have emerged which beside supplying methods for solving actual problems of certain sciences (such as physics, chemistry, biology, sociology etc.) aim at giving assistance to formulate their basic notions and relations in an exact way. Thereby the spectrum of applications of mathematics has broadened a great deal. Mathematical systems theory, for instance, has also been developing in this spirit.

Keeping in mind the background from which game theory has grown out we cannot help discussing briefly a few problems of modelling real-world conflict situations.

We first reconsider the concept of *strategy*. There is no doubt that the concept used in game theory is important from the theoretical point of view. However its practical significance is questionable.

Apart from very simple and well-structured games one cannot choose and carry out a particular strategy. In such a relatively simple (from game theoretical point of view) game as chess, because of the complexity of the game the players cannot choose a strategy which tells them in every conceivable situation that can arise during the game which move to make in order to guarantee a certain "pay-off". This can only be done in certain endings where the number of possible situations is not that large.

As we have already mentioned the players may have different information in a game about the strategy sets and the pay-off function. They may also differ concerning experience gathered during previous runs of the play. (This is precisely the situation in chess.)

As an illustration let us formalize how a two-person game take place in reality. Denote $J_1(t)$, $J_2(t)$ the information sets of the players at time t (we do not define precisely now what we mean by information, we use it in the common sense of the word). Let $C_1(t)$, $C_2(t)$ denote the sets of feasible "actions" at time t. (An "action" is not necessarily a strategy though in certain cases it can be).

The real-valued functions $K_1: C_1 \times C_2 \to \mathbb{R}$, $K_2: C_1 \times C_2 \to \mathbb{R}$ measure the consequences of the pair of actions (c_1, c_2) $c_1 \in C_1$, $c_2 \in C_2$. When performing their next action the players have new sets of informations $J_1(1) \neq J_1(0)$, $J_2(1) \neq J_2(0)$ since $c_2(0)$, $K_1(c_1(0), c_2(0))$ and $c_1(0)$, $K_2(c_1(0), c_2(0))$ may generate new information. (We treat time as a discrete variable with values $0, 1, 2, \ldots$). Thus $J_1(0) \subseteqq J_1(1)$, $J_2(0) \subseteqq J_2(1)$. Of course, from time k on the players might perform the same actions $c_1(t) = c_1(k)$, $c_2(t) = c_2(k)$, $(t \geqq k)$ based on the information $J_1(t)$, $J_2(t)$. The pair of actions $c_1(k)$, $c_2(k)$ mean a certain kind of "stabilization" and can be considered an "equilibrium point" of the game.

This very general model of a two-person game has been meant to emphasize the importance of the time parameter in the practical applications of game-theory, i.e., the significance of a "dynamic approach". If C_1, C_2, K_1, K_2 are not assumed to be time-dependent, then the above game is determined by the tuple $\{J_1(t), J_2(t); C_1, C_2; K_1, K_2\}$.

The problem discussed so far cannot always be put in the category of "games with incomplete information" defined in Chapter 28, since the "set of information" available for the players may change during the play of the game. In reality, apart from very simple games, this is exactly the way things go.

Coming to *Nash equilibrium points* their nature as a "safety strategy" has already been criticized by many authors [11], [142]. Choosing Nash safety strategies seems to be a reasonable goal for players engaged in a two-person zero-sum game. In other games aspirations of players may lead to situations other than those characterized by Nash equilibrium points. This is more so in case of a cooperative game though we are only dealing with noncooperative games now.

We will consider six basic patterns of behaviour in an n-person noncooperative game $\Gamma = \{\Sigma_1, \ldots, \Sigma_n; K_1, \ldots, K_n\}$ and call them, in turn, *rational, antisocial, social, mazochist, malevolent and philanthropic.* As a basic assumption we suppose that in the game the players think in the same way, i.e., their goals and behaviour patterns are identical. Various concrete realizations of these attitudes can be conceived of. Out of these we mention two for each type under paragraphs (a) and (b) resp. The resulting n-tuple of strategies we simply call a "*solution point*" in the first case and "*equilibrium point*" in the second.

1. Rational behaviour. The players are concerned only with their own pay-off without taking any risk and are indifferent as to the interests of others

(a) A solution point $(\bar{\sigma}_1, \ldots, \bar{\sigma}_n)$ can be defined by the following inequalities

$$\text{(1)} \qquad \inf_{\sigma_1, \ldots, \sigma_{i-1}, \sigma_{i+1}, \ldots, \sigma_n} K_i(\sigma_1, \ldots, \sigma_{i-1}, \bar{\sigma}_i, \sigma_{i+1}, \ldots \sigma_n) \geqq$$
$$\geqq \inf_{\sigma_1, \ldots, \sigma_{i-1}, \sigma_{i+1}, \ldots, \sigma_n} K_i(\sigma_1, \ldots, \sigma_n)$$

for all $\sigma_i \in \Sigma_i$ and $i = 1, \ldots, n$. In case of two-person games these are "defensive strategies" as defined on page 111.

(b) The equilibrium point under the assumption of this kind of "rationality" is the Nash equilibrium point. The noncooperative part of our book is almost entirely committed to this concept and it is discussed in great detail.

2. Antisocial behaviour. The players do not care about their own pay-offs, they want to hurt everybody (including themselves) as badly as they can.

(a) This attitude can result in a "solution point" $(\bar{\sigma}_1, \ldots, \bar{\sigma}_n)$ satisfying the system of inequalities

$$\text{(2)} \qquad \sup_{\sigma_1,\ldots,\sigma_{i-1},\sigma_{i+1},\ldots,\sigma_n} \sum_{j=1}^{n} K_j(\sigma_1, \ldots, \sigma_{i-1}, \bar{\sigma}_i, \sigma_{i+1}, \ldots, \sigma_n) \leqq$$

$$\leqq \sup_{\sigma_1,\ldots,\sigma_{i-1},\sigma_{i+1},\ldots\sigma_n} \sum_{j=1}^{n} K_j(\sigma_1, \ldots, \sigma_n)$$

for all $\sigma_i \in \Sigma_i$ and $i=1, \ldots, n$.

Each player minimizes the maximal achievable total pay-off on the grounds of a way of thinking: "I do harm to the common interest allowing others for being wise and cooperative".

(b) As an equilibrium based on antisocial behaviour we define an n-tuple of strategies $(\sigma_1^*, \ldots, \sigma_n^*)$ satisfying the inequality system

$$\text{(3)} \qquad \sum_{j=1}^{n} K_j(\sigma_1^*, \ldots, \sigma_n^*) \leqq$$

$$\leqq \sum_{j=1}^{n} K_j(\sigma_1^*, \ldots, \sigma_{i-1}^*, \sigma_i, \sigma_{i+1}^*, \ldots, \sigma_n^*)$$

for all $\sigma_i \in \Sigma_i$ and $i=1, \ldots, n$.

3. Social behaviour. Now the players try to take into consideration "everybody's interest" measured by the sum of the pay-off functions.

(a) One way for doing so is picking "solution strategies" $(\bar{\sigma}_1, \ldots, \bar{\sigma}_n)$ to satisfy

$$\text{(4)} \qquad \inf_{\sigma_1,\ldots,\sigma_{i-1},\sigma_{i+1},\ldots,\sigma_n} \sum_{j=1}^{n} K_j(\sigma_1, \ldots, \sigma_{i-1}, \bar{\sigma}_i, \sigma_{i+1}, \ldots, \sigma_n) \geqq$$

$$\geqq \inf_{\sigma_1,\ldots,\sigma_{i-1},\sigma_{i+1},\ldots,\sigma_n} \sum_{j=1}^{n} K_j(\sigma_1, \ldots, \sigma_n)$$

for all $\sigma_i \in \Sigma_i$ and $i=1, \ldots, n$.

This means that each player maximizes the lower bound of the total common pay-off. This pattern of behaviour can best be characterized by the following train of thought: "I do my best for the community's interest but allow that others might act against it (e.g. through ignorance)".

(b) The equilibrium point $(\sigma_1^*, \ldots, \sigma_n^*)$ in this situation satisfies the inequalities

$$(5) \qquad \sum_{j=1}^{n} K_j(\sigma_1^*, \ldots, \sigma_n^*) \geqq$$
$$\geqq \sum_{j=1}^{n} K_j(\sigma_1^*, \ldots, \sigma_{i-1}^*, \sigma_i, \sigma_{i+1}^*, \ldots, \sigma_n^*)$$

for all $\sigma_i \in \Sigma_i$ and $i = 1, \ldots, n$.

4. Mazochist behaviour. This is the opposite of rational behaviour which can best be formalized through multiplying the original pay-off functions by -1.

5. Malevolent behaviour. This is the same as antisocial except that summation in (2) and (3) is taken over all $j \neq i$ which means that each player wants to hurt everybody else's interests. (For two-person games this leads to "attacking strategies", see page 111).

6. Philanthropic behaviour. It is the same as social save for the summation in (4) and (5) taken now over all $j \neq i$ which means that each player intends to enhance public welfare without considering his own benefit.

In general we cannot say much about the existence of "equilibrium points" of the above kind. "Solution points" do exist under very general conditions but they are not "stable" with respect to the particular behavioural pattern.

Of course other behavioural patterns can also be set up (e.g. mixtures of those discussed above, inclusion of subjective assessment of other players' expected behaviour etc.). Our sole purpose by listing a few has been to emphasize the need for a theory in which other than "rational" behaviour can also be incorporated.

For two-person games similar "philosophical" aspects of game theory are discussed in [98] and [142].

References

[1] Ackoff, R. L.; Sasieni, M. W.: *Fundamentals of Operations Research*. J. Wiley and Sons, N. Y. 1968.

[2] Aggarwal, V.: On the generation of all equilibrium points for bimatrix games through the Lemke–Howson algorithm. *Mathematical Programming*, **4**, 1973, pp. 233–234.

[3] Alexandrov, P. S.: *Introduction to the Theory of Sets and Functions*. (Russian), Ogiz, Gostehizdat, Moscow–Leningrad, 1948.

[4] Ambrosenko, V. V.: Acceleration of the convergence of Brown's method for solving matrix games. (Russian). *Ekonom. i Mat. Metody*, **1.**, No. *4*, 1965. pp. 570–575.

[5] Anderson, B. D.; Arbib, M. A.; Manes, E. G.: *General Systems Theory. Finitary and Infinitary Conditions*. Lecture Notes in Economics and Mathematical Systems. Springer, Berlin, 1976.

[6] Appelgren, L.: An attrition game. *Operations Research*, **15**, 1967, pp. 11–31.

[7] Aumann, R. J.: Almost strictly competitive games. *SIAM J. on Appl. Math.*, **9**, 1961. pp. 544–550.

[8] Aumann, R. J.: Markets with a continuum of traders. *Econometrica*, **32**, 1964, pp. 39–50.

[9] Aumann, R. J.: Existence of competitive equilibria in markets with a continuum of traders. *Econometrica*, **34**, 1966, pp. 1–17.

[10] Aumann, R. J.; Maschler, M.: The bargaining set for cooperative games in [41].

[11] Aumann, R. J.; Maschler, M.: Some thoughts on the minimax principle. *Management Science*, **18**, 1971/72, pp. 54–63.

[12] Baligh, H. H.; Richartz, L. E.: Variable-sum game models of marketing problems. *J. Market Res.*, **4**, 1967, pp. 173–183.

[13] Baum, L. E.; Ferguson, J. D.; Katz, M.: Infinitely repeated matrix games for which pure strategies suffice. *Bull. Amer. Math. Soc.*, **69**, 1963, pp. 467–470.

[14] Behzad, M.; Harary, F.: Which directed graphs have a solution? *Mathematica Slovaca*, **27**, 1977, pp. 37–41.

[15] Belenkij, V. Z.; Volkonszkij, V. A.; Ivankov, S. A.: A certain general approach to the study of the convergence of iterative processes. (Russian). *Ekon. i Mat. Metody*, **10**, 1974, pp. 140–158.

[16] Bellman, R.: On "Colonel Blotto" and analogous games *SIAM Rev.* **11**, 1969, pp. 66–68.

[17] Berkovitz, L. D.; Dresher, M.: A game-theory analysis of tactical air-war. *Operations Research*, **7**, 1959, pp. 599–620.

[18] Billera, L. J.: Existence of general bargaining sets for cooperative games without side payments. *Bull. Amer. Math. Soc.*, **76**, 1970, pp. 375–379.

[19] Billera, L. J.: Some theorems on the core of an *n*-person game without side-payments. *SIAM J. on Appl. Math.*, **18**, 1970, pp. 567–579.

[20] Billera, L. J.: Some recent results in *n*-person game theory. *Mathematical Programming*, **1**, 1971, pp. 58–67.

[21] Blackwell, D.: Minimax and irreducible matrices. *Journal of Math. Anal. Appl.*, **3**, 1961, pp. 37–39.

[22] Blackwell, D.; Girschick, M. A.: *Theory of Games and Statistical Decisions*. J. Wiley and Sons, New York, 1945.

[23] Bondareva, O. N.: Same applications of the methods of linear programming to the theory of cooperative games. (Russian), *Problemy Kibernetiky*, No. *10*, 1963, pp. 119–130.

[24] Borch, K. H.: *The Economics of Uncertainty*. Princeton Univ. Press, Princeton, N. J., 1968.

[25] Borisova, E. P.; Magarik, I. V.: On two modifications of Brown's method for solving matrix games. (Russian). *Ekon. i Mat. Metody*, No. *5*, 1966, pp. 732–738.

[26] Boudwin, K.; Rosendahl, R.; Wilson, R.: *Computation of equilibria of extensive games*. Stanford University, Tech. Rep. No. *69–12*.

[27] Bregman, L. M.; Fokin, I. N.: The structure of the optimal strategies in certain matrix games. (Russian). *Dokl. Akad. Nauk SSSR.*, **188**, 1969, pp. 974–977.

[28] Bubjalis, V.: The connection between noncooperative *n*-person and three-person games. In "*Current trends in game theory*". (Russian). Izdat. "Mosklas", Vilnius, 1976, pp. 18–24.

[29] Buck R. C.: Preferred optimal strategies. *Proc. Amer. Math. Soc.* **9**, 1958, pp. 312–314.

[30] Bulatov, V. P.: Numerical methods for the solution of certain game problems (Russian). *Optimization methods and their applications* (All-Union Summer Sem., Khakusy, Lake Baikal, 1972), 183. Sibirsk. Energet. Inst. Sibirsk. Otdel. Akad. Nauk SSSR, Irkutsk, 1974, pp. 164–178.

[31] Burger, E.: *Einführung in die Theorie der Spiele*. de Gruyter, Berlin, 1966.

[32] Butterworth, R. W.: A set theoretic treatment of coherent systems. *SIAM J. on Appl. Math.*, **22**, 1972, pp. 590–598.

[33] Cassidy, R. G.; Field, C. A.; Kirby, M. J.: Random pay-off games with partial information: one person games against nature. *Revue. franç. Informatique et de Recherche Operationelle*, **3**, 1971, pp. 3–18.

[34] Cooper, J. N.; Restrepo, R. A.: Some problems of attack and defense. *SIAM Review*, **9**, 1967, pp. 680–691.

[35] Cover, T. M.: The probability that a random game is unfair. *Ann. Math.*, **37**, 1966, pp. 1796–1799.

[36] Danskin, J. M.: A game over spaces of probability distributions. *Naval Research Logistics Quarterly*, **11**, 1964, pp. 157–189.

[37] Dantzig, G. B.: *Linear Programming and Extensions*. Princeton Univ. Press, Princeton, 1963.

[38] Davis, M.; Maschler, M.: The kernel of a cooperative game. *Naval Research Logistics Quarterly*, **12**, 1965, pp. 223–259.

[39] Dresher, M.: *Games of Strategy: Theory and Applications*. Prentice Hall Applied Mathematics Series. Prentice Hall, Englewood Cliffs, N. J. 1961.

[40] Dresher, M.: Probability of a pure equilibrium point in n-person games. *J. of Combinatorial Theory*, **8**, 1970, pp. 134–145.

[41] Dresher, M.; Tucker, A. W.; Wolfe, P. (eds): Contributions to the Theory of Games, **III**. *Annals of Math. Studies*, **39**, Princeton University Press, Princeton, 1957.

[42] Dresher, M.; Shapley, L. S.; Tucker, A. W. (eds): Advances in Game Theory, *Annals of Math. Studies*, **52**. Princeton University Press, Princeton, 1964.

[43] Dubey, P.: *On the uniqueness of the Shapley value*. Technical Report, Cornell University, June, 1974.

[44] Dubin, G. N.: The set of games on the unit square with unique solution. (Russian). *Dokl. Akad. Nauk. SSSR*, **184**, 1969, pp. 267–269.

[45] Eaves, B. C.: Polymatrix games with joint constraints. *SIAM J. on Appl. Math.*, **24**, 1973, pp. 418–423.

[46] Edlefsen, L. E.; Millham, C. B.: On a formulation of discrete n-person noncooperative games. *Metrika*, **18**, 1971/72, pp. 31–34.

[47] Egerváry, J.: On hypermatrices consisting of piecewise interchangeable blocks and their application to grid dynamics. (Hungarian). *MTA Alk. Mat. Közl.*, **3**, 1954, pp. 31–47.

[48] Eichhorn, B. H.: On sequential minimax. *J. Math. Anal. Appl.*, **14**, 1966, pp. 31–37.

[49] Everett, H.: *Recursive games*. Princeton Univ. Press. Princeton, 1954.

[50] Fan, K.: *Minimax theorems*. Proc. of Nat. Acad. Sci. **39**, pp. 42–47.

[51] Fenstad, J. E.: Good strategies in general games. *Math. Zeitschrift*, **101**, 1967, pp. 322–330.
[52] Forgó, F.: *The relationship between continuous zero-sum two-person games and linear programming*. Dep. of Math. Karl Marx Univ. of Economics, Budapest, 1969.
[53] Forgó, F.: *A SUMT-method and its application to the solution of certain continuous games*. Dep. of Math. Karl Marx University of Economics. Budapest, 1970.
[54] Forgó, F.: *Dynamic processes and games with limited information about the pay-off function*. Dep. of Math. Karl Marx University of Economics, Budapest, 1971.
[55] Forgó, F.: Multicriteria decision making: a game theoretical approach. (Hungarian) *SZIGMA*, **XIV**, 1981, pp. 29–38.
[56] Fox, M.; Kimeldorf, G. S.: Noisy duels. *SIAM J. on Appl. Math.*, **17**, 1969, pp. 353–361.
[57] Frick, H.: *Spieltheoretische Behandlung mehrfacher Inventur-probleme*. Dissertation, Universität Fridericiana Karlsruhe, 1976.
[58] Friedman, J. W.: *Oligopoly and the Theory of Games*. North-Holland Publ. Co., Amsterdam, 1977.
[59] Germeyer, J. B.: *Games with Nonantagonistic Interests*. (Russian). Nauka, 1976.
[60] Glicksman, A. M.: *An Introduction to Linear Programming and the Theory of Games*. J. Wiley and Sons, 1963.
[61] Goldberg, K.; Goldman, A. J.; Newman, M.: The probability of an equilibrium point. *J. Res. Nat. Bur. Stnds. USA*, **72B**, 1968, pp. 93–101.
[62] Groemer, H.: On the min-max theorem for finite two-person zero-sum games. *Zeitschrift für Wahrscheinlichkeitstheorie und verwandte Gebiete*, **9**, 1967, pp. 59–61.
[63] Goldman, A. J.: The probability of a saddle-point. *American Math. Monthly*, **60**, 1957, pp. 729–730.
[64] Hansen, T.: On the approximation of a Nash equilibrium point in an n-person noncooperative game. Norwegian School of Econ. and Bus. Adm. Bergen 1970.
[65] Hansen, T.: On the approximation of Nash equilibrium points in an N-person noncooperative game. *SIAM J. on Appl. Math.*, **26**, 1974, pp. 622–637.
[66] Harsanyi, J. C.: Rationality postulates for bargaining solutions in cooperative and noncooperative games. *Management Science*, **9**, 1962, pp. 141–153.
[67] Harsanyi, J. C.: A general solution for finite noncooperative games, based on risk-dominance. *Ann. Math. Studies*, **52**, 1964, pp. 651–679.

[68] Harsanyi, J. C.: A general theory of rational behaviour in game situations. *Econometrica*, **34**, 1966, pp. 613–634.

[69] Harsanyi, J. C.: Games with incomplete information played by "Bayesian" players. *Management Science*, **14**, 1967, No. *3–5–7*, pp. 159–182, 320–334, 486–502.

[70] Harsanyi, J. C.: An equilibrium point interpretation of stable sets and a proposed alternative definition. *Management Science*, **20**, 1973/74, pp. 1472–1495.

[71] Harsanyi, J. C.: The tracing procedure: a Bayesian approach to defining a solution for n-person noncooperative games. Parts I–II. University of California, Berkeley and University of Bielefeld, May, 1974.

[72] Harsanyi, J. C.: *Advances in understanding rational behaviour*. University of California, Berkeley, July 1975.

[73] Harsanyi, J. C.: Analysis of a family of two-person bargaining games with incomplete information. Working papers in Management Science, University of California, Berkeley, CP-406, April, 1978.

[74] Harsanyi, J. C.; Selten, R.: A generalized Nash-solution for two-person bargaining games with incomplete information. *Management Science*, **18**, 1971/72.

[75] Hoffman, A. J.; Karp, R. M.: On nonterminating stochastic games. *Management Science*, **12**, 1966, pp. 359–370.

[76] Isaacs, R.: Differential Games: A Mathematical Theory, with Applications to Warfare and Pursuit, Control and Optimization. J. Wiley and Sons, N. Y., 1965.

[77] Jacob, J. P.; Polak, E.: On finite dimensional approximations to a class of games. *J. Math. Anal. Appl.*, **21**, 1968, pp. 287–303.

[78] Kalman, R.: *Lectures on Algebraic Systems Theory*. Lecture Notes in Mathematics. Springer, Berlin, 1969.

[79] Karlin, S.: *Mathematical Methods and Theory in Games, Programming and Economics*, Addison Wesley Publ. Co., Reading, Mass., 1959.

[80] Kiruta, A. J.: Equilibrium points in non-atomic noncooperative games in "*Mathematical Methods in Social Sciences*", (Russian). No. *6*. Vilnius, 1975, pp. 18–71.

[81] Kohlberg, E.: On the nucleolus of a characteristic function game. *SIAM J. on Appl. Math.*, **20**. 1971, pp. 62–66.

[82] Kohlberg, E.: The nucleolus as a solution of a minimization problem. *SIAM J. on Appl. Math.*, **23**, 1972, pp. 34–39.

[83] Krelle, W.; Coenen, D.: Das nichtkooperative Nichtnullsummen-Zwei-Personen-Spiel I–II. *Unternehmensforschung*, **9**, 1965, pp. 57–59, 137–163.

[84] Kuhn, H. W.: An algorithm for equilibrium points in bimatrix games. *Proc. Nat. Acad. Sci. USA.*, **47**, 1961. pp. 1657–1662.

[85] Kuhn, H. W.; Tucker, A. W. (eds): Contributions to the Theory of Games, I–II. *Annals of Math. Studies*, 24–28. Princeton University Press, Princeton, 1950–1953.

[86] Kuhn, H. W.; Szegő, G. P. (eds): *Differential Games and Related Topics*. North-Holland Publ. Co., Amsterdam, 1971.

[87] Kuznecov, I. N.: *Minimax problem for an arbitrary pay-off function*. (Russian). Izv. Akad. Nauk SSSR Techn. Kibernet., 1967, No. *2*, pp. 30–37.

[88] Lebediev, V. N.: *On the equivalence of convex-concave antagonistic games with mathematical programming problems*. (Russian). Izv. Akad. Nauk SSSR Techn. Kibernetik, 1966, No. *2*, pp. 32–38.

[89] Lemaire, J.: A new value for games without transferable utilities. *Internat. J. Game Theory*, **2**, 1973, pp. 205–213.

[90] Lemke, C. E.; Howson, J. T. Jr.: Equilibrium points of bimatrix games. *SIAM J. on Appl. Math.*, **12**, 1964, pp. 413–423.

[91] Lemke, C. E.: Bimatrix equilibrium points and mathematical programming. *Management Science*, **11**, 1965, pp. 681–689.

[92] Lihterov, J. A.: *On processes that are solutions of a matrix game*. (Russian). Izv. Akad. Nauk. SSSR Techn. Kibernet., 1965, No. *5*, pp. 23–41.

[93] Lipatov, E. P.: On products of matrix games. *Problemy Kibernet.*, 1965, No. *19*, pp. 259–261.

[94] Littlechild, S. C.: A simple expression for the nucleolus in a special case. *Internat. J. Game Theory*, **3**, 1974, pp. 21–29.

[95] Lucas, W. F.: A game with no solution. *Bull. Amer. Math. Soc.*, **74**, 1968, pp. 237–239.

[96] Lucas, W. F.: An overview of the mathematical theory of games. *Management Science*, **18**, 1972, pp. 3–19.

[97] Luce, R. D.: k-stability of symmetric and quota games. *Annals of Mathematics*, **62**, No. *3*, 1955, pp. 517–527.

[98] Luce, R. D.; Raiffa, H.: *Games and Decisions*, J. Wiley and Sons, N. Y., 1957.

[99] Maitra, A.; Parthasarathy, T.: On stochastic games. *J. Optimization Theory Appl.*, **5**, 1970, pp. 289–300.

[100] Maitra, A.; Parthasarathy, T.: On stochastic games II. *J. Optimization Theory Appl.*, **8**, 1971, pp. 154–160.

[101] Majthay, A.: A lexicographic complementary pivot algorithm for the solution of bimatrix games. *Studia Sci. Math. Hungar.*, **7**, 1972, pp. 181–188.

[102] Mangasarian, O. L.: Equilibrium points of bimatrix games. *SIAM. J. on Appl. Math.*, **12**, 1964, pp. 778–780.

[103] Mangasarian, O. L.; Stone, H.: Two-person nonzero-sum games and quadratic programming. *J. Math. Anal. Appl.*, **9**, 1964, pp. 348–355.

[104] Marchi, E.: Simple stability of general *n*-person games. *Naval Research Logistics Quarterly*, **14**, 1967, pp. 163–171.

[105] Maschler, M.; Peleg, B.: A characterization, existence proof and dimension bounds for the kernel of a game. *Pacific J. Math.*, **18**, 1966, pp. 289–328.

[106] Maschler, M.; Peleg, B.; Shapley, L. S.: The kernel and the nucleolus of a cooperative game as locuses in the strong ε-core. Research Memorandum No. *60*. Department of Mathematics, The Hebrew University of Jerusalem, Jerusalem. May, 1970.

[107] Martos, B.: *Nonlinear Programming Theory and Methods*. Akadémiai Kiadó, Budapest, 1976.

[108] McKinsey, J. C. C.: *Theory of Games*. McGraw-Hill, N. Y., 1952.

[109] Meggido, N.: Nucleoluses of compound simple games. *SIAM J. on Appl. Math.*, **26** (1974), pp. 607–621.

[110] Meggido, N.: On the nonmonotonicity of the bargaining set, the kernel and the nucleolus of games. *SIAM J. on Appl. Math.*, **27**, 1974, pp. 355–358.

[111] Menges, G.: On the "Bayesification" of the minimax principle. *Unternehmensforschung*, **10**, 1966, pp. 81–91.

[112] Mesarovic, M. D.: *General Systems Theory. Mathematical Foundations*. Acad. Press. N. Y. 1975.

[113] Millham, C. B.: On the structure of equilibrium points in bimatrix games. *SIAM Review*, **10**, 1968, No. *4*.

[114] Mills, H.: Equilibrium points in finite games. *SIAM J. on Applied Math.*, **8**, 1960, pp. 397–402.

[115] Milnor, J.; Shapley, L. S.: *On games of survival*. Princeton University Press, Princeton, 1957.

[116] Mine, H.; Yamada, K.; Asaki, S.: On terminating stochastic games. *Management Science*, **16**, 1970, pp. 560–571.

[117] Miyashawa, K.: On the convergence of the learning process in a 2 by 2 nonzero sum two-person game. Princeton, University Research Memorandum, No. *33*, 1961.

[118] Mon, G. R.: A dynamic theory of zero-sum two-person games. *J. Math. Anal. Appl.*, **29**, 1970, pp. 392–411.

[119] Mond, B.: On the direct sum and tensor product of matrix games. *Naval Research Logistics Quarterly*, **11**, 1964, pp. 205–215.

[120] Morozov, V. V.: A certain approach to cooperative games (Russian). *Ž. Vyčisl. Mat. i. Mat. Fiz.* **13**, 1973, pp. 781–787.

[121] Nash, J.: The bargaining problem. *Econometrica*, **18**, 1950, pp. 155–162.

[122] Nash, J.: Equilibrium points in finite games. *Proc. Nat. Acad. Sci. USA*, **36**, 1950, pp. 48–49.

[123] Nash, J.: Noncooperative games. *Ann. of Math.*, **54**, 1951, pp. 286–295.

[124] Nash, J.: Two-person cooperative games. *Econometrica*, **21**, 1953. pp. 128–140.

[125] von Neumann, J.: A numerical method to determine optimal strategy. *Naval Research Logistics Quarterly*, **1**, 1954, pp. 109–115.

[126] von Neumann, J.; Morgenstern, O.: *Theory of Games and Economic Behavior*. Princeton University Press, Princeton, 1944, 1947, 1953.

[127] Newman, D. J.: Another proof of the minimax theorem. *Proc. Amer. Math. Soc.*, **11**, 1960, pp. 692–693.

[128] Nietsche, J.: Das Problem der Dominanz bei Zwei-Personen Spielen. *Zeitschrift für Wahrscheinlichkeitstheorie und verwandte Gebiete*, **6**, 1966, pp. 119–124.

[129] Orkin, M.: Recursive matrix games. *J. Appl. Probability*, **9**, 1972, pp. 813–820.

[130] Owen, G.: An elementary proof of the minimax theorem. *Management Science, Ser. A.*, **13**, 1967, p. 765.

[131] Owen, G.: *Game Theory*. Phil. Saunders, 1968.

[132] Owen, G.: On the core of linear production games. *Mathematical Programming*, **9**, 1975, pp. 358–370.

[133] Parthasarathy, T.: On games over the unit square. *SIAM J. on Appl. Math.*, **19**, 1970, pp. 473–476.

[134] Parthasarathy, T.: Discounted and positive stochastic games. *Bull. Amer. Math. Soc.* **77**, 1971, pp. 134–136.

[135] Parthasarathy, T.: Raghavan, S.: *Some Topics in Two-person Games*. American Elsevier Publ. Comp. Inc. N.Y., 1971.

[136] Peisakoff, M. P.: *More on games of survival*. Research Memorandum RM-884, The RAND Corporation, Santa Monica, 1952.

[137] Peleg, B.: On the bargaining set M_0 of m-quota games. in [41].

[138] Peleg, B.: Existence theorem for the bargaining set $M_1^{(i)}$. *Bull. Am. Math. Soc.*, **69** (1963), pp. 109–110.

[139] Ponstein, J.: An extension of the min-max theorem. *SIAM Review*, **7**, 1965, pp. 181–188.

[140] Raghavan, T. E. S.: Completely mixed strategies in bimatrix games. *J. London Math. Soc.* (2). **2**, 1970, pp. 709–712.

[141] Rao, A. G.; Shakun, M. F.: A quasi-game theory approach to pricing. *Management Science*, **18**, 1972. pp. 110–124.

[142] Rapoport, A.: *Two-person Game Theory*. (The Essential Ideas). The University of Michigan Press, Ann Arbor. 1969.

[143] Reinganum, J. F.: On the diffusion of new technology: a game theoretical approach. *Review of Economic Studies* XLVIII, 1981, pp. 395–405.

[144] Réti, J.: A game-theoretic solution of the von Neumann model of general economic equilibrium. (Hungarian). *Szigma*, **5**, 1972, pp. 97–105.

[145] Robinson, J.: An iterative method of solving a game. *Ann. Math.* **154**, 1951, pp. 296–301.

[146] Rogers, Ph.: *Nonzero-sum stochastic games*. Operation Research Center, University of California, Berkeley, April, 1969.

[147] Romanovsky, I. V.: Exploitation of the convexity of the pay-off function in the solution of matrix games. (Russian). *Vestnik Leningrad. Univ.*, **20**, 1965, No. *13*, pp. 161–163.

[148] Rosen, J. B.: Existence and uniqueness of equilibrium points for concave n-person games. *Econometrica*, **33**, 1965, pp. 520–534.

[149] Rosenmüller, J.: On a generalization of the Lemke-Howson algorithm to noncooperative N-person games. *SIAM J. Appl. Math.*, **21**, 1971, pp. 73–79.

[150] Rosenmüller, J.: *Kooperative Spiele und Märkte*. Lecture Notes in Operations Research and Mathematical Systems. Springer Verlag, Berlin–Heidelberg–New York, 1971.

[151] Roy, G.: Problems and methods with multiple objective functions. *Mathematical Programming*, **1**, 1971, No. 2.

[152] Scarf, H. E.; Shapley, L. S.: Games with partial information. *Ann. Math. Studies*, **39**, 1957, pp. 213–229.

[153] Scarf, H. E.: The approximation of fixed points of a continuous mapping. *SIAM J. on Appl. Math.*, **15**, 1967, pp. 1328–1343.

[154] Scarf, H.: The core of an n-person game. *Econometrica*, **35**, 1967, pp. 50–69.

[155] Schlaifer, R.: *Introduction to Statistics for Business Decisions*. McGraw-Hill, New York, 1961.

[156] Schmeidler, D.: The nucleolus of a characteristic function game. *SIAM J. on Appl. Math.*, **17**, 1969, pp. 1163–1170.

[157] Schroeder, R. G.: Linear programming solutions to ratio games. *Operations Research*, **18**, 1970, pp. 300–305.

[158] Shapiro, H. N.: Note on computation method in the theory of games. *Communs. Pure and Appl. Math.*, **11**, 1958, No. *4*.

[159] Shapley, L. S.: Stochastic games. *Proc. Nat. Acad. Sci. USA*, **39**, 1953, pp. 1095–1100.

[160] Shapley, L. S.: Equilibrium points in games with vector payoffs. *Naval Research Logistics Quarterly*, **6**, 1959, pp. 57–63.

[161] Shapley, L. S.: Some topics in two-person games. in [41].

[162] Shapley, L. S.: On balanced sets and cores. *Naval Research Logistics Quarterly*, **14**, 1967, pp. 453–461.

[163] Shapley, L. S.; Shubik, M.: On market games. *Journal of Economic Theory*, **1**, 1969, pp. 9–25.

[164] Shapley, L. S.; Shubik, M.: Pure competition, coalitional power, and fair division. *International Economic Review*. **10**, No. *3*, 1969.

[165] Shenoy, P. P.: On game theory and coalition formation. Technical Report No. *342*. College of Engineering, Cornell University, Ithaca, New York. July, 1977.

[166] Shenoy, P. P.: On committee decision making: a game theoretical approach. *Management Science*, **26**, 1980, *4*, pp. 387–400.

[167] Shenoy, P. P.: A two-person non-zero-sum game model of the world oil market. *Appl. Math. Modelling*, **4**, 1980, pp. 295–300.

[168] Shmadich, K.: The existence of graphic solutions. (Russian). *The Leningrad University Herald*, **1**, 1976, pp. 88–92.

[169] Shubik, M.: *Strategy and Market Structure. Competition, Oligopoly, and the Theory of Games*. J. Wiley and Sons, N. Y. 1959.

[170] Shubik, M.: Some experimental non-zero sum games with lack of information about the rules. *Management Science*, **8**, 1961/62, pp. 215–234.

[171] Shubik, M.; Thomson, G. L.: Games of economic survival. *Naval Research Logistics Quarterly*, **6**, 1959, pp. 111–123.

[172] Smith, P. E.: A game theoretic interpretation of a Leontief input-output system. *Metroeconomica*, **15**, 1963. pp. 47–54.

[173] Sobel, M. J.: Noncooperative stochastic games. *Ann. Math. Statistics*, **42**, 1971, pp. 1930–1935.

[174] Sobolev, A. I.: The characterization of optimality principles in cooperative games by functional equations in "Mathematical Methods in Social Sciences" No. *6*, Vilnius, 1975, pp. 18–71.

[175] Stearns, R. E.: On the axioms for a cooperative game without side payments. *Pr. Am. Math. Soc.*, **15**, 1964, pp. 82–86.

[176] Suzuki, M.; Nakayama, M.: The cost assignment of the cooperative water resource development: a game theoretical approach. *Management Science*. **22**, 1976, pp. 1081–1086.

[177] Szép, J., Hegedüs, M.: On equilibrium systems I. Dep. of Math., Karl Marx Univ. of Economics, Budapest, 1970.

[178] Szép, J.; Forgó, F.: *Introduction to the Theory of Games*. (Hungarian). Közgazdasági és Jogi Könyvkiadó, Budapest, 1974.

[179] Szidarovszky, F.: *Theory of Games*. (Lecture Notes). (Hungarian). Tankönyvkiadó, Budapest, 1977.

[180] Szidarovszky, F.: On the multiproduct oligopoly game. (Hungarian). *SZIGMA*, **IX**, 1976, pp. 243–248.

[181] Szidarovszky, F.: The nondifferentiable oligopoly problem. (Hungarian). *SZIGMA*, **X**, 1977.

[182] Szidarovszky, F.: A generalization of Nash's concept for cooperative games. (Hungarian). *SZIGMA*, **IX**, 1978, *1–2*, pp. 69–74.

[183] Szidarovszky, F.; Bogárdi, I.: An application of game theory in water management. (Hungarian). *Hidr. Közlöny*, 1976, No. *10*, pp. 456–462.

[184] Thomas, D. R.; David, H. T.: Game value distributions I–II. *Ann. Math. Statistics*, **38**, 1967, pp. 242–260.

[185] Thompson, G. L.; Weil, R. L. Jr.: Further relations between game theory and eigensystems. *SIAM Review*, **11**, 1969, pp. 597–602.

[186] Thrall, R. M.; Lucas, W. F.: *n*-person games in partition function form. *Naval Research Logistics Quarterly*, **10**, 1963. pp. 281–298.

[187] Tucker, A. W.; Luce, R. D. (eds).: Contribution to the Theory of Games, *IV Annals of Math. Studies*, **40.** Princeton University Press, Princeton, 1959.

[188] Uzawa, H.: The stability of dynamic processes. *Econometrica*, **29**, 1961, pp. 617–630.

[189] Vajda, S.: *Theory of Games and Linear Programming*. Methuen, London, 1956.

[190] Vilkas, E. I.: Axiomatic definition of the value of a matrix game (Russian). *Teor. Veroyatn. i Prim.*, **8**, 1963, pp. 324–327.

[191] Vilkas, E. I.: Decision regions of a parametric matrix game. (Russian). *Litovsk. Mat. Sb.*, **4**, 1964. pp. 31–35.

[192] Vilkas, E. I.: An axiomatic definition of equilibrium situation and value for *n*-person coalitionfree game. *Theor. Prob. Appl.*, **13**, 1968, pp. 523–527.

[193] Vilkas, E. I.: Games with variable payoffs. (Russian). *Litovsk. Mat. Sb.*, **10**, 1970, pp. 693–703.

[194] Vilkas, E. I.; Korbut, A. (ed): *Current trends in game theory*. (Russian). Izdat. "Mosklas", Vilnius, 1976.

[195] Volkovsky, V. A.: Optimal planning under conditions of large-scale systems. (Russian). *Ekon. i Math. Meth.*, **1**, 1965, pp. 195–219.

[196] Vorobyev, N. N.: Equilibrium points in bimatrix games. *Prob. Theory and Appl.* (Russian), **3**, (1958) pp. 318–331.

[197] Vorobyev, N. N.: The present state of game theory. (Russian). *Uspehi Mat. Nauk.*, **25**, 1970, pp. 81–140.

[198] Vorobyev, N. N.: *Game theory*. Lectures for Economists and Systems Scientists. Springer, Berlin, 1977.

[199] Wald, A.: *Statistical Decision Functions*. J. Wiley and Sons, N. Y., 1950.

[200] Weber, R. J.: Bargaining solutions and stationary sets in *n*-person games. Technical Report. No. *223*. Department of Operations Research, College of Engineering, Cornell University, Ithaca, New York. July, 1974.

[201] Westphal, L. C.; Stubberud, A. R.: The associated maximization problem for separable games. *Mathematical Programming*, **5**, 1973, pp. 374–380.

[202] Westphal, L. C.: An associated maximization problem for two-person nonzero-sum separable games. Mathematical Programming, **10**, 1976, pp. 124–129.

[203] Wernick, R. J.: Parametrized games and the eigenproblem. Linear Algebra and Appl., **3**, 1970, pp. 311–346.

[204] Williams, J. D.: The Compleat Strategist. McGraw-Hill, N. Y., 1954.

[205] Wilson, R.: Computing equilibria of two-person games from the extensive form. Stanford University, Working paper No. *176*, May, 1970.

[206] Wilson, R.: Computing equilibria of n-person games. *SIAM J. on Appl. Math.*, **21**, 1971. No. *1*.

[207] Woodbury, M.: On games whose matrices are the sums of the matrices of other games. *Bull. Amer. Math. Soc.*, **56**, 1950, 122.

[208] Wu Wen-Tsun: On noncooperative games with restricted domains of activities. *Sci. Sinica.* **10** (1961), pp. 387–409.

[209] Yanovskaya, E. B.: Minimax theorems for games on the unit square. *Teor. Ver.* **9**, 1964, pp. 554–555.

[210] Yanovskaya, E. B.: Antagonistic games played in function spaces. (Russian). *Litovsk. Mat. Sb.* **7**, 1967, pp. 547–557.

[211] Zeleny, M.: Linear multiobjective programming. Lecture notes in economics and mathematical systems 95. Springer. Berlin–Heidelberg–New York, 1976.

[212] Zuhovitsky, S. I.; Polyak, R. A.; Primak, M. E.: The concave n-person game and a certain production model. (Russian). Dokl. Akad. Nauk SSSR, **191**, 1970, pp. 1220–1223.

[213] Zuhovitsky, S. I.; Polyak, R. A.; Primak, M. E.: Concave n-person games (numerical methods). (Russian). *Ekonom. i. Mat. Metody*, **7**, 1971, pp. 888–900.

Name index

Subject index

www.ingramcontent.com/pod-product-compliance
Ingram Content Group UK Ltd.
Pitfield, Milton Keynes, MK11 3LW, UK
UKHW041831200726
13854UKWH00002BA/990

* 9 7 8 9 4 0 0 9 5 1 9 4 5 *